Climate Hazards, Disasters, and Gender Ramifications

This book focuses on the challenges of living with climate disasters, in addition to the existing gender inequalities that prevail and define social, economic and political conditions.

Social inequalities have consequences for the everyday lives of women and girls where power relations, institutional and socio-cultural practices make them disadvantaged in terms of disaster preparedness and experience. Chapters in this book unravel how gender and masculinity intersect with age, ethnicity, sexuality and class in specific contexts around the globe. They look at the various kinds of difficulties for particular groups before, during and after disastrous events, such as typhoons, flooding, landslides and earthquakes. Research focuses on how issues of gender hierarchies, patriarchal structures and masculinity are closely related to gender segregation, institutional codes of behaviour and to a denial of environmental crisis. This book stresses the need for a gender-responsive framework that can provide a more holistic understanding of disasters and climate change. A critical feminist perspective uncovers the gendered politics of such disaster and climate change.

This book will be useful for practitioners, students, and researchers working within the areas of climate change response, gender studies, disaster studies and international relations.

Catarina Kinnvall is Professor at the Department of Political Science, Lund University, Sweden.

Helle Rydstrom is Professor at the Department of Gender Studies, Lund University, Sweden.

Routledge Studies in Hazards, Disaster Risk and Climate Change

Series Editor: Ilan Kelman

Reader in Risk, Resilience and Global Health at the Institute for Risk and Disaster Reduction (IRDR) and the Institute for Global Health (IGH), University College London (UCL)

This series provides a forum for original and vibrant research. It offers contributions from each of these communities as well as innovative titles that examine the links between hazards, disasters and climate change, to bring these schools of thought closer together. This series promotes interdisciplinary scholarly work that is empirically and theoretically informed, with titles reflecting the wealth of research being undertaken in these diverse and exciting fields.

Community Engagement in Post-Disaster Recovery
Edited by Graham Marsh, Iftekhar Ahmed, Martin Mulligan, Jenny Donovan and Steve Barton

Climate, Environmental Hazards and Migration in Bangladesh
Max Martin

Governance of Risk, Hazards and Disasters
Trends in Theory and Practice
Edited by Giuseppe Forino, Sara Bonati and Lina Maria Calandra

Disasters, Vulnerability, and Narratives
Writing Haiti's Futures
Kasia Mika

Climate Change Impacts and Women's Livelihood
Vulnerability in Developing Countries
Salim Momtaz and Muhammad Asaduzzaman

Risk Communication and Community Resilience
Edited by Bandana Kar and David M. Cochran

Climate Hazards, Disasters, and Gender Ramifications
Catarina Kinnvall and Helle Rydstrom

For more information about this series, please visit: www.routledge.com/ Routledge-Studies-in-Hazards-Disaster-Risk-and-Climate-Change/book-series/HDC

Climate Hazards, Disasters, and Gender Ramifications

Edited by Catarina Kinnvall
and Helle Rydstrom

Routledge
Taylor & Francis Group

LONDON AND NEW YORK

First published 2019
by Routledge
2 Park Square, Milton Park, Abingdon, Oxon OX14 4RN

and by Routledge
605 Third Avenue, New York, NY 10017

First issued in paperback 2020

Routledge is an imprint of the Taylor & Francis Group, an informa business

British Library Cataloguing-in-Publication Data
A catalogue record for this book is available from the British Library

Library of Congress Cataloging-in-Publication Data
A catalog record for this book has been requested

ISBN 13: 978-0-367-72789-5 (pbk)
ISBN 13: 978-1-138-35436-4 (hbk)
ISBN 13: 978-0-429-42486-1 (ebk)

Typeset in Times New Roman
by Apex CoVantage, LLC

Contents

Illustrations

Figures

Tables

Contributors

Sara Bondesson, the Swedish National Defense University, Sweden, holds a PhD in Political Science from the Department of Government, Uppsala University. She is Assistant Professor at the Department of Security, Strategy and Leadership and is also affiliated with the Centre for Natural Hazards and Disaster Science (CNDS). Her research interests include disaster risk reduction, disaster management, gender studies, social movements, grassroots organising and social vulnerability. Current projects include an article combining democratic theory with ethnographic field studies of Occupy Sandy, a relief network active after Hurricane Sandy in New York City, and a book project about disaster collaboration that rests on story-telling from individuals involved in disaster management in Sweden.

Arya Gautam is a Planning, Monitoring and Social Safeguard Specialist for Small Irrigation Programme, Nepal, a joint initiative between the Government of Nepal and the Swiss Agency for Development and Cooperation (SDC). She holds a master's degree from the London School of Economics and Political Science in NGOs and Development. Arya has extensive research experience on security and justice issues, governance, livelihoods and resilience, working on projects funded by the UK's DFID. She has expertise in qualitative and participatory methods and is also a trained Reality Check Approach researcher.

Nicole George is Associate Professor of Peace and Conflict Studies in the school of Political Science and International Studies at the University of Queensland, Australia. Her work responds to theoretical debates on gender peace and security, gender and institutional change, and the gendered impacts of pluralised systems of regulatory authority. She conducts empirical studies of these issues in the Pacific Islands. This work has recently sparked her interest in the ways that conflict-affected communities are navigating the impacts of environmental degradation and climate change in this part of the world. Her recent publications appear in *International Affairs*, *International Feminist Journal of Politics*, *Policing and Society*, *Third World Thematics*, *International Political Science Review* and the *Australian Journal of International Affairs*.

Sidsel Hansson holds a PhD and is a South Asia specialist. She has been lecturing and tutoring at the Centre for East and South-East Asian Studies at Lund

University, Sweden, since 2003. Her research has focused on religion and environmentalism, and rural women's community organisation in South Asia. She is currently writing on gendered violence and masculinity in Pakistan.

Martin Hultman is Associate Professor at Chalmers University of Technology, Sweden. He has published widely on energy, climate and environmental issues in journals such as *Environmental Humanities, NORMA: International Journal for Masculinity Studies, History & Technology* and *Hydrogen Energy*. Especially notable are the articles "The Making of an Environmental Hero: A History of Ecomodern Masculinity, Fuel Cells and Arnold Schwarzenegger", "A Green Fatwā? Climate Change as a Threat to the Masculinity of Industrial Modernity" and the books *Posthumanistiska Nyckeltexter, Discourses of Global Climate Change* and *Ecological Masculinities*. As part of his academic work he writes chronicles in a wide range of newspapers and gives public lectures commenting on contemporary politics.

Catarina Kinnvall is Professor at the Department of Political Science, Lund University, Sweden. Her research interests involve issues of political psychology, globalisation, gender, religion and nationalism, with a particular focus on South Asia and Europe. She is especially concerned with the gendered dimension of extremist movements and has, together with Jennifer Mitzen, initiated and developed the field of ontological security studies. Her most recent book is *Governing Borders and Security: The Politics of Connectivity and Dispersal* (co-ed., Routledge, 2014). Her latest articles have appeared in *Review of International Studies*; *Journal of International Relations and Development*; *Cambridge Review of International Affairs*; *Cooperation and Conflict*; *European Security*; *Humanity and Society*; and *Nature & Human Behavior*.

Phu Doma Lama is a researcher at the Division of Risk Management and Societal Safety at Lund University, Sweden. Her doctoral thesis focuses on the "Consequence of Adaptation on Risks: Case Study of Maldives and Nepal". She was trained in development studies at the Tata Institute of Social Sciences, Mumbai, India, and has an MPhil in law and governance from Jawaharlal Nehru University, Delhi, India. Her expertise is in the area of community-based adaptation, gender issues in climate change and disaster research, and environmental governance. Lama's current research focuses on understanding transformative adaptation through the lens of power and participation in environmental governance settings. She is engaged as evaluator in a number of projects involving Swedish and other international agencies. She is also coordinator of the annual Gender, Disaster and Climate Risk Summer School at the Division of Risk Management and Societal Safety, Lund University.

Matthew Scott leads the People on the Move thematic area at the Raoul Wallenberg Institute of Human Rights and Humanitarian Law, Sweden, coordinating the Institute's activities relating to refugees and other displaced persons across Africa, Asia-Pacific, Europe and MENA regions. He defended his doctoral thesis in 2018, entitled "Refugee status determination in the context of 'natural' disasters and climate change: A human rights-based approach". His academic work has been published in the *International Journal of Refugee Law*, *Refugee Survey Quarterly*, *Forced Migration Review* and in specialist edited volumes. Matthew sits on the advisory committee of the Platform on Disaster Displacement and the editorial committee of the *Yearbook of International Disaster Law*.

Claudia Merli is Associate Professor of Cultural Anthropology at Uppsala University, Sweden. She is a medical anthropologist specialising in bodily practices, body politics and reproductive health, primarily in the context of medical care and demographic control of minorities in Southeast Asia (Thailand). Her second field of research is critical risk research; she has conducted ethnographic fieldwork during and after the 2004 Indian Ocean tsunami in Thailand, focusing on theological discourses, disaster victim identification (DVI) and identity politics. She has written several articles and chapters on disaster and religion, and was main contributor to the *World Disaster Report 2014: Focus on Culture and Risk* for the International Federation of Red Cross and Red Crescent Societies.

Huong Thu Nguyen holds a PhD from the University of Amsterdam, the Netherlands, and is Lecturer at the Department of Anthropology of Vietnam National University, Hanoi. Previously, she was affiliated with Department of Gender Studies at Lund University, Sweden, as a Postdoctoral Research Fellow (2015–2017). Her research interests centre on the intersection of sexual violence, gender, diversity, ethnicity, climate disasters and humanitarianism in the Philippines and Vietnam. Her latest work has appeared in *Culture, Health and Sexuality*; *Violence Against Women*; *American Anthropologist*; *Women's Studies International Forum*; and *NORMA: International Journal for Masculinity Studies*.

Katie Oven is Research Associate in the Department of Geography, Durham University, UK. Her research focuses on environmental processes including landslides and earthquakes and human vulnerability, with a particular focus on South and more recently Central Asia. Recent research has explored issues of development, policy and governance in the context of disaster risk drawing on fieldwork undertaken in Nepal and Northern India (Bihar State). Related publications have sought to highlight the everyday lived experience of householders, and have taken a critical view of the institutionalised concepts of vulnerability and resilience.

Paul Pulé is an Australian freelance scholar, social entrepreneur and social and environmental justice activist specialising in men's lives and their impacts on planet, people and self. After receiving his doctorate in 2013, Paul founded MenAlive Australia, a consultancy that provides research and leads workshops, seminars, public talks and keynotes about men and masculinities, as well as profeminist mentoring for boys and men. His research and experiences formed the basis of his first book, an Earth-inspired and profeminist university text titled *Ecological Masculinities: Theoretical Foundations and Practical Guidance* (2018). *Ecological Masculinities* explores the costs of toxic masculinities on us all and seeks alternatives to help point us in the direction of a deep green future for life.

Emmanuel Raju is Assistant Professor at the University of Copenhagen, Denmark. Raju is affiliated with the Copenhagen Centre for Disaster Research (COPE) linking disaster research and education. Raju's research interests include disaster risk reduction, disaster recovery and governance. He also works on issues of disasters and memory; international frameworks on disasters; and integration of climate change adaptation and disaster risk reduction. He has conducted research in India, Sri Lanka, Bangladesh, South Africa and Denmark. Raju holds a PhD from Lund University, Sweden. His PhD focused on disaster recovery coordination post-tsunami in India.

Shubheksha Rana was the former Director of Programs at the Kathmandu-based research organisation Foundation for Development Management. She has a master's degree in social work from Hawaii Pacific University, USA, and more than 12 years' experience working with national and international organisations in Nepal, with a particular focus on the rights of women and girls. Shubheksha's recent research has focused on disaster resilience and vulnerability and social justice issues, including issues of gender and ethnicity, particularly in the context of the 2015 Gorkha earthquake. This includes research for the INGO Mercy Corps and a DFID-funded review of community-based disaster risk reduction initiatives in collaboration with Durham University.

Jonathan Rigg holds a Chair in Human Geography at the School of Geographical Sciences at the University of Bristol, UK. He has been working on issues of human vulnerability and exposure in connection with earthquakes, tsunamis and climate change since the 1980s, undertaking fieldwork in Laos, Nepal, Sri Lanka, Thailand and Vietnam. His most recent books are *More than Rural: Textures of Thailand's Agrarian Transformation* (Hawaii University Press, 2019) and *People and Climate Change: People, Adaptation, and Social Justice* (OUP, 2019), edited with Lisa Reyes Mason.

Helle Rydstrom is Professor at the Department of Gender Studies at Lund University, Sweden and has a background in social anthropology and international development studies. Asia is at the fore of her research with a special focus on Vietnam, where she explores how gender and masculinity inform crises, violences and precariousness. She currently coordinates the Pufendorf Institute for Advanced Studies Theme on CRISIS (with 10 colleagues) as well as a research project on climate disasters and gendered violence in Asia (with Kinnvall and Nguyen) and one on gendered precariousness at the Nordic workplace outsourced to Asia (with Kinnvall, Eklund and Tonini) (both funded by the Swedish Research Council). She has edited/co-edited six volumes including *Gendered Inequalities in Asia* (2010) and *Gender Practices in Contemporary Vietnam* (with Drummond, 2004), authored the award winning *Embodying Morality* (2003) and published, e.g., in *Ethnos*; *European Journal of Women's Studies*; *Gender, Place and Culture*; *Men and Masculinities*; and *Signs*.

Toran Singh was a former Program Coordinator for the Poverty Alleviation Fund, funded by the World Bank and the International Fund for Agriculture Development. He holds a bachelor's degree in rural development from Tribhuwan University, Nepal. Toran has more than four years' experience working on a range of research projects in the sectors of education, rural livelihoods, climate change, community-based disaster risk reduction, security and social justice, funded by the World Bank, the UK's Department for International Development and the Swiss Agency for Development and Cooperation. He has expertise in quantitative, qualitative and participatory research methods.

Maria Tanyag is a Postdoctoral Research Fellow at Monash University's Centre for Gender, Peace and Security, Australia. In 2018, she was awarded her PhD in politics and international relations. Her doctoral thesis investigated sexual and reproductive health in crisis settings of armed conflicts and environmental disasters from a feminist political economy lens. Her article "Resilience, Female Altruism and Bodily Autonomy: Disaster-induced Displacement in Post-Haiyan Philippines" was published in *Signs: Journal of Women in Culture and Society* (2018). Twitter: @maria_tanyag.

Jacqui True, FASSA, is Professor of Politics and International Relations and Director of the Gender, Peace and Security Centre in the School of Social Sciences at Monash University, Australia. Her research interests include feminist political economy, conflict-related sexual and gender-based violence in international relations, gender and global governance and feminist research methodologies. Her book, *The Political Economy of Violence Against Women* in the Oxford Gender and International Relations Series (2012), won the American Political Science Association's 2012 biennial prize for the best book in human rights and the British International Studies Association 2013 International Political Economy book prize. She is co-editor of *The Oxford Handbook of Women, Peace and Security* (Oxford, 2018).

Misse Wester is currently a Lise Meitner Visiting Professor at the Division of Risk Management and Societal Safety at Lund University, Sweden. She holds a PhD in psychology from Orebro University and is Associate Professor at the Royal Institute of Technology in Stockholm, Sweden. During her three-year visiting professor appointment, she has worked with broadening her research on gender, risk and disasters. Her research interests relate to two areas of study: first, how do we think about different types of risk, and how do citizens and authorities communicate about risks? Second, how do people behave in times of disaster and how can it inform the identification of specific needs among citizens and groups?

Acknowledgements

This volume is one of many outcomes of an ongoing research project on "Climate Disasters and Gendered Violence in Asia: A Study on the Vulnerability and (In)Security of Women and Girls in the Aftermath of Recent Catastrophes in Pakistan, the Philippines, and Vietnam". The research project has been generously funded by the Swedish Research Council (*Vetenskapsrådet*) and has provided a unique opportunity to focus on the ways in which catastrophes are inherently gendered. Participants of the research project include the editors of this volume, Catarina Kinnvall, Department of Political Science, and Helle Rydstrom (PI), Department of Gender Studies, both at Lund University, Sweden; as well as Huong Thu Nguyen, previously affiliated with the Department of Gender Studies at Lund University but now at the Department of Anthropology at Hanoi University in Vietnam. Within the project framework, we have explored the intersections between climate hazards, disasters and gender ramifications in the region of Asia. With this book, we bring together a rich variety of project findings and research that speak to a global research field and provide a platform for the insightful studies conducted by the excellent scholars and colleagues contributing to this volume.

1 Introduction

Climate hazards, disasters, and gender ramifications

Helle Rydstrom and Catarina Kinnvall

Prologue

In the fall of 2013, a tropical storm took form in Micronesia and soon accelerated into a Category 5 storm, a super typhoon. Heading for the Philippines, super typhoon Haiyan (locally called Yolanda) made landfall on the island of Leyte on November 8, as one of the strongest storms ever recorded. Initial reports coming out of the Philippines described the horrors erupting in the maelstrom of the pounding winds of nearly 315 kilometres per hour: more than 12 million people were affected, including 4 million displaced persons; more than 6,000 persons were killed; and at least 300,000 homes were damaged or destroyed (NASA 20 November 2013; *Reliefweb* 14 November 2013; *Weather Underground*, 10 November 2013).

Sustento, a young woman from Leyte, luckily survived the disaster but she tragically lost her mother, father, eldest brother, sister-in-law and young disabled nephew. The typhoon was an extremely hazardous event which brutally took lives and drastically shaped the future of survivors like Sustento (*Ecologist*, 9 November 2017). In the aftermath of the typhoon, many people could not be reached and they did not receive any assistance for weeks, something which especially affected pregnant women and children as well as the elderly, sick and disabled. Of the displaced persons, 1.7 million were children, who, after being separated from their families, were at risk of falling victim to abuse and trafficking. Women and girls were particularly susceptible to be subjected to violence. For instance, men in uniform (i.e., police and military), who were stationed to protect inhabitants and guarantee their safety at such tumultuous times, reportedly sexually abused female survivors (IDMC 2013; Nguyen and Rydstrom 2018; *Philippine Star*, 2014; *Washington Post*, 4 January 2014).

The reconstruction process after the destructive disaster has been slow and the suffering and agony persistent across the country. Four years after the typhoon, more than 1,000 people were still missing; one of the missing was Sustento's father, whose body was never found after he was swept away by the extreme inland waters. The human and societal recovery process has been prolonged, not least because donations earmarked for rehabilitation allegedly remain in a bank account of the implementing branch of the Philippine National Disaster Risk Reduction and Management Council (*Ecologist*, 9 November 2017; see also Nguyen 2018).

As a devastating and horrifying crisis of emergency which caused high death tolls, destroyed homes, damaged infrastructure, disrupted means of communication, set aside social order, jeopardised the security of survivors and fostered political torpor, the Haiyan typhoon provides a window to the dire hazards and differentiated ramifications brought on by a climate disaster.[1] The experiences and politics related to storms, flooding, landslides, droughts and earthquakes are informed by gender and its intersections with sexuality, ethnicity, age, class and bodyableness (Butler, Gambetti, and Sabsay 2016; Crenshaw 1989; Godfrey and Torres 2016; UNDP 2007, 2013). In this volume, we engage with climate change disasters from a gender perspective by examining first policies, techniques and strategies employed by various agencies and governments to mitigate climate change related ramifications and then by exploring the rupturing impacts of disasters which enforce the redefinition of lifeworlds and the reinvention of livelihoods after a catastrophe.

Climate disasters

A disaster unfolds societal dynamics at the structural level and a community's relation to its environment, as well as the capability to adapt and the extent to which local knowledge can be infused to reduce vulnerability and harm (Fordham et al. 2013; Oliver-Smith 1999). The United Nation's International Strategy for Disaster Reduction (UNISDR 2009, 9) defines a disaster as "a serious disruption of the functioning of a community or a society involving widespread human, material, economic or environmental losses and impacts, which exceeds the ability of the affected community or society to cope using its own resources". One of the following criteria must be met for a disaster to be registered in the UNISDR database: (1) a report of 10 or more people killed; (2) a report of 100 people affected; (3) a declaration of a state of emergency by the relevant government; or (4) a request by the national government for international assistance (cited in Bradshaw 2013, 3).

According to Tom Cowton and colleagues (2018), a clear connection can be identified between warmer air temperature, retreating glaciers in Greenland, rising sea levels and climatic warming (see also IPCC 2018). In a similar vein, Victor Galaz (2017, 3) demonstrates how ocean acidification, modification of freshwater, deforestation, mineral flows and releases of carbon dioxide and methane gasses all contribute to climate changes. Such changes result in rising sea levels, extreme weather and changing biodiversity which impact freshwater resources, infrastructure, terrestrial and marine ecosystems, agriculture and coastal systems and in turn living conditions across the globe. The United Nations Framework Convention on Climate Change (UNFCCC 2013) thus defines climate change as "a change of climate which is attributed directly or indirectly to human activity that alters the composition of the global atmosphere and which is in addition to natural climate variability observed over comparable time periods" (cited in UNISDR 2009, 7).

Some places, however, are more disposed to disasters than others and some people's lifeworlds and livelihoods are more precarious to climate hazards than others. The extent to which the Global South struggles with climate-related disasters compared to the Global North is striking (Crutzen 2000; Galaz 2017; Ogden et al. 2015; Sternberg 2019; Wisner et al. 2003). Enforced monocultures designed to extract wealth from the Global South to the Global North during colonialism have translated into irrigation systems, exploitation of land, deforestation, hydroelectric plants, mining and industrial enterprises which thwart ecological stability and provoke environmental alterations (Cowton et al. 2018; *Guardian*, 9 January 2017; Oliver-Smith 1996, 1999).

Historically constituted socio-ecological relations between the Global North and the Global South, as James Ferguson (1990) argues, are concealed by the application of technologies designed to intensify agricultural and industrial production and revenue for transnational developers and investors. Launched as apolitical, as the 'anti-politics machine', development techniques contribute to the masking of a colonial legacy and the ways in which this legacy has transmuted into politics of power, conflicts of distribution, marginalisation of livelihoods and social inequalities in the present postcolonial era (see also Momtaz and Asaduzzaman 2019; Ogden et al. 2015; Sternberg 2019; Wolf 1982).

The époque of the Anthropocene

Climate change indicates a new type of reciprocity between the Earth and human beings. According to Paul Crutzen and Eugene Stoermer (2000), human interventions in nature have propelled us into the era of the Anthropocene; into the Geological Age of Man (see also Crutzen 2000). In this époque, the Earth has left the "interglacial state called the Holocene" (Crutzen and Stoermer 2000, 17) and is now facing the perils of planetary *terra incognita* (see also Crutzen 2006; Galaz 2017; Steffen, Crutzen, and McNeill 2007, 614).[2]

As outlined in the UN Intergovernmental Panel on Climate Change (IPCC) report (2018, 61), "human influence has become a principal agent of change on the planet, shifting the world out of the relatively stable Holocene period into a new geological era, often termed the Anthropocene". Founded on an *a priori* division between the things usually referred to as 'nature' and 'culture', the notion of the Anthropocene might be like a 'poisonous gift' (Latour 2014) due to the paradoxical way in which it simultaneously embraces and eschews a differentiation of matter and meaning (see also Barad 2007, 2014).

A typical divide between 'nature' and 'culture' emerges as a result of the ways in which each of these is apprehended as a separate phenomenon composed of particular properties and faculties. Our knowledge applies to things, or objects, which are empirically known to us, as those of which we can have knowledge. Perceived as familiar, a thing comes into existence as a recognisable object, with a quality for itself. Restricted epistemic insight, however, fixates the materiality lurking behind what we apprehend as 'nature', and its qualities in itself, by dichotomising the materiality of 'nature' to the construction of 'culture' (Rydstrom forthcoming; see also Guyer and Wood 1998; Heathwood 2011; Kreines 2017).

Plastic epitomises the problem captured by the notion of the Anthropocene in revoking our understandings and definitions of 'nature' and 'culture' (see also Buckingham and Masson 2017). Heaps of plastic are burgeoning on land and in the oceans. Microplastics are absorbed by plants, insects and animals while entire fragments are consumed by whales and other marine biology. By saturating our environment and pervading our diets, microplastics are eventually infused into the human body (Eriksen et al. 2014; *Newsweek*, 22 October 2018).

The imagined boundaries between artefacts, the environment, humans and other existences (animals, marine biology, insects and plants) are not only transgressed but even blurred by the immanent forces of plastic (Eriksen et al. 2014). Increasingly sophisticated knowledge about the ways in which plastic engages as a mediating power between the entities of 'nature' and 'culture' recasts established assumptions about their internal demarcations and reciprocal relations (Rydstrom forthcoming).

The ways in which plastic is imposing new insights about the 'nature' and 'culture' divide might reveal that "the very stuff of 'matter' is ubiquitous and chameleon and 'the natural order' is essentially sociological, errant, and always 'out of place', or 'out of sync' with itself" (Kirby 2017, x; see also Barad 2007, 2014; Deleuze and Guattari 2002; MacGregor 2017). The notion of the Anthropocene, we would suggest, captures these intricate alternations of previously agreed stable entities. It therefore emerges as a powerful analytical entry point to advance our knowledge about the complex ways in which climate change obscures and frustrates the internal lines as regards ideas about nature and materiality vis-à-vis culture and humanity (Rydstrom 2019; see also Geerts and van der Tuin 2016; Haraway et al. 2016; Hoffman and Oliver-Smith 2002).[3]

Gendered intersections

In her article "A Stranger Silence Still: The Need for Feminist Social Research on Climate Change", Sherilyn MacGregor (2010, 137) calls for "more case studies and more evidence that will contribute to a thorough understanding of gender differences in perceptions, impacts, and responses". When the article by MacGregor was published, studies of disasters were mainly focused on emergency issues in a pretended 'gender neutral' way. However, even more recently, as Susan Buckingham and Virginie Le Masson (2017) observe, gender perspectives tend to be circumvented in the making of global climate change policies. International sustainable development planning has been reluctant to incorporate gender into environmental protective strategies, such as the Kyoto Protocol; the Paris Agreement; Agenda 21; and the Beijing Conference. According to Buckingham and Le Masson, a more complex understanding of gender needs to be translated into macro-political-economic climate change and environmental policies in order to support the introduction of inclusive and sustainable climate strategies (see also Arora-Jonsson 2017).

As a notion and organising principle of systems and lifeworlds, the conceptualisation of gender, as introduced by Simone de Beauvoir in 1949 [1972], refers to sex as a bodily materiality upon which a socially constructed sex, called gender, is formatted. Judith Butler (1993, 2), however, rescinds the Beauvoiran sex and gender dichotomy, arguing that there is "no way to understand 'gender' as a cultural construct which is imposed upon the surface of matter, understood either as 'the body' or its given sex. Rather, once 'sex' itself is understood in its normativity, the materiality of the body will not be thinkable apart from the materialisation of that regulatory norm". If both gender and sex are the result of socio-economic and historical configurations, as Butler (1993) argues, they become indistinguishable as gender is rendered sexed and sex is rendered gendered through the ways in which corporeal properties and faculties are apprehended and instantiated in particular socio-economic and political contexts of power (Rydstrom 2002, 2003).

While sex in Butler's optic is absorbed by gender, becoming a being in terms of a woman, man or non-binary person emerges as materially conditioned, not in any essentialist senses, but in terms of being-in-the world (Merleau-Ponty 1996; see also Barad 2007, 2014; Derrida 1970; Grosz 1994; Gunnarson, Martinez Dy, and van Ingen 2018). Gender is consolidated at particular sites through intersections with other defining resources and parameters, as discussed by Kimberlé Crenshaw (1989). Like one stem of a rhizome (Deleuze and Guattari 2002), gender engages with other stems such as age, sexuality, ethnicity/race, class and bodyableness to produce images and realities of people and their powers, roles, status and abilities (Braidotti 2002; Rydstrom 2016).

As Nancy Fraser (1996, 16) notes, gender "codes pervasive cultural patterns of interpretation and evaluation, which are central to the status order as a whole". Such patterns shape and are shaped by androcentrism: a forceful power, which Fraser (1996, 16) defines as "the authoritative construction of norms that privilege traits associated with masculinity and the pervasive devaluation and disparagement of things coded as 'feminine', paradigmatically – but not only – women" (see also Nussbaum 2000). A nuanced, yet operational definition of gender useful for the development of risk reduction and coping and mitigation strategies prior to and in the wake of a climate disaster is provided by the UNDP Bureau for Crisis Prevention and Recovery (2010, 1):

> Gender determines what is expected, allowed and valued in a woman or a man in a given context. It determines opportunities, responsibilities and resources, as well as powers associated with being male and female. Gender also defines the relationships between women and men and girls and boys, as well as the relationships between women and those between men. These attributes, opportunities and relationships are socially constructed and are learned through socialization processes. They are context- and time-specific, and changeable.

In pinpointing the ways in which societal antecedents frame and restrict pre-disaster life, this definition reminds us that a catastrophe does not land in a socio-economic and political void, but rather in a place where the autonomy and rights of women and girls might be severely limited. A damaging event interlocks with gender specific inequalities which underpin ordinary times and might, in doing so, not only fuel but even exacerbate gendered susceptibility to short-term and long-term disaster-related hazards and ramifications (see Enarson and Chakrabarti 2009; Enarson and Pease 2016; Ginige, Amaratunga, and Haigh 2014; Nguyen and Rydstrom 2018). Climate disasters thus tap into gendered asymmetries which are already saturating social and political life, as described by Madhavi Malalgoda Ariyabandu (2009, 7):

> Gender-based prejudices and divisions in many societies mainly affect girls and women and as these are based on views of them as physically and emotionally weak, inferior in comparison to men and boys, dependent, subordinate and a burden to family such gender-based perceptions form the nature of interactions at personal, family and institutional levels and contribute towards the formulation of gendered attitudes leading to observations, decisions and actions within family and society, as well as the informal institutions of different kinds including the state. In crisis, these pre-established views are extended to girls and women to identify them as passive and incapacitated victims who are in need of rescue and help.

While females might be construed as if they were more emotional and even closer to nature than men, the chapters in this volume show that women and girls are not passive victims in times of disasters (see also Arora-Jonsson 2017; Ortner 1972). Yet, societal inequalities make them disproportionally susceptibility to various kinds of harm prior to, during and after a climate catastrophe meaning that "women always tend to suffer most from the impact of disasters", according to the United Nations Asian Disaster Preparedness Centre (UN/ADPC) (2010, 8).

Women's lack of access to resources and dependency upon livelihood systems sensitive to climate changes make female life precarious in connection with a climate disaster, as does limited access to media information and electronic devices; legal restrictions for women; reduced influence on decision-making at the level of society and the family; major responsibility for children and the household; and constraining cultural practices such as not teaching girls and women how to read (i.e., preventing them from accessing information), and not teaching them how to swim (i.e., jeopardising their escape from rising waters) (Bradshaw and Fordham 2013; Enarson and Chakrabarti 2009; UNFPA 2010).

Those who experience the greatest negative impacts after a climate disaster tend to be groups who were disadvantaged prior to the destructive event and gender is pivotal as regards the distribution of exposure and insecurity in the era of the Anthropocene. In the aftermath of a disaster, insufficient measures to rehabilitate and reconstruct societies deteriorate the safety of women and girls. Turmoil might break out in the wake of a disaster; family members might be separated and populations run the risk of being forced to resettle far from home. Public safety can derail after an urgent event and foster lawlessness, as seen in the aftermath of super typhoon Haiyan, which exposes women and girls to the perils of violence perpetrated by men in the public sphere, in shelters and in ruined homes (Aoláin 2011; First, First, and Houston 2017; True 2013; UNFPA 2010).

Men, masculinities and precariousness

Conflating gender with women might mean that the struggles faced by men and boys in connection with a climate disaster are overlooked (Ariyabandu 2009; Bradshaw 2013). Boys and young men are ascribed new responsibilities in the era of the Anthropocene such as being the saviours of crucial livelihood properties (e.g., a family's tractor, boat, ox, carriage and tools) when a disaster strikes (Enarson and Pease 2016; Mudavanhu et al. 2015; Rydstrom 2019). Hence, some international aid organisations have voiced concern about the ways in which boys are rendered vulnerable in emergency situations due to stereotypically defined male roles which in general value strength and even more so in times of danger when males are expected to take a protective role.[4]

Focusing on men's experience of the horrifying tsunami which slammed into Sri Lanka in 2004, Malathi de Alwis (2016) describes how the disaster was devastating for everyone, regardless of gender, and how the suffering of men was rendered invisible in the aftermath. International aid projects were reluctant to engage with local men due to narratives about a pervasive type of masculinity defined by alcohol consumption and wife beating. Critically scrutinising predominant assumptions about men and masculinities circulated in and by the international aid community, De Alwis (2016) shares testimonies from the field which highlight male vulnerabilities that run counter to influential images projected of men in post-tsunami Sri Lanka.

The 'ethnographic moment' of critical studies of men and masculinities has elucidated the heterogeneity of men in regard to power, vulnerability and pain (Enarson and Pease 2016). The roles and precariousness of men and boys are important to take into account for gaining nuanced insights about the dynamics between genders and disasters (Rydstrom 2019). While one particular type of maleness might dominate as 'hegemonic masculinity', to use the notion coined by R. W. Connell (1995), the 'ethnographic moment' in critical studies of men and masculinities has provided in-depth studies and detailed accounts of the heterogeneity of men, maleness and masculinity in the Global North and Global South (e.g., Horton and Rydstrom 2011; Jensen 2015; Ruspini et al. 2011).

"The double complexity that men are both a social category formed by the gender system and collective and individual agents, often dominant agents, of social practices" (Hearn, Blagojević, and Harrison 2013, 6) remind us of the ways in which a sociological axiom allows for individual men and men as a group to gain and hold gender specific power and privileges. Dominance is demonstrated and rendered intelligible as a gender-intersected instantiation within a context-specific societal fabric of power relations (Connell and Messerschmidt 2005). Ethnographic studies, however, draw our attention to the differences amongst men, as these are informed by age, sexuality, ethnicity, class and bodyableness and how men cannot be clustered into one and the same group despite holding large-scale androcentric privileges.

Men's violence against women and girls

A gender-intersected life framed by economic and social difficulties is likely to be wrought into an even more challenging state in catastrophic times. On a world scale, 35 percent of women have experienced either physical and/or sexual intimate partner violence (IPV) or non-partner sexual violence (i.e., 7% of women of the 35%). Thirty percent of women who have experienced violence are abused by their partner and 38 percent of all murders of women are committed by intimate partners (WHO 2017). As an underpinning risk and/or reality of daily life, men's abuse of women and girls is rendered mundane as a 'banal' condition of female life (Arendt 1970; see also Hearn 2004). Men's violence against women and girls, we approach as a means for the "struggle for the maintenance of certain fantasies of identity and power" (Moore 1994, 70), as a tool used by men to demarcate a space of impunity in which a battered woman is treated as if she were nothing but naked corporeality (Rydstrom 2012, 2015, 2017), or 'bare life' (Agamben 1998).

Studies find that gender-based violence, including sexual, tends to increase in the aftermath of a climate disaster (First, First, and Houston 2017; International Federation of Red Cross and Red Crescent Societies 2015).[5] Violence against women and girls who have just survived a disruptive event such as a typhoon or an earthquake is particularly gruesome in taking advantage of those who are traumatised and exceptionally vulnerable. Violence causes harm and pain at a personal level and can even prevent survivors from engaging in the rebuilding of their life, family and society (Aoláin 2011; First, First, and Houston 2017; Fordham et al. 2013). Women's experiences of being subjected to male abuse tend to be met by silence in ordinary pre-disaster times and even so in the aftermath of a catastrophe. Violence against the female population in connection with a disaster is not uncommonly reduced to a secondary problem when 'first things first' protocols are implemented to save lives (Bradshaw 2013; Ginige, Amaratunga, and Haigh 2014; True 2012, 2013).

Violence manifests itself as direct physical violence but also takes more abstract forms. Nguyen and Rydstrom (2018) conceptualise gendered inequalities justified by the nation-state and the state apparatus as systemic harm, or 'structural violence' (Farmer 2004; Galtung 1969). Men's physical abuse of women in the home or in the public sphere is endorsed by larger socio-political and economic structures, including insufficient legal mechanisms to enforce laws to protect females from men's violence. Systemic violence, in terms of an androcentric politics of inequity and misrecognition, Nguyen and Rydstrom (2018) show, is dialectically related to men's direct physical violence against women and girls on the ground and is due to the fostering of a political discourse of permissiveness and a culture of impunity.

A surge in physical violence inflicted upon women by known – and unknown – men in disastrous times, is explained in various studies with references to a change in the dynamics between partners in extreme situations; aggressive solutions implemented in times of crisis; stress due to the destruction of homes and livelihoods; frustration due to unemployment and lack of income; prolonged waiting in shelters; insufficient or impaired support systems; and trauma (see Bradshaw 2013; Denton 2002; Ginige, Amaratunga, and Haigh 2014; Fordham et al. 2013; True 2013). Yet, men's violence against women needs to be put in its right context by connecting the dots between pre-disaster violence, violence during a disaster and post-disaster violence to understand how gendered harm crosses time and space. In catastrophic times, men's violence against women capitalises upon already existing androcentric gender and power asymmetries, which foster and are fostered by 'structural violence'. These dynamics of violence might be fortified by disaster specific factors that cause stress and the perpetration of violence either by first time or habitual abusers (Nguyen and Rydstrom 2018; Rydstrom 2019).

Risk, vulnerability and (in)security

A risk and hazard paradigm in disaster studies, as Kenneth Hewitt (1995) argues, tends to distract from pre-disaster injustice and misrecognition (see also Valdés 2009). Mitigation strategies and the development of disaster reduction techniques chiefly focus on what is thought to be the 'natural' component of a disaster and therefore tend to overlook the ways in which the living conditions of impoverished groups, their autonomy and legal protection were compromised before a crisis of emergency (Momtaz and Asaduzzaman 2019; Wisner, Gaillard, and Kelman 2012).

Collapsing houses, soil erosion and inland flooding are examples of climate-related hazards which inevitably raise concerns about the level of security and the risks to which people are exposed (Ginige, Amaratunga, and Haigh 2014; Hamza 2015; Kelman, Mercer, and Gaillard 2017). Risk studies (e.g., Beck 1992; Giddens 2004) have contributed to the development of policies designed to protect populations but have not however been particularly gender sensitive (MacGregor 2017; Kinnvall and Mitzen 2017).

A sufficiently supple analytical language is needed to unravel the risks, insecurities and precariousness with which various groups must cope to deal with the havoc erupting in the path of a climate disaster (Kulick and Rydstrom 2012). Disaster Risk Reduction (DRR) designed by international aid organisations need to mainstream gender to identify how survival and recuperation take shape in regard to women, men, girls, boys and non-binary persons as well as people of different ethnicities/races, sexualities, classes, ages and bodyableness (Bradshaw 2013; Cruz-Torres and McElwee 2012; Enarson and Chakrabarti 2009; Fordham et al. 2013).

Jeopardised gendered security in ordinary times, during a disaster and when a society is recovering from a catastrophe indicate that security issues cannot be captured by a traditional state-centric model (Kent 2006; Kinnvall and Mitzen 2017; Tripp, Ferree, and Ewig 2013). Social and human security matters go beyond an international politics level, as traditionally associated with securitisation (Bilgin

2010; Buzan and Hansen 2009; Kinnvall 2004, 2013), risk (Douglas 1991; Enloe 2016), vulnerability (Butler, Gambetti, and Sabsay 2016), resilience (Wisner, Gaillard, and Kelman 2012), rights (Bradshaw 2013) and empowerment (Kabeer 2005; Sylvester 2010).

The identification of risk and the study of vulnerability overlap in the sense that increased hazards foster new types of gendered vulnerability (Kulick and Rydstrom 2012). We approach vulnerability in connection with a climate disaster as a matter of exposure to ecological and socio-economic and political anteced-ents and the specific ways in which these interlock with disaster hazards and rami-fications (Cutter 1996; Fordham et al. 2013; Momtaz and Asaduzzaman 2019). In this spirit, Ben Wisner and colleagues (2003, 11; italics removed) define vulner-ability as follows:

> [Vulnerability refers to] the characteristics of a person or group and their situ-ation that influence their capacity to anticipate, cope with, resist and recover from the impact of a natural hazard (an extreme natural event or process). It involves a combination of factors that determine the degree to which some-one's life, livelihood, property and other assets are put at risk by a discrete and identifiable event (or series or 'cascade' of such events) in nature and in society.

Recognising that some groups are more prone to suffer from the damaging forces of a climate disaster than others means to identify the unequal level of disaster vulnerability distributed amongst and endured by particular groups. In disaster studies, vulnerability has been used as an avenue to estimate the extent to which ecologies, livelihoods and populations are exposed to risk, hazards, insecurity and stress as well as the adaptive capacity, sensitivity and resilience of places and peo-ple (Aoláin 2011; Aradau and van Munster 2011; Kinnvall 2006; Yamin, Rahman, and Huq 2005). Such parameters have been applied to measure the susceptibility to damage in a specific place and particular groups' vulnerability to various kinds of harm. A vulnerability perspective thus helps to identify exposure of people and places (Adger 2006; Fordham et al. 2013; Momtaz and Asaduzzaman 2019; Wisner 2016).

Maureen Fordham (1999) relates complex and different degrees of vulnerabil-ity in the aftermath of a disaster with larger visions of social justice for all (see Fraser 1996). Vulnerability and social justice, Fordham (1999) argues, encourage a mapping of what Hewitt (1997, 164) defines as 'geographies of vulnerability' to unfold the differentiation of vulnerability and how it hampers social equality prior to and after a climate disaster (see also Momtaz and Asaduzzaman 2019). Wisner and colleagues (2003) suggest that a number of simple questions should be asked to identify differentiated climate disaster vulnerability. Paraphrased, these include questions such as "to which kinds of hazards which kind of people are vulnera-ble"; "how particular groups are able to cope with and resist a disaster"; and "who is able to recover in which ways from which kinds of disastrous ramifications" (see also Enarson, Fothergill, and Peek 2007).

Resilience and agency

An emphasis on vulnerability identifies exposure, while a resilience perspective sheds light on local agency and the extraordinary human capabilities mustered by survivors in situations of danger and urgency as "the social capacity to absorb and recover from the occurrence of a hazardous event" (Smith 1992, 25 cited in Wisner et al. 2003). According to Sarah Bradshaw (2013), an explicit focus on vulnerability could imply assumptions about an inability to manage life under devastating circumstances while a resilience approach brings to the fore that survivors can reorganise life after a disaster (Gamburd 2013; Kelman, Mercer, and Gaillard 2017; Rigg et al. 2008; Rydstrom 2012). Thus, while the disastrous realities of the Anthropocene are encompassing, the era also conditions agency, resistance, empowerment and hope (IPCC 2018; Scott 1992; Spivak 1988). Resilience is a goal which pervades various frameworks such as the Millennium Development Goals and the Hyogo Framework for Action (HFA) 2005–2015 International Strategy for Disaster Reduction (ISDR 2007) as an agenda aimed at "systematically linking and integrating risk reduction and crises management" (FAO 2019). A resilience agenda, though, is underpinned by politics and hidden agendas, as explained by Pauline Eadie (2017):

> Survivors may state that they are resilient when they are not (for fear of being seen as weak) and accept material resources that they have no need for (for fear of appearing ungrateful or compromising access to future goods and services). Government and non-governmental agencies may use the term resilience to congratulate themselves on a job well done without due consideration of what it actually means to be resilient.

Vigilance to what Eadie (2017) labels the 'tyranny of resilience' is significant to critically examine how the idea of resilience plays into the politics of governments and aid organisations. The work of various agencies and organisations is not *per se* guided by noble ideals and objectives concerned with gendered equality, recognition and justice for all regardless of whether 'resilience' is promoted as a guiding principle for recovery after a disaster (cf. Carothers 1999/2000; Howell 2007; Kaldor 2003; Kinnvall and Mitzen 2017). The rebuilding after a profoundly dangerous and destructive climate disaster, such as super typhoon Haiyan, is a path fraught with difficulties and challenges (Nguyen 2018; ILO 2018). In a University of Nottingham Asia Research Institute study on the recovery of the Philippines after Haiyan, Eadie (2017) thus observes:

> [T]he term resilience was being thrown around somewhat carelessly in the media and policy circles. It was particularly worrying that negative coping strategies can be equated to positive connotations of resilience. Reducing food intake, living in inadequate housing without adequate sanitation, the increase of excessive working hours are not positive adaption strategies or real indicators of 'strength'.

The potential for local and international associations, agencies and organisations to negotiate the divisions of power and responsibilities between citizens and the nation-state through the possible expansion of gendered rights and equality thus needs to be identified by research as well as mitigation and coping strategies prior to a catastrophic event. When society and people suffer from the serious hazards and dire ramifications of a climate catastrophe, strategic means to support and empower survivors like Sustento, introduced in the opening of this chapter, are critical to develop and implement on the ground (Bradshaw 2013; Fraser 1996).

In this volume, scholars of anthropology, engineering, gender studies, geography, human rights and political science draw on quantitative as well as in-depth ethnographic findings to carefully examine how climate hazards, disasters and ramifications are intrinsically gendered. They explore how these forces act as conditions that frame lifeworlds, livelihoods and politics in Bangladesh, Cambodia, India, Kenya, the Maldives, Nepal, the Pacific Islands, Pakistan, the Philippines, Sri Lanka, Thailand, Vanuatu, Vietnam and the USA. In doing so, the volume provides critical insights into an urgent problem of relevance for scholars, students, local and international aid organisations, and for agencies dedicated to preventing and combatting gender-specific hazards and ramifications brought about by a climate disaster in the era of the Anthropocene.

Part I: aid, frameworks and policies

In Part I, climate disaster policies and frameworks are considered from a gender and masculinity perspective to critically examine how these correspond with various climate disaster coping and mitigation strategies both in the Global North and the Global South.

In Chapter 2, Maria Tanyag and Jacqui True map out the key elements for developing a framework that can strengthen women's voice and leadership in responding to climate change and related crises. Drawing from a collaborative research project with the global non-governmental organisation Action Aid, the chapter demonstrates how an integrated approach to gender-responsive climate change mitigation and adaptation begins from women's standpoints. These standpoints provide a bridge between the current fragmentation existing across humanitarian, security and development agendas.

Tanyag and True propose three elements to a gender-responsive framework. First, the recognition of women's and men's localised knowledges on 'early warning' indicators at household and community/village levels, and awareness of policies at national, regional and global levels. Second, women's direct representation at different levels of governance, especially in climate-related leadership positions from the community, national agencies/decision-making bodies and global climate negotiations. Third, institutional and strategic pathways for linking these localised knowledges and gender-inclusive political representations at all levels of governance. Bringing these elements together yields a more holistic understanding of climate change as it intersects with other forms of crisis, often causing or compounding them. Tanyag and True show that gender-responsive climate change solutions are those that encompass comprehensive security and varied forms of leadership, especially among the most marginalised groups of women and girls.

A focus on gender contingencies in relation to climate disaster risk reduction frameworks and the power with which these are imbued is also explored in Chapter 3. Sara Bondesson's chapter critically analyses the underlying politics of Disaster Risk Reduction (DRR) strategies to reveal whether a DRR rationale stimulates or hampers the incorporation of a gender perspective. More specifically, she examines the Sendai Framework for action, the central policy document in the global field of DRR policies aimed at reducing vulnerabilities and disaster risks by building resilient communities and mitigating negative consequences of climate disasters. Within this framework, a gender perspective tends to be side-stepped and Bondesson thus considers whether it is possible to mainstream gender into DRR strategies to reach the goal of gender equality.

Bondesson argues that a sufficient gender sensitive approach to disaster vulnerability should be developed in order for DRR strategies to consider how social and political inequalities influence women, men, boys and girls differently. Included here are their different capacities to withstand a disaster; the ways in which they are affected by a catastrophe; the extent to which they can recover from a disastrous event; and how they can prepare themselves and their families for future emergency incidents. DRR policies and strategies, Bondesson concludes, need to address gender contingencies and dynamics imbued in climate disasters to adjust to the needs of specific groups.

Seeking more comprehensive gender security strategies, which include all women while also acknowledging contextual specifics, is also at the fore of Chapter 4. Here, Wester and Lama argue that if women are to be empowered, they need to be included in all aspects of life prior to a disaster and when a damaged society is being rebuilt. The negative effects of climate change manifested as flooding, droughts, heatwaves and land erosion are, they maintain, experienced around the globe as abrupt changes to everyday life and livelihoods. The consequences of the Anthropocene and the ways in which the Global North is impacted and can handle climate challenges compared to the Global South are skewed due to unequal distributions of economic resources which favour the Global North. However, as Wester and Lama show, the climate change coping strategies employed by women in the Global North are not necessarily more efficient or adequate compared with those implemented in the Global South.

In environmental projects, women are often recognised as champions. Women in the Global South, Wester and Lama say, are particularly targeted in empowerment projects aimed at increasing preparedness to cope with climate hazards. However, even if discussions on climate change attempt to include gender, efforts rarely challenge the status quo, but rather reinforce stereotypical perceptions of gender or result in even greater vulnerability. This is also the reason, Wester and Lama emphasise, why empowerment is always dependent on inclusion in all phases of the pre- and post-disaster process.

Similar to Wester and Lama, the authors of Chapter 5, Paul Pulé and Martin Hultman, draw our attention to climate disaster-related developments in the

Global North. In their chapter, they consider how the Paris Agreement (2015) has been seen as a landmark for how to deal with global climate change. Despite the consensus reached by the vast majority of researchers about the peril that is upon us, achieving and actioning mitigating responses to anthropogenic climate change have been far from the smooth start to a new beginning that was the great promise of the Paris Agreement. Rather, a number of countries with heavy investments in fossil fuels, including Saudi Arabia, Kuwait, Russia and the United States (with tacit support from Australia), have recently combined forces to obstruct international proceedings that are designed to commit nations to strict carbon pollution controls.

Such obstructions support corporate attempts to expand and extend fossil-fuel development, spurred on by lobby groups and researchers who identify as unapologetic and vocal climate change deniers, their cumulative efforts assuring humanity's collision course with these apocalyptic scale concerns. Pulé and Hultman discuss how hyper-masculinities are at the heart of this pressing climate emergency. Accordingly, they reflect on an industrial/breadwinner typology and in doing so expose an alliance between the masculine identities of industrial elites and working-class workers at the 'coal face' of industrial productivity and corporatisation throughout the Global North; a typology of gender identity whose impulses are being met by tepid government regulations seeking systemic compromise and reform at best. The authors conclude the chapter by discussing how climate change denialism epitomises white male effect, providing a portent example of the intersections between race, power and resource inequities that have reasserted white men's primacy precisely because 'malestream' norms persist and shape some men's values and actions in overtly uncaring directions.

Part II: mitigation and coping strategies

Moving from overarching discussions of climate disaster, gender and masculinity and how these inform climate policies, in Part II we direct our attention towards particular climate disaster coping strategies and the implementation of projects aimed at reliving human suffering caused by cataclysmic events such as earthquakes and tsunamis.

Pacific Islanders are at the 'front line' of global climate change. Nicole George examines in Chapter 6 the vulnerability of Pacific Islanders to a global future of more frequent extreme weather events, sea-level rise and other forms of ecological damage. To uncover the gendered politics of climate change in the Pacific Islands, George develops the concept of an 'architecture of entitlement' to analyse where and how women feature in these accounts. In line with this architecture, she shows that women are politically rewarded when their responses to climate change related phenomena emphasise gendered tropes of maternal nurturing that link care for dependents with care for the environment and survival of indigenous culture more broadly.

These tropes, she argues, sit in interesting tension with the Orientalising and eroticising gendered representations of Pacific peoples that were a common aspect of the colonial encounter. Conversely, this same architecture of entitlement tends to frustrate the ambitions of women who respond to climate change-related insecurity in ways that push beyond representations of women as bearers of culture and/or maternal duty. George shows that while Pacific Island communities are generally keen to demonstrate and promote their capacities for adaptation and resilience to the future climactic challenges that the region will face, gendered ideas of entitlement are strongly institutionalised and exert a powerful influence on where and how women might contribute to that future.

Examining a number of cases in South Asia in Chapter 7, Emmanuel Raju demonstrates that women are susceptible to the hazards and ramifications of climate disasters in gender specific ways. The chapter draws on the vulnerability paradigm in the context of disasters. Here disasters are viewed as root causes of vulnerability and connected to various processes and settings, such as development failures, larger structural agendas and patriarchal structural contexts. Disaster narratives from different communities in Bangladesh, India and Pakistan show the importance of engaging with vulnerability and the development of capacity assessment methods to overcome the challenges these may encounter in particular areas.

South Asia, Raju argues, has not only experienced major disasters but also been forced to live through and learn from disasters, yet the meaning of gender and power has not been brought to the forefront in post-disaster strategies. While gender sensitive methods are important to include when working with disaster-affected areas, engaging with the very notion of gender, Raju shows, is a process fraught with difficulties due to the potential critique of established structures imbued in a gender perspective. The difficulties of applying a gender perspective in 'normal' times are augmented in an emergency situation when gender specific needs and nuances take new shapes.

Katie Oven, Jonathan Rigg, Shubheksha Rana, Arya Gautam and Toran Singh similarly argue in Chapter 8 that gender is highly relevant to disaster risk reduction given the disproportionate number of women and girls who are impacted negatively by disasters. Similar to Bondesson, Oven and her colleagues proceed from the Sendai Framework for Disaster Risk Reduction to explore how the participation of women is critical to effective disaster risk management. Here they argue that community-based approaches to Disaster Risk Reduction (CBDRR) have become a focus of significant development funding in countries in the Global South and how such approaches reflect a shift towards more bottom-up planning and community empowerment.

In their chapter, Oven and her co-authors draw on research undertaken across 24 case study wards as part of a review of CBDRR policy and practice in Nepal to show how women shoulder a good deal of responsibility for CBDRR. For some female participants, their involvement had the desired effect: they felt empowered and better able to respond to disasters. For others, it was an additional burden. Such interventions, however, were concerned largely with disaster preparedness: first aid and search and rescue training, the construction of raised grain stores and water pumps to protect crops and drinking water from flooding, and the establishment of early-warning systems. The CBDRR projects did little to address the root causes of disasters that are often bound up in caste and gender relations with implications for citizenship and land rights. The authors conclude that gendered reflections on CBDRR must pay attention to the ways in which historical and inherited patterns of gendered relations and processes of marginalisation intersect with contemporary transformations in economy and society.

Gender disparities experienced in the wake of mass-scale disasters as death, suffering and displacement are also at the fore of Claudia Merli's chapter, Chapter 9. In the chapter, Merli emphasises how gender disparity has become an important focus in disaster research across disciplinary boundaries and how research on gender and disaster can contribute important knowledge to effective planning and to international emergency intervention and risk management. While gender features prominently as a category of analysis with regards to violence, trafficking and exploitation of primarily children and women in times of crisis, the bodies and embodied experiences of women and men in disasters remain opaque if not completely invisible.

Merli examines the embodied experiences of women and men during and in the aftermath of the 2004 Indian Ocean tsunami. By drawing on ethnographic material collected in Thailand's southernmost western province, she analyses how people have lived the disaster through their bodies as well as the bodies of others. Merli sheds new light on the ways in which tourists' bodies were discussed and perceived in certain interpretations that emerged in southern Thailand and how these imputed the disaster to specific behaviours in tourist areas. This also had implications in terms of how local inhabitants experienced changes in their own bodies after the tsunami and how well-intended aid interventions ignored such embodied realities.

Part III: contextual realities and experiences

In Part III, intersections between climate disasters, gender, masculinity and social life are central for analysing how insecurity, crises and ruination fostered in connection with climate disasters such as storms, landslides and flooding inflict differing ramifications upon various groups in particular contexts, one being men's violence against women and girls.

In Chapter 10, Sidsel Hansson and Catarina Kinnvall explore the impacts of climate disasters and reconstruction efforts on gender relations and gender-based violence in the context of Pakistan. The chapter focuses on situations in which reconstruction efforts, including displacement (where many are forced to live in makeshift tents and other 'camp-like' housing areas), have become recurrent and normalised due to recurring disasters in Pakistan. The authors highlight how such conditions shape and construct gender relations and gendered violence in relation to ongoing crises and traumas of reconstructing livelihoods (often requiring short-term migration and/or going into debt) in the face of armed conflict.

The ways in which particular norms of masculinity are strengthened in areas struck by disaster and how these multiply through religious organisations and militaries involved in disaster relief are often at the heart of these experiences, influencing the situation and the agency of women and girls, and making their lives particularly difficult. An influential factor, Hansson and Kinnvall argue, is the emergence of a particular tripartite form of security-seeking masculinity shaped by the co-existence of men's crisis; the consistent crises of religion, nationhood and the state; and the perpetuation of multiple large and small-scale crises in the country.

In Chapter 11, Helle Rydstrom explores the conceptualisation of crisis, arguing that a crisis perspective provides a useful analytical entry point to examinations of climate disasters as inherently gendered and masculinised. A sudden incident like a typhoon is ruining as an immediate crisis, yet it also interlocks with gender-specific socio-economic crises conditions which framed life prior to the disaster. Drawing on ethnographic data from central coastal Vietnam, Rydstrom highlights why disasters in terms of recurrent storms only can be seen as a bracketing of daily life for those with resources.

Elucidating the interactions between a disastrous crisis of emergency and a spectrum of crises antecedents such as gendered and masculinised livelihoods, patriarchal kinship hierarchies, male privileges and male-to-female violence, Rydstrom identifies various crisis temporalities, modalities and intensities in the Vietnamese context. A crisis of emergency caused by a storm might, for particularly precarious groups, morph into a new normalcy; into a crisis that takes shape as a chronicity which is encroaching ruination and slow harm upon lifeworlds, livelihoods and environments and, in doing so, impeding human recovery and the rebuilding of social worlds.

In Chapter 12, Huong Thu Nguyen applies a ruination perspective to her analysis and thus examines the notion's various dimensions. She conducts an ethnographic study of the lived experiences of women survivors directly affected by Haiyan in the provinces of Leyte and Eastern Samar of the Philippines. Nguyen highlights the human suffering, great losses and frustrations with what survivors saw as a failure of the authorities to provide adequate and timely relief aid and rehabilitation support. Nguyen also examines how relief and rehabilitation efforts were beset by conflicting interests among international and national stakeholders and by the dynamics of metropolitan and provincial politics set against the background of pre-existing armed conflict, all of which combine to render these women into, figuratively and literally, a state of *ruination*. By examining the webs of power which conditions female insecurity and precariousness, Nguyen argues that the concept of ruination seems to facilitate resistance, particularly in the form of collective efforts to overcome the condition of *being ruined*.

The chapter draws on a number of examples to show how local residents were able to manage their shattered lives with the resources available to overcome immediate hardship, thus indicating that survivors of a disaster do not collapse into passivity and victimhood in a wider sense. The possibilities for agency also carry a positive message to the effect that cultural, social and institutional factors that drive gendered injustice against women are not inevitable and can be transformed, making this a key moment to implement the intervention. Furthermore, this conceptualisation of agency as a possibility for action seems to be in line with the participatory community-based approach in many initiatives aimed at promoting actual gender equality in the contexts of the Philippines and prevent men's violence against the female population before, during and after a climate disaster.

Taking a human rights-based approach, Matthew Scott, in Chapter 13, examines the role that legal actors can play in promoting accountability for incidences of sexual assaults in situations of disaster displacement, particularly in evacuation centres and temporary centres. Drawing on the evolving principle of due diligence obligations of states to prevent sexual assault generally, his chapter argues that justice-sector actors can contribute to accountability for and a reduction in sexual assault in evacuation centres and temporary shelters by advancing legal arguments through litigation in domestic and regional courts, as well as in wider advocacy initiatives.

Scott considers how lawyers, legal assistance organisations and national human rights institutions might be able to contribute to the protection of human rights in times of climate disasters and disaster-related displacement by providing knowledge about, first, legal action aimed at preventing and combatting gender-based violence in evacuation shelters, and second, legal action aimed at addressing inadequate provision of the care for survivors of gender-based violence. The chapter concludes with the recognition that the articulation of relevant legal arguments concerning state responsibility for the prevention of sexual assault in evacuation centres and temporary shelters may have more or less leverage depending on the particular context in which it is invoked, and that a reduction in sexual assault in evacuation centres and temporary shelters requires local actors and vernacular strategies.

Finally, the concluding chapter, Chapter 14, brings together and critically analyses the patterns that have emerged in the amalgam of theoretical approaches and empirical cases throughout the book.

Notes

1 Disasters of a certain level of intensity are recognised as hazards, as dangerous conditions due to which lives are lost, people are injured and their health jeopardised. Hazards also mean that property is damaged, livelihoods are hampered or lost and societies are ruptured socio-economically (Bradshaw 2013; UNISDR 2009).
2 Claudia Aradau and Res van Munster (2011) contrast disaster, emergency and risk to a catastrophe. A catastrophe, they argue, refers to first, the induction of a sense of the limit and second, a radical overturn. This definition, we would contest, because of the ways in which a disaster ravages and inevitably provokes feelings that the world might be coming to an end, which for some sadly is the case. Thus, we do not distinguish between catastrophe and disaster.
3 See Maria Mies et al. (1988) and Vandana Shiva (1991) and the gender unique qualities they associate with women who interact with nature and environment.
4 Helle Rydstrom: personal communication with representatives of the UN Women, UNDP and WHO, amongst others.
5 Following the UN Declaration (1993) on the Elimination of Violence against Women, we use the term 'violence against women' as referring to "any act of gender-based violence that results in, or is likely to result in, physical, sexual or psychological harm or suffering to women, including threats of such acts, coercion or arbitrary deprivation

of liberty, whether occurring in public or in private life". The term 'intimate partner violence' (IPV) refers to "behaviour by an intimate partner or ex-partner that causes physical, sexual or psychological harm, including physical aggression, sexual coercion, psychological abuse and controlling behaviours" (WHO 2017).

References

Adger, Neil W. 2006. "Vulnerability." *Global Environmental Change* 16: 28–281.

Agamben, Giorgio. 1998. *Homo Sacer: Sovereign Power and Bare Life*. Stanford: Stanford University Press.

Alwis, Malathi de. 2016. "The Tsunami's Wake: Mourning and Masculinity in Eastern Sri Lanka." In *Men, Masculinities, and Disaster*, edited by Elaine Enarson and Bob Pease, 92–103. London and New York: Routledge.

Aoláin, Fiona. 2011. "Women, Vulnerability, and Humanitarian Emergencies." *Michigan Journal of Gender & Law* 18, no. 1: 1–23.

Aradau, Claudia, and Rens van Munster. 2011. *Politics of Catastrophe: Genealogies of the Unknown*. London and New York: Routledge.

Arendt, Hannah. 1970. *On Violence*. New York: Harcourt Brace and Company.

Ariyabandu, Madhavi Malalgoda. 2009. "Sex, Gender, and Gender Relations in Disasters." In *Women, Gender, and Disaster: Global Issues and Initiatives*, edited by Elaine Enarson and P. G. Dhar Chakrabarti, 5–18. Los Angeles: Sage.

Arora-Jonsson, Seema. 2017. "Gender and Environmental Policy." In *Routledge Handbook of Gender and Environment*, edited by Sherilyn MacGregor, 289–304. London and New York: Routledge.

Barad, Karen. 2007. *Meeting the Universe Halfway: Quantum Physics and the Entanglement of Matter and Meaning*. Durham: Duke University Press.

Barad, Karen. 2014. "Diffracting Diffraction: Cutting Together-Apart." *Parallax* 20, no. 3: 168–87.

Beck, Ulrich. 1992. *Risk Society: Towards a New Modernity*. Berkeley: California University Press.

Bilgin, Pinar. 2010. "The 'Western-Centrism' of Security Studies." *Security Dialogue* 41: 615.

Bradshaw, Sarah. 2013. *Gender, Development and Disasters*. Cheltenham and Northhampton: Edward Elgar.

Bradshaw, Sarah and Maureen Fordham. 2013. *Women and Girls in Disasters. Project Report*. UK: Department for International Development.

Braidotti, Rosi. 2002. *Metamorphoses: Towards a Materialist Theory of Becoming*. Cambridge: Polity Press.

Buckingham, Susan, and Virginie Masson. 2017. *Understanding Climate Change Through Gender Relations*. London and New York: Routledge.

Butler, Judith. 1993. *Bodies that Matter*. London and New York: Routledge.

Butler, Judith, Zeynep Gambetti, and Leticia Sabsay. 2016. *Vulnerability in Resistance*. Durham: Duke University Press.

Buzan, Barry, and Lene Hansen. 2009. *The Evolution of International Security Studies*. Cambridge: Cambridge University Press.

Carothers, Thomas. 1999/2000. "Civil Society", *Foreign Policy*, Winter 1999–2000: 18–29.

Connell, Robert W. 1995. *Masculinities*. Cambridge: Polity.

Connell, Robert W., and James Messerschmidt. 2005. "Hegemonic Masculinity: Rethinking the Concept." *Gender & Society* 19: 829–59.

Cowton, Tom R., Andrew J. Sole, Peter W. Nienow, Donald A. Slater, and Poul Christoffersen. 2018. "Linear Response of East Greenland's Tidewater Glaciers to Ocean/Atmosphere Warming." *Proceedings of the National Academy of Sciences* 115, no. 31: 7907–12; published ahead of print July 16, 2018 https://doi.org/10.1073/pnas.1801769115.

Crenshaw, Kimberlé. 1989. "Demarginalizing the Intersection of Race and Sex: A Black Feminist Critique of Antidiscrimination Doctrine, Feminist Theory and Antiracist Politics." *University of Chicago Legal Forum* Issue 1, Article 8.

Crutzen, Paul. 2006. "The Anthropocene." In *Earth System Science in the Anthropocene*, edited by Eckart Ehlers and Thomas Krafft, 13–18. Berlin and Heidelberg: Springer.

Crutzen, Paul, and Eugene F. Stoermer. 2000. "The 'Anthropocene'." *Global Change Newsletter* 41: 1–17.

Cruz-Torres, Maria Luz, and Pamela McElwee. 2012. *Gender and Sustainability: Lessons from Asia and Latin America*. Tucson: University of Arizona Press.

Cutter, Susan L. 1996. "Vulnerability to Environmental Hazards." *Progress in Human Geography* 20: 529–39.

De Alwis, Malathi. 2016. "The Tsunami Wake: Mourning and Masculinity in Eastern Sri Lanka." In *Men Masculinities and Disaster*, edited by Elaine Enarson and Bob Peace, 92–103. London and New York: Routledge.

de Beauvoir, Simone. 1972 [1949]. *The Second Sex*. Harmondsworth: Penguin Books.

Deleuze, Gilles, and Félix Guattari. 2002. *A Thousand Plateaus: Capitalism and Schizophrenia*. London and New York: Continuum.

Denton, Fatme. 2002. "Climate Change Vulnerability, Impacts, and Adaptation: Why Does Gender Matter?" *Gender & Development* 10, no. 2: 10–20.

Derrida, Jacques. 1970. *Om Grammatologi*. Copenhagen: Arena.

De Schutter, Olivier, and Emile Frison. 2017. "Modern Agriculture Cultivates Climate Change – We Must Nurture Biodiversity." *The Guardian*, January 9, 2017. www.theguardian.com/global-development/2017/jan/09/modern-agriculture-cultivates-climate-change-nurture-biodiversity-olivier-de-schutter-emile-frison.

Douglas, Mary. 1991. *Purity and Danger: An Analysis of the Concepts of Pollution and Taboo*. London and New York: Routledge.

Eadie, Pauline. 2017. Typhoon Yolanda and the 'Tyranny' of Resilience. Asia Dialogue: The Online Magazine of the University of Nottingham Asia Research Institute. http://theasiadialogue.com/2017/10/30/typhoon-yolanda-and-the-tyranny-of-resilience/.

The Ecologist. 2017. "Typhoon Haiyan Four Years Later: 'Everything was Taken Away by that Storm'." *Ecotopical*, November 9. https://ecotopical.com/the-ecologist/200205/typhoon-haiyan-four-years-later-everything-was-taken-away-by-that-storm/.

Enarson, Elaine, and P. G. Dhar Chakrabarti. 2009. *Women, Gender and Disaster*. New York: Springer.

Enarson, Elaine, and Bob Pease. 2016. "The Gendered Terrain of Disaster: Thinking About Men and Masculinities." In *Men Masculinities and Disaster*, edited by Elaine Enarson and Bob Peace, London and New York: Routledge.

Enarson, Elaine, Alice Fothergill, and Lori Peek. 2007. "Gender and Disaster: Foundations and Directions." In *Handbook of Disaster Research*, edited by Havidan Rodriguez, Enrico L. Quarantelli, and Russell Dynes, 130–46. New York: Springer.

Enloe, Cynthia. 2016. *Globalization and Militarism: Feminists Make the Link*. New York: Rowman & Littlefield Publishers.

Eriksen, Marcus, Laurent C. M. Lebreton, Henry S. Carson, Martin Thiel, Charles J. Moore, Jose C. Borerro, Francois Galgani, Peter G. Ryan, and Julia Reisser. 2014. "Plastic Pollution in the World's Oceans: More Than 5 Trillion Plastic Pieces Weighing over

250,000 Tons Afloat at Sea". *PLoS ONE* 9, no. 12: e111913. https://doi.org/10.1371/journal.pone.0111913.

FAO (Food and Agriculture Organization of the United Nations). 2019. "Sustainable Development Goals." www.fao.org/sustainable-development-goals/overview/fao-and-the-post-2015-development-agenda/resilience/en/.

Farmer, Paul. 2004. *Pathologies of Power*. Berkeley: California University Press.

Ferguson, James. 1990. *The Anti-Politics Machine: "Development", Depoliticization and Bureaucratic Power in Leshoto*. Cambridge: Cambridge University Press.

First, Jennifer M., Nathan L. First, and J. Brian Houston. 2017. "Intimate Partner Violence and Disasters." *Journal of Women and Social Work* 32, no. 3: 390–403.

Fordham, Maureen. 1999. "The Intersection of Gender and Social Class in Disaster: Balancing Resilience and Vulnerability." *International Journal of Mass Emergencies and Disasters* 17, no. 1: 15–36.

Fordham, Maureen, William E. Lovekamp, Deborah S. K. Thomas, and Brenda D. Phillips. 2013. "Understanding Social Vulnerability." In *Social Vulnerability to Disasters* (2nd ed.), edited by Deborah S. K. Thomas, Brenda D. Phillips, William E. Lovekamp, and Alice Fothergill, 1–32. London and New York: CRC Press.

Fraser, Nancy 1996. *Social justice in the age of identity politics: Redistribution, recognition and participation*. The Tanner Lectures on Human Values, Stanford University, April 30–May 2.

Galaz, Victor. 2017. "Planetary Terra Incognita." *Global Environmental Governance, Technology and Politics* 1–20.

Galtung, Paul. 1969. "Violence, Peace, and Peace Research." *Journal of Peace Research* 6, no. 3: 167–91.

Gamburd, Michele R. 2013. *The Golden Wave*. Bloomington and Indianapolis: Indiana University Press.

Geerts, Evelien, and Iris van der Tuin. 2016. "New Materialism: How Matter Comes to Matter Diffraction and Reading Diffractively." http://newmaterialism.eu/almanac/d/diffraction.

Giddens, Anthony. 2004. *The Consequences of Modernity*. Stanford: Stanford University Press.

Ginige, Kanchana, Dilanthi Amaratunga, and Richard Haigh. 2014. "Tackling Women's Vulnerabilities through Integrating a Gender Perspective into Disaster Risk Reduction in the Built Environment." *Procedia: Economics and Finance* 18: 327–35.

Godfrey, Phoebe, and Denise Torres. 2016. "World Turning: Worlds Colliding?" In *Systematic Crises of Global Climate Change: Intersections of Race, Class and Gender*, edited by Phoebe Godfrey and Denise Torres, 1–17. London and New York: Routledge.

Grosz, Elizabeth. 1994. *Volatile Bodies Volatile Bodies: Toward a Corporeal Feminism*. Bloomington: Indiana University Press.

Gunnarson, Lena, Angela Martinez Dy, and Michiel van Ingen, eds. 2018. *Gender, Feminism and Critical Realism*. London and New York: Routledge.

Guyer, Paul, and Allen W. Wood. 1998. "Introduction." In *Critique of Pure Reason*, edited by Immanuel Kant, 1–81. Cambridge: Cambridge University Press.

Hamza, Mo, ed., 2015. "World Disasters Report: Local Actors the Key to Humanitarian Effectiveness." International Federation of the Red Cross and Red Crescent Society (IFRC), Geneva.

Haraway, Donna, Noboru Ishikawa, Scott F. Gilbert, Kenneth Olwig, Anna L. Tsing, and Nils Bubandt. 2016. "Anthropologists Are Talking – About the Anthropocene." *Ethnos* 82, no. 3: 535–64.

Harlan, Chico. "After Typhoon, Philippines Faces One of the Most Profound Resettlement Crises in Decades." *Washington Post*, January 4, 2014. www.washingtonpost.com/

world/after-typhoon-philippines-faces-one-of-the-most-profound-resettlement-crises-in-decades/2014/01/04/2118c6a2–71f2–11e3-bc6b-712d770c3715_story.html?utm_term=.b1762fa03c49.

Hearn, Jeff. 2004. "From Hegemonic Masculinity to the Hegemony of Men." *Feminist Theory* 5: 49–72.

Hearn, Jeff, Marina Blagojević, and Katherine Harrison, eds. 2013. *Rethinking Transnational Men: Beyond, Between and Within Nations.* London and New York: Routledge.

Heathwood, Chris. 2011. "The Relevance of Kant's Objection to Anselm's Ontological Argument." *Religious Studies* 47: 345–57.

Hewitt, Kenneth. 1995. "Excluded Perspectives in the Social Construction of Disaster," *International Journal of Mass Emergencies and Disasters* 13, no. 3: 317–39.

Hewitt, Kenneth. 1997. *Regions of Risk: A Geographical Introduction to Disasters.* London and New York: Routledge.

Hoffman, Susanna M., and Antony Oliver-Smith, eds., 2002. *Catastrophe and Culture: The Anthropology of Disaster.* School of American Research Press, James Currey.

Horton, Paul, and Helle Rydstrom. 2011. "Heterosexual Masculinity in Contemporary Vietnam: Privileges, Pleasures, Protests." *Men and Masculinities* 14, no. 5: 542–64.

Howell, Jude. 2007. "Gender and Civil Society." *Social Politics* 14, no. 4: 415–36.

ILO (International Labor Organization) 2018. After Haiyan – The Philippines Builds Back. www.ilo.org/global/about-the-ilo/multimedia/features/haiyan/lang – en/index.htm.

Intergovernmental Panel on Climate Change (IPCC). 2018. "Global Warming of 1.5 C°." www.ipcc.ch/report/sr15/.

Internal Displacement Monitoring Centre (IDMC). 2013. "1 in 10 Filipinos Affected by Haiyan, as Picture of Mass Displacement Emerges." www.internal-displacement.org/expert-opinion/1-in-10-filipinos-affected-by-haiyan-as-picture-of-mass-displacement-emerges.

International Federation of Red Cross and Red Crescent Societies. 2015. *Unseen, Unheard: Gender-based Violence In Disasters.* Geneva: IFRC. www.ifrc.org/Global/Documents/Secretariat/201511/1297700_GBV_in_Disasters_EN_LR2.pdf.

International Strategy for Disaster Reduction (ISDR). 2007. "Hyogo Framework for Action (HFA) 2005–2015: Building Resilience of Nations and Communities to Disasters." Extract from the final report of the World Conference on Disaster Reduction (A/CONF.206/6). www.unisdr.org/we/coordinate/hfa.

Jensen, Steffen. 2015. "Between Illegality and Recognition: Exploring Sacrificial Violence in a Manila Brotherhood." *Critique of Anthropology* 35, no. 1: 64–77.

Kabeer, Naila. 2005. "Gender Equality and Women's Empowerment." *Gender and Development* 13, no. 1: 13–24.

Kaldor, Mary. 2003. "The Idea of Global Civil Society." *International Affairs* 79, no. 3: 583–93.

Kelman, Ilan, Jessica Mercer, and J. C. Gaillard, eds. 2017. *The Routledge Handbook of Disaster Risk Reduction Including Climate Change Adaptation.* London and New York: Routledge.

Kent, Alexandra. 2006. "Reconfiguring Security." *Security Dialogue* 37, no. 3: 333–61.

Kinnvall, Catarina. 2004. "Globalization and Religious Nationalism: Self, Identity and the Search for Ontological Security." *Political Psychology* 25, no. 4: 741–67.

Kinnvall, Catarina. 2006. *Globalization and Religious Nationalism in India: The Search for Ontological Security.* London: Routledge.

Kinnvall, Catarina. 2013. "Borders, Security and (Global) Governance." *Global Society* 27, no. 3: 261–68.

Kinnvall, Catarina, and Jennifer Mitzen. 2017. "Ontological Security in World Politics: An Introduction." *Cooperation & Conflict* 52, no. 1: 3–11.

Kirby Vicki. 2017. "Foreword: Where to Begin?" *What if Culture Was Nature All Along?*, edited by Vicky Kirby, viii–xii. Edinburgh: Edinburgh University Press.

Kreines, James. 2017. "Kant and the Laws of Nature: Laws, Necessation, and the Limitations of Knowledge." *European Journal of Philosophy* 17, no. 4: 527–58.

Kulick, Don, and Helle Rydstrom. 2012. "Hazardous Times and Exposed Lives: Security, Precariousness, and Risk." Unpublished. Program Outline.

Latour, Bruno. 2014. "Anthropology at the Time of the Anthropocene. A Personal View of What is to Be Studied." *Distinguished Lecture at the American Anthropologists Association Meeting in Washington*, December 2014.

MacGregor, Sherilyn. 2010. "A Stranger Silence Still: The Need for Feminist Social Research on Climate Change." *Sociological Review* 57: 124–40.

MacGregor, Sherilyn, ed. 2017. *Routledge Handbook of Gender and Environment*. London and New York: Routledge.

Mercy Corps. 2013. "Quick Facts: What you Need to Know about Super Typhoon Haiyan." *Reliefweb*, November 14. https://reliefweb.int/report/philippines/quick-facts-what-you-need-know-about-super-typhoon-haiyan.

Merleau-Ponty, Maurice.1996. *Phenomenology of Perception*. London: Routledge.

Mies, Maria, Veronika Bennholdt-Thomsen, and Claudia von Werlhof, eds. 1988. *The Last Colony*. London: Zed Books.

Momtaz, Salim, and Muhammad Asaduzzaman. 2019. *Climate Change Impacts and Women's Livelihood: Vulnerability in Developing Countries*. London and New York: Routledge.

Moore, Henrietta. 1994. "The Problem of Explaining Violence in the Social Sciences." In *Sex and Violence: Issues in Representation and Experience*, edited by Penelope Harvey and Peter Gow, 138–55. London and New York: Routledge.

Mudavanhu, Chipo, Siambalala Benard Manyena, Andrew E. Collins, Paradzayi Bongo, Emmanuel Mavhura, and Desmond Manatsa. 2015. "Taking Children's Voices in Disaster Risk Reduction a Step Forward." *International Journal of Disaster Risk Science* 6, no. 3: 267–81.

NASA (National Aeronautics and Space Administration). 2013. "Evidence of Destruction in Tacloban, Philippines." November 20. https://earthobservatory.nasa.gov/images/82420/evidence-of-destruction-in-tacloban-philippines.

Nguyen, Huong. 2018. "Gendered Vulnerabilities in Times of Natural Disasters: Male-to-Female Violence in the Philippines in the Aftermath of Super Typhoon Haiyan." *Violence Against Women* 25, no. 4: 421–40. DOI: 10.1177/1077801218790701.

Nguyen, Huong T., and Helle Rydstrom. 2018. "Climate Disaster, Gender, and Violence: Men's Infliction of Harm upon Women in the Philippines and Vietnam." *Women's Study International Forum* 71: 56–62.

Nussbaum, Marta. 2000. "Women's Capabilities and Social Justice." *Journal of Human Development* 1, no. 2: 219–47.

Ogden, Laura, Nik Heynen, Ulrich Oslender, Paige West, Karim-Aly Kassam, Paul Robbins, Francisca Massardo, and Ricardo Rozzi. 2015. "The Politics of Earth Stewardship in the Uneven Anthropocene." In *Earth Stewardship*, edited by Ricardo Rozzi, F. Stuart Chapin III, J. Baird Callicott, Steward Pickett, Mary E. Power, Juan J. Armesto, and Roy H. May Jr., 137–57. Switzerland: Springer International Publishing.

Oliver-Smith, Anthony. 1996. "Anthropological Research on Hazards and Disasters." *Annual Review of Anthropology* 25: 303–28.

Oliver-Smith, Anthony. 1999. "What Is a Disaster? Anthropological Perspectives on a Persistent Question." In *The Angry Earth*, edited by Anthony Oliver-Smith and Susanna Hoffman, 18–34. London and New York: Routledge.

Ortner, Sherry B. 1972. "Is Female to Male as Nature is to Culture?" In *Woman, Culture, and Society*, edited by Michelle. Z. Rosaldo and Louise Lamphere, 68–87. Stanford, CA: Stanford University Press.

Osborne, Hannah. 2018. "Microplastics Have been Found in Feces Across the World." October 22. www.newsweek.com/microplastics-found-human-feces-across-world-1181121.

The Philippine Star/Philstar Global (2014, November 20). *Gabriela: Cops Top Perpetrators of Violence vs Women, Kids*.

Rigg, Jonathan, Carl Grundy-Warr, Lisa Law, and May Tan-Mullins. 2008. "Grounding a Natural Disaster: Thailand and the 2004 Tsunami." *Asia Pacific Viewpoint* 49, no. 2: 137–54.

Ruspini, Elisabetta, Jeff Hearn, Bob Pease, and Keith Pringle. 2011. *Men and Masculinities Around the World: Global Masculinities*. Basingstoke and New York: Palgrave Macmillan.

Rydstrom, Helle. 2002. "Sexed Bodies/Gendered Bodies: Children and the Body in Vietnam." *Women's Studies International Forum* 25, no. 3: 359–372.

Rydstrom, Helle. 2003. *Embodying Morality: Growing Up in Rural Northern Vietnam*. Honolulu: University of Hawai'i Press.

Rydstrom, Helle. 2012. "Gendered Corporeality and Bare Lives: Local Sacrifices and Sufferings During the Vietnam War." *Signs* 37, no. 2: 275–301.

Rydstrom, Helle. 2015. "Politics of Colonial Violence: Gendered Atrocities in French Occupied Vietnam." *European Journal of Women's Studies* 22, no. 2: 191–207.

Rydstrom, Helle. 2016. "Vietnam Women's Union and the Politics of Representation." In *Gendered Citizenship and the Politics of Representation*, edited by Hilde Danielsen, Kari Jegerstedt, Ragnhild Muriaas, and Brita Ytre-Arne, 209–34. Basingstoke and New York: Palgrave Macmillan.

Rydstrom, Helle. 2017. "A Zone of Exception: Gendered Violence of Family 'Happiness' in Vietnam." *Gender, Place and Culture* 24, no. 7: 1051–70.

Rydstrom, Helle. 2019. "Disasters, Ruins, and Crisis: Masculinity and the Ramifications of Storms in Vietnam." *Ethnos*. DOI: 10.1080/00141844.2018.1561490.

Rydstrom, Helle. (forthcoming). "Episte-Ontology, Gender, and Masculinity: Corporeal Properties and Confrontations in Vietnam."

Scott, James. 1992. *Domination and the Arts of Resistance: Hidden Transcripts*. London and New Haven: Yale University Press.

Shiva, Vandana. 1991. *Staying Alive: Women, Ecology and Survival in India*. London: Zed Books.

Spivak, Gayatri C. 1988. "Can the Subaltern Speak?" In *Marxism and the Interpretation of Culture*, edited by Cary Nelson and Lawrence Grossberg, 66–111. London: Macmillan.

Steffen, Will, Paul J. Crutzen, and John R. McNeill. 2007. "The Anthropocene: Are Humans Now Overwhelming the Great Forces of Nature?" *Royal Swedish Academy of Sciences, Ambio* 36, no. 8: 614–21.

Sternberg, Troy, ed., 2019. *Climate Hazards Crises in Asian Societies and Environments*. London and New York: Routledge.

Sylvester, Christine. 2010. "Tensions in Feminist Security Studies." *Security Dialogue* 41: 607.

Tripp, Aili Mari, Myra Marx Ferree, and Christina Ewig, eds. 2013. *Gender, Violence, and Human Security: Critical Feminist Perspectives*. New York: New York University Press.

True, Jacqui. 2012. *The Political Economy of Violence against Women*. Oxford: University of Oxford Press.

True, Jacqui. 2013. "Gendered Violence in Natural Disasters: Learning from New Orleans, Haiti and Christchurch." *Aotearoa New Zealand Social Work* 25, no. 2: 78–89.

UNDP's Bureau for Crisis Prevention and Recovery. 2010. "2010 Annual Report." www.undp.org/content/dam/undp/library/crisis%20prevention/2010report/full-annual-report-2010-lowres.pdf.

UNISDR, UNDP, and IUCN. 2009. *Making Disaster Risk Reduction Gender-Sensitive-Policy and Practical Guidelines*. Geneva: UNISDR, UNDP, and IUCN. www.unisdr.org/files/9922_MakingDisasterRiskReductionGenderSe.pdf.

United Nations Asian Disaster Preparedness Centre (UN/ADPC). 2010. "Disaster Proofing the Millennium Development Goals (MDGs)." UN Millennium Campaign and the Asian Disaster Preparedness Center.

United Nations Development Program (UNDP). 2007. *Human Development Report 2007/8 – Fighting Climate Change: Human Solidarity in a Divided World*. Houndmills, Basingtoke, Hampshire and New York: Palgrave Macmillan. http://hdr.undp.org/sites/default/files/reports/268/hdr_20072008_en_complete.pdf.

United Nations Development Programme (UNDP). 2013. "Overview of Linkages between Gender and Climate Change." *Policy Brief*. www.undp.org/content/dam/undp/library/gender/Gender%20and%20Environment/PB1-AP-Overview-Gender-and-climate-change.pdf.

United Nations Framework Convention on Climate Change (UNFCCC). 2013. "Warsaw International Mechanism for Loss and Damage Associated with Climate Change Impacts." http://unfccc.int/adaptation/workstreams/loss_and_damage/items/8134.php.

United Nations Population Fund (UNFPA). 2010. "Natural Disasters: Gender-based Violence Scenarios. Gender-based Violence and Natural Disasters in Latin America and the Caribbean." https://lac.unfpa.org/sites/default/files/pub-pdf/UNFPA%20version%20ingles%201.pdf.

Valdés, Helena M. 2009. "A Gender Perspective on Disaster Risk Reduction." In *Women, Gender and Disaster: Global Issues and Initiatives*, edited by Enarson, Elaine and P. G. Dhar Chakarabarti, 18–28. New Delhi: Sage Publications.

Weather Underground (2013, November 10). "Category 1 Typhoon Haiyan Hitting Vietnam; Extreme Damage in the Philippines". https://www.wunderground.com/blog/Jeff Masters/category-1-typhoon-haiyan-hitting-vietnam-extreme-damage-in-the-phili.html

Wisner, Ben. 2016. "Vulnerability as Concept, Model, Metric, and Tool." *Natural Hazard Science*. DOI: 10.1093/acrefore/9780199389407.013.25.

Wisner, Ben, Piers Blaikie, Terry Cannon, and Ian Davis. 2003. *At Risk: Natural Hazards, People's Vulnerability and Disasters*. Follow up to the Hyogo Framework for Action 2005. Routledge and UNDP.

Wisner, Ben, J. C. Gaillard, and Ilan Kelman. 2012. "Framing Disaster: Theories and Stories Seeking to Understand Hazards, Vulnerability and Risk." In *The Routledge Handbook of Hazards and Disaster Risk Reduction*, edited by Ben Wisner, J. C. Gaillard, and Ilan Kelman. London: Routledge.

Wolf, Eric. 1982. *Europe and the People Without History*. Berkeley: University of California Press.

World Health Organization (WHO). 2017. "Violence Against Women." November 20. www.who.int/news-room/fact-sheets/detail/violence-against-women.

Yamin, Farhana, Atiq Rahman, and Saleemul Huq. 2005. "Vulnerability, Adaptation and Climate Disasters: A Conceptual Overview." *IDS Bulletin* 36, no. 4: 1–14.

Part I

2 Gender-responsive alternatives on climate change from a feminist standpoint

Maria Tanyag and Jacqui True

Introduction

How do women contending with multiple climate hazards and disaster risks understand climate change? Scientific research predicts catastrophic impacts of climate change on humanity unless emissions of greenhouse gases are drastically reduced *now*. By 2100 climate change will constitute a major threat to "human health, water, food, economy, infrastructure and security" (Mora et al. 2018, 1062, see also Cramer et al. 2018). What can women's everyday lives tell us about what security means at the frontlines of cumulative crises such as those related to climate, disaster and conflict? Importantly, how can climate governance at national, regional and global levels become more responsive to women's knowledge and more able to replenish their contributions to mitigating crises?

This chapter addresses these questions by mapping out the key elements of a framework that aims to strengthen women's voices and leadership in climate change and related crisis responses. The complex consequences of climate change demand such an approach that encompasses experienced-based as well as scientific knowledge about the interplay of different risks and hazards. Focusing on one or a few hazards runs the risk of "mask[ing] the changes and impacts of other hazards, [and] giving an incomplete or misleading assessment of the consequences of climate change" (Mora et al. 2018, 1062). Drawing from a collaborative, multi-country research project with the global non-governmental organisation, ActionAid, we demonstrate how an integrated and gender-responsive approach to climate change begins from the standpoint of women on the margins.

Feminist scholarship contributes to an interdisciplinary analysis of climate change for it has built an expertise, the relevant tools and lenses, for revealing multiple and overlapping forms of insecurity by starting from women's everyday life on the margins of global politics. A feminist standpoint allows us to advance and deepen scientific knowledge by drawing out the gendered interconnections among multiple consequences of climate change. Consistent with standpoint theory, we define being 'on the margins' as *geographic* (remote, rural and displaced), *spatial* (invisibility and exclusion from decision-making spaces), *epistemic* (devaluation or denial of particular ways of knowing and sources of knowledge) and *structural* (inequality or discrimination in access to resources).

Situatedness in these various margins often simultaneously imbues particular groups of women with a distinct understanding of the multi-layered causes and effects of climate change in our global ecosystem. Importantly, as Cynthia Enloe reminds us, "[N]o individual or social group finds themselves on the 'margins' of any web of relationships . . . without some other individual or group having accumulated enough power to create the 'centre' somewhere else" (1996, 186). The Anthropocene is the fullest representation of human accumulation and concentration of power. Climate change is and will be even more so a struggle for justice and equality because the worst of its impacts are borne by people who have contributed the least in its creation and hastening. Yet, these very people particularly women and girls are *kept* in the margins so as not to have meaningful and global impact in reshaping a new vision of how humans ought to relate with the environment.

We propose three main elements required for a gender-responsive framework that addresses the challenges of climate change and its community impacts. First, such a framework needs to recognise localised and traditional knowledge within policies at national, regional and global levels. This type of knowledge can provide novel warning indicators as well as long-term adaptation plans at household and community/village levels. Second, women require direct representation at different levels of governance especially in climate-related leadership positions within the community, national agencies/decision-making bodies and global climate negotiations. Third, institutional and strategic pathways need to be identified to facilitate connections with localised knowledge about climate factors and gender-inclusive political representation at all levels of climate governance. Bringing these elements together will enable a more holistic understanding of climate change and its intersection with other crises that may cause or compound climate events and natural disasters. Climate change solutions that are gender-responsive require a comprehensive approach to security and different types of leadership to encompass the needs of the most marginalised groups of women and girls.

Our research partnership with ActionAid has allowed us to design and implement an action-research project which involves direct and indirect participation from members of climate change-affected communities in Cambodia, Kenya and Vanuatu.[1] These communities participate in ActionAid's different gender equality, disaster risk reduction and climate change programmes particularly through initiatives that promote women's leadership and activism in their respective communities and nationally (see, for example, ActionAid 2016a; 2016b; The Economist Intelligence Unit 2014). We draw evidence from our case studies in the three countries to demonstrate that if we start from women's lives and their knowledge as legitimate sources for climate governance and decision-making, then we will arrive at both a more holistic understanding of the problem of climate change and its contextualised impacts as well as a more durable and transformative agenda for mitigation, adaptation and resilience. As we discuss further in this chapter, ensuring gender-equal representation across all levels of climate governance matters because many of the barriers to knowledge on climate risks are inseparable from this representation. The achievement of gender balance is also important for equitable resource distribution and facilitating synergies between women-led community-level efforts and the international community, in order that international climate benchmarks align with the needs of the most affected by climate change.

Our chapter has three parts. The first section provides an overview of existing research on gender, climate change and related crises to contextualise the impetus for our collaborative research with ActionAid. The second section outlines the three key elements for a more gender-responsive climate change agenda. In particular, we examine how feminist standpoint theory enables us to re-examine the multiple realities of climate change through the analysis of how unequal gender relations originate, evolve and are sustained in the natural and social environments. Finally, the third section considers future scenarios and applications of this gender-responsive framework as well as suggestions for policy and community-level changes engendered by our framework.

Gender, climate change and related crises

There have been crucial strides in mainstreaming gender within global climate change frameworks. In 2018, General Recommendation No. 37 from the Committee on the Elimination of Discrimination against Women (CEDAW) emphasised states' obligations in addressing the gender-related dimensions to disaster risks and climate change. According to the Committee, "[S]tates parties should ensure that all policies, legislation, plans, programmes, budgets and other activities related to disaster risk reduction and climate change are gender responsive and grounded in human-rights based principles" (2018, 7). Moreover, states have a responsibility vis-à-vis non-state actors, to create an environment conducive for "gender-responsive investment in disaster and climate change prevention, mitigation and adaptation" across which different groups of women should equally participate and lead (2018, 14). This landmark CEDAW recommendation comes at a time of more than four decades of gender research in climate change and environmental policy (Arora-Jonsson 2014). It also comes together with other major advancements in global climate policy and governance.[2] For instance, the Green Climate Fund which is a financial mechanism for the UN Framework Convention on Climate Change (UNFCC)[3] launched its pioneering manual on mainstreaming gender into climate finance in 2017. The manual is a recognition that programmes and initiatives require the necessary resourcing for them to effectively address the gendered ramifications of climate change. In the same year, at the annual Conference of the Parties (COP23) to the Paris Agreement, the first ever Gender Action Plan was adopted. This milestone is in large part due to the strong mobilisation and lobbying of civil society groups representing women's, indigenous and environmental rights.[4]

An unequivocal and consistent commitment to gender equality is also present within disaster risk reduction (DRR) agendas such as the global Hyogo Framework for Action (2005–2015) and the Sendai Framework for Disaster Risk Reduction (DRR), 2015–2030. In 2011, the UN Office for Disaster Risk Reduction (UNISDR) developed a 20-point checklist on making DRR gender sensitive.

The checklist serves to facilitate and coordinate DRR by governments at national and local levels to reflect a gender perspective across five main priority processes: technical, political, social, developmental and humanitarian.[5] It explicitly states that gender is integral to the whole cycle of DRR beginning with how problems are understood, how resources are mobilised and allocated, what issues are prioritised and how decisions are made, to the practice and implementation of DRR in development and humanitarian programmes.

The state of gender research has also grown across various disciplines and geographic locations with case studies from the Global North and the Global South where the consequences of climate change and disasters are most apparent and severe. Scholars have contributed to the development of rich and complex analysis of the relationship between gender and climate change (Arora-Jonsson 2014; CEDAW 2018; Pearse 2017), and gender and disasters (Bradshaw 2014; Enarson 1998; True 2013; Tanyag 2018).

Research has shown that women through their roles as primary caregivers are socialised from a very young age about food provisioning. This means that they can develop a distinct or gender-specific knowledge on natural resources compared to men (Mies and Shiva 2014; Agarwal 2010). They may have first-hand experiences of how climate change manifests and intersects with other everyday insecurities for their household and communities. For instance, in our research with a remote, pastoralist community in Kenya, we find that women are tasked with the collection of water and firewood. The women we met reported that they know where the water sources are and increasingly, how far and how much time of their day, it takes to get there and back. The women have been passed on knowledge by their mothers and grandmothers on alternative food sources which require specific methods of food preparation in order to either remove poison or make them edible. They rely on this knowledge even more during times of drought when their households or communities face threats of food insecurity. Gender roles therefore can translate into women's capacity to identify context-specific 'warning signs' and heightened awareness that can further strengthen governance measures regarding the availability and management of accessing natural resources (see ActionAid 2017).

Women, as an extension of their caregiving roles, often take up volunteer or unpaid work in the community in order to ensure the well-being of their household. Women often serve as community workers and are involved in groups such as in churches, and advocating around development issues such as health and education (Elson 2012; Maathai 2006). They are therefore at the forefront of responding to communities in the aftermath of disasters and related consequences of climate change (ActionAid 2016a). However, despite strong knowledge of their community's needs and dynamics, their contributions especially leadership may not always be supported nor recognised by governments and aid organisations. For example, natural and economic resource shortages arise in the aftermath of disasters, health pandemics and economic crises. Communities have to survive through 'self-help' strategies precisely because state and global interventions are inadequate or fall short of meeting all immediate needs of affected populations.

Consequently, climate-induced disasters lead to intensified demands on care provisioning as health and welfare needs increase. In the process, women end up experiencing multiple burdens as they need to provide basic needs as well as attend to the care of others to the point of depleting their own health and well-being (Tanyag 2018). Especially in remote areas, where access to state relief assistance may be even more constrained or delayed, women's unpaid care labour serves as the safety net that ensures survival in times of crisis (Elson 2012).

At its most extreme, resource scarcity is directly linked to certain types of violence against women (True 2012). In the case of disasters, women and girls are more vulnerable to death (Neumayer and Plumper 2007) and violence due to socially constructed gender relations and pre-existing patterns of gender inequality and discrimination. As disasters become frequent and more intense due to climate change, eliminating gender inequalities is vital. For example, climate change intensifies the condition of poverty for families and pre-existing gender norms determine the kind of stop-gap, coping measures particularly negative adaptation undertaken to survive. A palpable manifestation of climate change documented in Malawi and Mozambique is that ever-younger girls are married off by their families as an economic survival strategy. Such a strategy produces a new generation of child brides, which in turn leads to dangerous, early pregnancies and associated complications (Chamberlain 2017).[6]

Similarly, research in a drought-prone area in Bangladesh suggests that the impacts of climate changes are increasingly serving as a further impetus for dowry practices to continue (Alston et al. 2014). This occurs as economic pressures intensify and as a result of the high value placed on family honour of the families of girls and young women [to] condone forced marriage or arrange early marriages for their daughters "because of the perceived economic risks associated with the continuance of education and the increasing costs of dowry as girls mature" (Alston et al. 2014, 137). There is a need to fully account for these gendered costs of climate change to include the long-term and gradual harms that women and girls may disproportionately bear as a result of these marriages: including the loss of personal autonomy and severe physical and mental stressors over time. Climate change policy interventions thus need to be multidimensional in order to respond to how communities are coping negatively in ways that may curtail women's and girls' rights.

Another way in which gender research has contributed to our knowledge of climate change is in highlighting how women occupy special cultural roles tied to the protection of land and other natural resources (Mies and Shiva 2014). For certain indigenous and ethnic groups, women may traditionally serve as stewards or custodians of the environment (Tuhiwai Smith 2013). Women, like men, acquire a deep understanding of how climate change manifests and impact upon their communities by accessing cultural and everyday knowledge. However, this knowledge is considered sacred and protected such that to be granted the knowledge, one must be specifically inducted through ceremonies and community investiture of responsibility as custodians of traditional lands and knowledge. In our research in Tanna Island of Vanuatu, the women we met described how their ancestors past and present, regard the active volcano on their island as supernatural. In fact, in their language the word for 'god' is Yasur which is also the name of the volcano, Mount Yasur.[7] Consequently, according to them, all the communities on this island have a profound relationship with the active volcano which then gives structure and meaning to their daily lives especially since they rely on agriculture for livelihood. An example is how they believe the behaviour of the volcano, especially the colour of fumes that come out of the crater, is an indicator for the success or loss of their harvest.

Through cultural stewardship and sacred beliefs, indigenous women and men have a direct stake in protecting the environment from unjust and extractive industries such as in mining and other development projects (Global Witness 2016). Land is not just a resource, it provides meaning and belonging to indigenous people. As such, this has meant that indigenous peoples are frequently most at risk of experiencing physical violence and extrajudicial killings as a result of their struggles to protect the environment. For women activists, the particular threat of sexual and gender-based violence is intensified and violence may be sexualised for women (True and Tanyag 2018). Their experiences can inform a more nuanced understanding of how climate governance and decision-making are part of a complex network of political and economic interests (Agarwal 2010). As Nayak (2009) suggests, development for some, is violence for others. Climate-friendly solutions for managing renewable sources of energy or in the sustainable reliance on land and natural resources therefore require an understanding of the environment's historical, religious and cultural significance. This may mean ensuring that pre-existing indigenous claims to land and group rights are fully redressed rather than omitted in the development of climate change policies. At the same time, these solutions need to comprehend how the intersection of climate-induced violence, new forms of gendered cultural restrictions and pre-existing gender discrimination within their own indigenous or ethnic groups affect certain groups of women and girls.

While there is now deliberate incorporation of gender in climate and disaster governance, and theoretical developments in gendering climate and disaster research, key challenges remain. Among them is that gender research "appears to have had a marginal effect on environmental practice on the ground", and that "the link between research and every day work appears to be more elusive" (Arora-Jonsson 2014, 296). For example, gender gaps in climate governance is evident in that women's localised knowledges and everyday experiences of climate change continue to be marginalised from global climate negotiations. A key barrier is in the continued low levels of direct participation from women leaders in global conferences such as the UNFCCC COP. In 2014, women comprised only 25 percent of government chief negotiators and this underrepresentation is also reflected at regional and national levels of environmental decision-making because most negotiators are also often leaders of environmental ministries in their home countries (IUCN GGO 2015, 10; see also Hemmati and Rohr 2009). Women face less barriers and have fared better when it comes to indirect participation in climate negotiations. COP participant lists show near equal representation of men and women among NGO delegations. As a result, "women occupy a larger share of NGO representatives to each COP than their government delegate counterparts" (IUCN GGO 2015, 15). However, where crucial and actual decision are made, usually by governments, women do not have equal numbers of seats at the table.

While women may have had access to community-level governance structures often by virtue of their traditional, social and cultural roles as mothers, women's participation overall remains low across national and global levels (see, for example, George 2014; Chandra et al. 2017). Ensuring that women representing communities at the margins of society and geopolitics are able to participate at the highest levels in climate governance is crucial to advancing gender equality and climate justice. Raising gender issues within climate negotiations begins at the national level where involving women at the highest political level "in state environmental protection is a necessary prerequisite to overcoming the gender gap in adaptation and mitigation actions, particularly on issues such as per capita greenhouse emissions reduction, food security, urbanisation, industrialisation and democracy" (Chandra et al. 2017, 46).

In our research, gender-responsiveness, in effect, means being able to bridge localised, 'everyday' knowledge with technical expertise on climate science to deepen and broaden global and national climate decision-making (Backstrand 2003; Harding 1991). By starting from women's lived experiences, we are able to map environmental realities that have been obscured from top-down climate change governance. The latter tend to be masculinised spheres of decision-making in that they are prevalently male dominated or biased towards behaviours that are considered masculine (see also Buckingham 2015). Hence, research has shown that even when there is a critical mass of women in climate policymaking, this does not automatically equate to gender-responsive climate policies (Magnusdottir and Kronsell 2015). This is even more striking in the Scandinavian context where there is a higher level of institutionalisation of gender equality issues and gender-balanced representation in climate policymaking bodies (Magnusdottir and Kronsell 2015, 314). Path-dependence in environmental governance privileges technical fixes and there is less understanding or appreciation of how a gender lens can help to both explain and solve climate change problems (Magnusdottir and Kronsell 2015, 317). In the case of Sweden, Annica Kronsell argues that equally including women "in the processes of deliberation on climate transitions does not guarantee gender sensitivity". Mainstreaming a gender perspective into climate governance "requires the input from those actors who are knowledgeable about gender aspects on climate issues. Those actors interested in change and transformation, and those who represent groups who have previously been excluded from climate governance should also be included" (2013, 12).

Gender is certainly now a mainstay in climate governance and across local and international policies to mitigate and adapt to climate change (Arora-Jonsson 2014). Yet, the debates are far from settled and questions abound about the relevance gender analysis in the practice of climate change and disaster policymaking. In response, an important step highlighted in gender research is citizen participation in the production, validation and application of scientific knowledge

(Backstrand 2003). We would add that it is also about broadening the sources of knowledge that are considered valid and legitimate in the understanding of climate change as well as the solutions developed to address its myriad consequences. In the next section, we propose a framework for developing more gender-responsive alternatives to climate change.

A framework for developing gender-responsive alternatives

The policy shift toward gender-responsiveness in climate governance represents a significant momentum. We are at a critical juncture to assess the conditions required for forms of climate governance that directly alleviate gender and other social inequalities along with climate justice. Our starting point for gender-responsiveness is the lived experience of women from/on the margins. We refer to marginalisation in an encompassing way. Women may be on the margins as in they are internally displaced as a result of disaster and climate change impacts. They may also be excluded in actual spaces where decisions are held to be deliberately far from where they are; or they may lack access to the resources that allow them to participate in climate policymaking in the first place. Importantly, women are on the margins and continue to be so because women from their community or group are either not considered legitimate sources of knowledge or what they can contribute is not valued (see also Guijt and Shah 1998).

Our framework and analysis is informed by feminist standpoint theory. Feminist standpoint theory emerged in the 1980s from feminist scholarship and activism (Hartsock 1983; Haraway 1988; Harding 1991). Like other emancipatory approaches in the social sciences, it stresses the distinct 'view from below'. It is emancipatory because a feminist standpoint approach argues that women's lived experiences provide a more critical vantage point on social reality precisely because of patriarchal structures and subordination. Women, by virtue of being in marginalised positions of power, understand the unequal nature of current systems and how they work to maintain inequality. Consequently, women have a standpoint on both what is wrong with the current system and how to change it based on their lived experience. Those in positions of power imbued by a 'view from the top' cannot conceive of alternatives to the status quo because they substantially benefit from – and have vested interests in maintaining – it. Therefore, women's standpoints especially at the intersections of multiple inequalities can inform strategies to collectively transform the situation for women, men, boys and girls. In gender and climate research this means understanding the causes and effects of climate degradation from the perspective of women's everyday perspectives and experiences. Women from the margins acquire different knowledge relating to the material impacts of climate change on households and communities because the consequences of climate change directly affect their social, economic and cultural roles. They can speak to and help advocate solutions that bridge mitigation, adaptation and long-term transformation.

Understanding gender in disaster contexts through women's eyes means investigating "the material conditions of women's everyday lives, focusing on the situated knowledge of those outside the dominant power structures but assuming no unified identity or set of experiences" (Enarson 1998, 157). This is similar to Chandra Mohanty's call for understanding women's agency in historically and context-specific ways such that we arrive at "a conceptualisation of agency which is multiple and often contradictory but always anchored in the history of specific struggles" (Mohanty 1991, 38). The researcher too has his or her own situated and partial knowledge. Feminist research ethics acknowledges that the researcher's own subjectivity shapes his or her relationships and inquiries throughout the research process. His or her own standpoint informs the kind of questions he or she asks, and the extent to which he or she is able to attend to which margins and silences. It is his or her role to develop an account of situatedness, how it is maintained and for what benefit; and to then be able to translate this knowledge into transformation (Ackerly and True 2010; Hooks 2000).

The framework we propose is anchored to the notion that gender-responsiveness is about analysing climate change from women's standpoints as well as understanding that gender inequalities intersect with class, marital status, race/ethnicity, sexuality, religion and geography (see also Kaijser and Kronsell 2014; Van Aelst and Holvoet 2016). By this we do not mean justifying women's capacity to participate and contribute in climate governance on the basis of natural, essential or inherent attributes. Women do not necessarily behave in 'climate friendly' ways. Like men, they may be agents in fuelling resource depletion when their immediate survival is threatened during times of resource scarcity. However, because of their marginal status within overlapping hierarchies of power, women often have specific knowledge to contribute. As Agarwal points out, "[P]eople's relationship with nature, their interest in protecting it, and their ability to do so effectively are significantly shaped by their material reality, their everyday dependence on nature for survival, and the social, economic, and political tools at their command for furthering their concerns". She adds further that "[I]deological constructions of gender, of nature, and of the relationship between the two would impinge on how people respond to an environmental crisis, but cannot be seen as the central determinants of their response" (2010, 42). People cannot promote gender equality and climate justice when they face material constraints to both.

Our framework considers the relational aspect of gender identities and structures which have both material and ideological dimensions. How people view the world stems from their material conditions, and at the same time, their material conditions are sustained or underpinned by context-specific beliefs, values and norms. In addition, gender is not synonymous with women. Our analysis of gender-responsiveness incorporates men's experiences and masculinities as vital to providing a fuller contextualisation of how and why women experience marginalisation in climate governance. Gender-responsiveness in climate change governance does not merely 'add women and stir'. Gender research has shown that such an approach ultimately fails or causes more harm to women than good in the long run (see for example Hozic and True 2016). For example, in Resurreccion's (2013) research, she identified that an entry point for women in climate change discussions has been through tying women's identity to nature and in flattening out differences among women to emphasise vulnerability to the consequences of climate change. Done this way, "problematic outcomes usually emerge when the simplifications that fuel politics segue into policy and programming" (2013, 40). Similarly, some discourses around women and climate change and disasters have tended to frame a binary of either women's virtues or vulnerability such that women are only visible or intelligible within climate policymaking when they conform to these stereotypes (Arora-Jonsson 2011; Jordan 2019; Smyth and Sweetman 2015).

Gender-responsiveness in climate governance aims to challenge siloes in climate policymaking where women are relegated to 'feminised' areas of expertise. Gender-responsive alternatives are about ensuring that the needs of the most affected are incorporated into long-term plans for adaptation to climate change. Gender-responsiveness is one measure that can inform concrete planning of where and how resources need to be allocated across all areas of climate policymaking such as in health, land and agriculture, as well as in monitoring changing weather patterns. It does not offer explanations to everything or disregard processes that equally impact all human beings. At the same time, security and development agendas need to be integrated from a gender perspective. This is vital because research has shown how climate, disasters and conflict risks are interrelated and that solutions must attend to both (Asia Pacific Forum on Women, Law, and Development [APWLD] 2015; Chandra et al. 2017; Lee-Koo 2012; True 2016). Research by Chandra et al. (2017) on smallholder men and women farmers in Mindanao, Philippines, suggest that for conflict-prone rural agrarian communities, conflict and climate change overlap to increase gender-specific vulnerability. Mindanao, the southernmost part of the archipelago country, has been the site of several decades of resource and identity-based conflicts.

Female-headed households are common in conflict-affected areas because many women have been widowed as a result of conflict. Poverty and high dependence on agricultural livelihood expose these households to food insecurity, as well as sexual and gender-based violence when they are displaced by both conflicts and disasters (as experienced in the Asian tsunami for instance, see True 2012). As such they stress the importance of connecting climate change with inclusive peacebuilding and development because "vulnerability due to climate change is symptomatic of broader gender justice and inequality issues left unaddressed by peace, post-conflict reconciliation and resettlement efforts by government and aid agencies" (Chandra et al. 2017, 54). Strategic connections in policy are needed to link up programmes and resourcing that emphasise women's capacity for leadership in peacebuilding and climate change. Coordination across security and development agendas makes sense because peace and environmentalism are not separate goals. This integrated approach is embodied by the Green Belt Movement in Kenya (Maathai 2006) and the UN 2030 Sustainable Development Goals.[8] Wangari Maathai, Nobel Peace Laureate and founder of the movement, firmly believed that acting locally have global repercussions. Furthermore, her knowledge of holistic peace stems from her experience of how local conflicts came hand in hand with environmental degradation. Promoting environmental conservation was a way to enhance security within localised and internal conflicts in which women suffer distinct consequences.

Given this discussion, the gender-responsive framework we propose for addressing climate change has three main and interconnected elements: (1) deliberation and democratisation of climate science; (2) gender-inclusive representation in climate governance; (3) rethinking norms and practices in climate governance.

Deliberation and Democratisation of Climate Science: This first element refers to women's and men's localised knowledge of warning indicators at household and community/village levels, and awareness of policies at national, regional and global levels. Developing alternatives to climate change governance thus begins with accounting for women's and men's localised knowledge on what climate change means in their everyday lives. Deliberation and democratisation of climate science ensure that all forms and sources of knowledge are recognised as contributing to climate responses, and that knowledge production is translated into information, assistance and services that benefit rather than exclude the communities most affected by climate change (see for a similar point on water management, Moraes and Perkins 2009). Women need access to the technical or scientific knowledge in order to be able to communicate their everyday experiences in the main language of climate governance. Democratising climate science is therefore a fundamental aspect to ensuring gender-responsiveness because "scientific knowledge can be conceived as a global public good in which the citizens have a stake" (Backstrand 2003, 25).

Moreover, from a feminist standpoint, all knowledge is partial and the types of knowledge that are predominantly engaged with as legitimate sources of information are shaped by dominant power relations. Women have been working to generate a more inclusive form of climate governance in localised and community-focused ways. Women-led initiatives such as *Women i Tok Tok Toketa* (Women Talk Together) in Vanuatu create spaces for women to share information around disaster risks, response and mitigation. They are also being used to directly support and mentor women's community leadership during elections and with respect to awareness-raising on violence against women. Through women's gatherings, climate change is understood not as a separate issue from the insecurities women experience such as election-related violence and political disenfranchisement as well as domestic violence. Technical and programmatic solutions to gender and climate change must grapple with these interrelated barriers. The first step is to understand how climate change as an abstract concept translates on the ground in the form of early-warning or everyday indicators.

In addition, nature or the environment cannot only be interpreted from a scientific lens especially given current moves to grant personhood to ancestral domains and bodies of nature based on human rights law. For example, in New Zealand, the promotion of Maori indigenous rights over ancestral domains have led to the *Te Urewera Act* 2014 which grants personhood to the Whanganui River. Te Urewera Act stipulates "Te Urewera is a place of spiritual value, with its own mana and mauri. . . [it] has an identity in and of itself, inspiring people to commit to its care" (see Kauffman and Martin 2018).[9] Other sources of knowledge need to be incorporated and recognised formally within resource management and sustainability programmes to encompass the non-tangible meaning ascribed to nature. Linda Tuhiwai Smith (2013), a Kaupapa Maori scholar, argues that environmental knowledge is earned and that the process itself of gaining knowledge is sacred for indigenous peoples. Moreover, expanding the conditions of justice for indigenous peoples involves enabling their alternative ways of knowing. For holistic climate solutions to respond to their marginalisation in geographic, spatial, structural and epistemic ways, efforts must be in place to foster long-term dialogue among all stakeholders in crafting climate solutions. This is most in need for those with deep connection to land and historically devalued worldviews because the consequences of climate change for them are both visible as in the loss of natural bodies of land and water, and invisible as in the loss of their culture, heritage and sense of purpose.

Gender-inclusive representation in climate governance

This second element refers to women's direct representation in different levels of governance especially in climate-related leadership positions from the community, national agencies/decision-making bodies and global climate negotiations. Gender representation is informed by an intersectional lens on women's and men's experiences. Climate governance must create spaces for diverse groups of women, not just those from dominant or elite groups. This means that climate change frameworks must take stock of the differences that create diversity among women's standpoints (Kaijser and Kronsell 2014). This can be achieved in part through ensuring that gaps among women's leadership and skills are progressively bridged so that representation is substantive and able to influence actual policy agendas. Climate change affects women in different ways depending on their economic status, geographic location, sexuality, age, religion, ethnicity and citizenship among others. For instance, women and girls internally displaced by environmental disasters are at a disadvantage because they are faced with higher mortality and long-term health complications as a result of displacement. Such barriers to their full human development also impinge upon their ability to politically participate in climate governance. This means that gender-responsive climate change frameworks are underpinned by gender equality goals because in order to achieve gender-equal representation across all levels of climate change, the various pre-existing material barriers to their human development need to be addressed.

At the same time, a cultural transformation in leadership is needed. For example, challenging male-dominance and the embeddedness of traditional masculinities in climate governance will require broader representation from different groups of men including spaces for caring masculinities. In our research in selected disaster-prone communities in Kampot and Pursat, Cambodia, we found that in areas where democratic spaces have historically been constrained due to authoritarian rule, political participation for both men and women has been limited. Our focus group discussions revealed that both men and women in these remote areas do not seek help from the government when there is a disaster because they are used to coping on their own because they often fear government control. However, DRR and climate change programmes can open spaces and help develop a culture of public engagement via new more open and deliberative forums for disaster preparedness, prevention and response policymaking. Women have actively participated and are even leading as 'DRR champions' in their communities through the assistance of ActionAid Cambodia.[10] This public sphere activity has ironically been allowed by the state ruling party, which is perhaps an indication that it is not viewed as politically threatening.

Importantly, the direct participation and substantive representation of women in key decision-making bodies must broaden to include leadership at all levels from the community, national agencies beyond designated women's ministries, and in global climate negotiations (Arora-Jonsson 2013). Research on gender and peace processes already make a case for how the direct inclusion of female peace negotiators make peace agreements more likely to be signed, and for the resulting peace agreement to be more durable (UN Women 2015; True and Riveros-Morales 2018). This may be because agreements are more likely to bring into the peace process different and often marginalised needs and perspectives of communities and to be regarded as fair and inclusive. Applying the same principle to climate governance, direct participation from women especially those at the frontline of mitigating the consequences of climate change can open decision-making to a more diverse set of perspectives. It can potentially address the critique levelled against the 'feminisation of responsibility' such that women end up taking on compounded responsibilities without guarantee that they will equally benefit from initiatives rolled-out, or if their efforts will be matched by resources that sustain their rights and well-being.

Rethinking norms and practices in climate governance

The third element of our framework involves re-visioning institutional and strategic pathways for linking these localised knowledges and gender-inclusive political representation at all levels of governance. This last element explicitly identifies institutional and strategic pathways for gender-inclusive political representation at all levels of governance (see also Hemmati and Rohr 2009). Women's presence is not enough when prevailing norms and codes in climate governance spaces continue to privilege masculinised forms of activities, behaviours and solutions (Magnusdottir and Kronsell 2015). Part and parcel to achieving gender-responsive alternatives is rethinking the very ideological and material structures that underpin climate governance (Kaijser and Kronsell 2014; Kronsell 2013). This is exemplified by the incorporation of women's unpaid care labour and the multiple burdens women will increasingly face as climate change unfolds in the future if unequal gender divisions of labour remain unchanged. There is a need to avoid divesting more and more responsibility on women and communities without the matching economic support from the state and global community (Tanyag 2018). Rethinking the very structure of climate governance can begin strategically such that there is gender balance in all major decision-making bodies especially around the allocation of climate financing, the development of climate solutions and who gets to benefit from climate adaptation projects (for example, Green Climate Fund, n.d.).

Another pathway is by changing the 'rules of the game' in climate negotiations. Interestingly, at COP23 held in Fiji, the distinct Pacifica approach of *Talanoa* was adopted as an overarching theme and principle.[11] The concept of Talanoa emphasises the relational aspect of knowledge and the importance of narratives or drawing from everyday experiences for understanding the full extent of climate change. Through the initiative of Fiji, talanoa is mainstreamed as an alternative approach to climate change because:

> Blaming others and making critical observations are inconsistent with building mutual trust and respect, and therefore inconsistent with the Talanoa concept. Talanoa fosters stability and inclusiveness in dialogue, by creating a safe space that embraces mutual respect for a platform for decision making for a greater good.[12]

Talanoa aligns with core principles reflected in feminist methodology and care ethics for example, mutual dependence, trust and empathy (Vaioleti 2014; see for example, Robinson 2011). The rules of the game must change to be more conducive to different perspectives and approaches and how such strategic pathways can be made, need to be made accessible to women especially at the community level. At the same time, women require opportunities to acquire the institutional and scientific knowledge including awareness of existing national policies in order to be in the same level playing field to be able articulate and share their narratives and solutions. To be truly gender-responsive, climate governance practices and norms that determine how climate decision-making is undertaken need to be challenged and transformed.

Conclusions

Climate change and disaster risk reduction (DRR) solutions cannot be developed and scaled up while glossing over existing gender inequalities around access to resources and decision-making, as well as whose knowledge is valued. Realising peace, justice and security in the context of climate change governance requires us to take seriously how women's standpoints can transform our economic, political and ecological systems. These standpoints have the transformative potential to bridge the current fragmentation of humanitarian, security and development agendas. In this chapter, we have outlined a gender-responsive framework, the three elements of which can together yield a more holistic understanding of climate change as it intersects with and is compounded by food insecurity, resource-based conflicts and gender-based violence. Crucially, our framework suggests that the distinct contribution of a gender lens for climate governance is in revealing how the consequences of climate change cannot be addressed by technological fixes alone. In our research in Cambodia, Kenya and Vanuatu, women's lived experience on the margins draws our attention to the many impacts that are already being felt as multidimensional insecurities. When left unaddressed these insecurities will likely result even further in negative coping mechanisms among communities in the face of future climate risks and hazards – that will themselves accelerate climate change and heighten climate-induced disasters.

Future research needs to examine the context-specific ways in which gender-responsive climate governance can be implemented depending on the nature of the climate change risks, and the social cleavages that define women's experiences and capacity to adapt and take on leadership. Some key areas for investigation include where and how traditional, localised knowledges and climate science can join together to improve women's participation within climate governance? Another avenue to be explored concerns the extent to which women's climate change leadership can be used as an entry point for women's participation in other security concerns such as conflict/peacebuilding, economic crises and health pandemics, and vice versa. In order to generate a comprehensive picture of the multiple dynamics engendered by climate change, alternative solutions need to be gender-responsive to the diversity of women's and men's experiences and the kind of mitigation and adaptation strategies that they inform. Climate change solutions that are gender-responsive are also those that encompass comprehensive security to recognise that environmental justice and peace incorporating mental, health and spiritual well-being are necessary components to a gender-responsive climate change agenda.

Notes

1 The project is funded by a grant from the Australian Department of Foreign Affairs and Trade, 2017–2019. The research partnership consists of ActionAid Australia, ActionAid Global, ActionAid Cambodia, ActionAid Kenya, ActionAid Vanuatu, Huairou Commission and Monash University's Gender Peace and Security Centre, where both authors are affiliated. Primary data collection was done in all three countries from April to October 2018 with ethics clearance from the Monash University Human Research Ethics Committee, project no. 11825.
2 See also, UN Women's "In Focus: Climate action by, and for, for, women." www. unwomen.org/en/news/in-focus/climate-change.
3 For further details see "Mainstreaming Gender," *Green Climate Fund*, www.greencli mate.fund/how-we-work/mainstreaming-gender.
4 The UNFCCC Subsidiary Body for Implementation outlined the provision of a gender action plan and its priority areas for implementation in 2017 (FCCC/SBI/2017/L.29). The full text is available via http://unfccc.int/resource/docs/2017/sbi/eng/l29.pdf.
5 For a comprehensive list of publications and programs relating to DRR and gender mainstreaming within UNISDR, please see www.unisdr.org/files/42360_20pointcheck listforgendersensitived.pdf.
6 See for further details "Brides of the Sun," www.bridesofthesun.com/.
7 Field notes, 1–2 October 2018, Tanna, Vanuatu.
8 The movement continues on the legacy of Maathai on land and conservation issues to this day; www.greenbeltmovement.org/.
9 A full copy of the Act can be retrieved from www.legislation.govt.nz/act/public/ 2014/0051/latest/whole.html.

10 In Cambodia, the work by ActionAid has also shown how women's leadership in DRR is extending to women's leadership in community. However, we note that the clear limitations to their leadership is in that it remains not politically threatening to the strongman rule of Hun Sen and his ruling party. For a video profile of the women DRR champions in Cambodia, see www.actionaid.org/cambodia/videos/promoting-womens-leadership-disaster-risk-reduction-drr.

11 The Talanoa approach was adopted after extensive consultations throughout 2017 and under the direction of the COP 22 (Morocco) and 23 (Fiji) presidencies. The informal note by the Presidencies provides several features of Talanoa in relation to COP 'preparatory and political phases'. See Annex II to 1/CP.23, http://unfccc.int/files/bodies/cop/application/pdf/approach_to_the_talanoa_dialogue.pdf.

12 UN Climate Change, *2018 Talanoa Dialogue Platform*, http://unfccc.int/focus/talanoa_dialogue/items/10265.php.

References

Ackerly, Brooke, and Jacqui True. 2010. *Doing Feminist Research in Political and Social Science*. London: Palgrave Macmillan.

ActionAid. 2016a. *On the Frontline: Catalysing Women's Leadership in Humanitarian Action*. Report. Johannesburg: ActionAid. www.actionaid.org/sites/files/actionaid/on_the_frontline_catalysing_womens_leadership_in_humanitarian_action.pdf.

ActionAid. 2016b. "Through a Different Lens: ActionAid's Resilience Framework." Policy Document. Johannesburg: ActionAid. https://actionaid.nl/wp-content/uploads/2017/12/2016_through_a_different_lens_-_actionaid_resilience_framework_0.pdf.

ActionAid. 2017. *Agroecology, Empowerment and Resilience: Lessons from ActionAid's Agroecology and Resilience Project*. Report. Johannesburg: ActionAid. www.actionaid.org/sites/files/actionaid/agroecologyempowermentresilience-lessons_from_aer.pdf.

Agarwal, Bina. 2010. *Gender and Green Governance: The Political Economy of Women's Presence Within and Beyond Community Forestry*. Oxford: Oxford University Press.

Alston, Margaret, Kerri Whittenbury, Alex Haynes, and Naomi Gooden. 2014. "Are Climate Challenges Reinforcing Child and Forced Marriage and Dowry as Adaptation Strategies in the Context of Bangladesh?" *Women's Studies International Forum* 47: 137–44.

Arora-Jonsson, Seema. 2011. "Virtue and Vulnerability: Discourses on Women, Gender and Climate Change." *Global Environmental Change* 21, no. 2: 744–51.

Arora-Jonsson, Seema. 2013. *Gender, Development and Environmental Governance Theorizing Connections*. New York: Routledge.

Arora-Jonsson, Seema. 2014. "Forty Years of Gender Research and Environmental Policy: Where Do We Stand?" *Women's Studies International Forum* 47 (Part B): 295–308.

Asia Pacific Forum on Women, Law, and Development (APWLD). 2015. *Climate Change and Natural Disasters Affecting Women Peace and Security*. Report. March 13, Chiang Mai.

Bäckstrand, Karin. 2003. "Civic Science for Sustainability: Reframing the Role of Experts, Policy Makers and Citizens in Environmental Governance." *Global Environmental Politics* 3, no. 4: 24–41.

Bradshaw, Sarah. 2014. "Engendering Development and Disasters." *Disasters* 39, no. S1: S54–S75.

Buckingham, Susan. 2015. "The Institutionalisation and Masculinisation of Environmental Knowledge." In *Why Women Will Save the Planet*, edited by Jenny Hawley, 49–57. London: Zed Books.

Chamberlain, Gethin. 2017. "Why Climate Change is Creating a New Generation of Child Brides." *The Guardian*, November 26. www.theguardian.com/society/2017/nov/26/climate-change-creating-generation-of-child-brides-in-africa.

Chandra, Alvin, Karen McNamara, Paul Dargusch, Ana Maria Caspe, and Dante Dalabajan. 2017. "Gendered Vulnerabilities of Smallholder Farmers to Climate Change in Conflict-Prone Areas: A Case Study from Mindanao, Philippines." *Journal of Rural Studies* 50: 45–59.

Committee on the Elimination of Discrimination Against Women (CEDAW). 2018. "General Recommendation No. 37 on Gender-Related Dimensions of Disaster Risk Reduction in the Context of Climate Change." Policy Report, CEDAW/C/GC/37. Office of the United Nations High Commissioner for Human Rights (OHCHR). http://tbinternet.ohchr.org/Treaties/CEDAW/Shared%20Documents/1_Global/CEDAW_C_GC_37_8642_E.pdf.

Cramer, Wolfgang, Joël Guiot, Marianela Fader, Joaquim Garrabou, Jean-Pierre Gattuso, Ana Iglesias, Manfred A. Lange, Piero Lionello, Maria Carmen Llasat, Shlomit Paz, Josep Peñuelas, Maria Snoussi, Andrea Toreti, Michael N. Tsimplis, and Elena Xoplaki. 2018. "Climate Change and Interconnected Risks to Sustainable Development in the Mediterranean." *Nature Climate Change* 8, no. 11: 972–80.

The Economist Intelligence Unit. 2014. The South Asian Women's Resilience Index: Examining the Rose of Women in Preparing for and Recovering from Disasters. Report. www.actionaid.org/sites/files/actionaid/the_south_asia_womens_resilience_index_dec8_1.pdf.

Elson, Diane. 2012. "Social Reproduction in the Global Crisis: Rapid Recovery or Long-Lasting Depletion?" In *The Global Crisis and Transformative Social Change*, edited by Peter Utting, Shahra Razavi, and Rebecca Varghese Buchholz, 63–80. Basingstoke: Palgrave Macmillan.

Enarson, Elaine. 1998. "Through Women's Eyes: A Gendered Research Agenda for Disaster Social Science." *Disasters* 22, no. 2: 157–73.

Enloe, Cynthia. 1996. "Margins, Silences and Bottom Rungs: How to Overcome the Underestimation of Power in the Study of International Relations." In *International Theory: Positivism and Beyond*, edited by Steve Smith, Ken Booth, and Marysia Zalewski, 186–202. Cambridge: Cambridge University Press.

George, Nicole. 2014. "Promoting Women, Peace and Security in the Pacific Islands: Hot Conflict/Slow Violence." *Australian Journal of International Affairs* 68, no. 3: 314–32.

Global Witness. 2016. *On Dangerous Grounds*. London: Global Witness. Report. www.globalwitness.org/en/reports/dangerous-ground/.

Guijt, Irene, and Meera Kaul Shah, eds. 1998. *The Myth of Community: Gender Issues in Participatory Development*. London: ITDG Publishing.

Green Climate Fund. n.d. "About the Fund." https://www.greenclimate.fund/who-we-are/about-the-fund.

Haraway, Donna. 1988. "Situated Knowledge: The Science Question in Feminism and the Privilege of Partial Perspective." *Feminist Studies* 14, no. 3: 575–99.

Harding, Sandra. 1991. *Whose Science? Whose Knowledge? Thinking from Women's Lives*. Ithaca: Cornell University Press.

Hartsock, Nancy. 1983. "The Feminist Standpoint: Developing the Ground for a Specifically Feminist Historical Materialism." In *Feminism and Methodology: Social Sciences Issues,* edited by Sandra Harding, 157–80. Bloomington: Indiana University Press.

Hemmati, Minu, and Ulrike Rohr. 2009. "Engendering the Climate-Change Negotiations: Experiences, Challenges, and Steps Forward." *Gender & Development* 17, no. 1: 19–32.

Hooks, Bell. 2000. *Feminist Theory: From Margins to Center* (2nd ed.). London: Pluto Press.

Hozic, Aida, and Jacqui True, eds. 2016. *Scandalous Economics: The Politics of Gender and Financial Crises*. New York: Oxford University Press.

IUCN Global Gender Office (IUCN GGO). 2015. "Women's Participation in Global Environmental Decision Making." EGI Supplemental Report. IUCN: Washington, DC. http:// genderandenvironment.org/resource/womens-participation-in-global-environmental-decision-making-an-egi-supplemental-report/.

Jordan, Joanne C. 2019. "Deconstructing Resilience: Why Gender and Power Matter in Responding to Climate Stress in Bangladesh." *Climate and Development* 11, no. 2: 167–79. DOI: 10.1080/17565529.2018.1442790.

Kaijser, Anna, and Annica Kronsell. 2014. "Climate Change Through the Lens of Intersectionality." *Environmental Politics* 23, no. 3: 417–33.

Kauffman, Craig, and Pamela Martin. 2018. "Constructing Rights of Nature Norms in the US, Ecuador, and New Zealand." *Global Environmental Politics* 18, no. 4: 43–62.

Kronsell, Annica. 2013. "Gender and Transition in Climate Governance." *Environmental Innovation and Societal Transitions* 7(June): 1–15.

Lee-Koo, Katrina. 2012. "Gender at the Crossroad of Conflict: Tsunami and Peace in Post-2005 Aceh." *Feminist Review* 101, no. 1: 59–77.

Maathai, Wangari. 2006. *The Green Belt Movement: Sharing the Approach and the Experience*. New York: Lantern Books.

Magnusdottir, Gunnhildur Lily, and Kronsell, Annica. 2015. "The (in)Visibility of Gender in Scandinavian Climate Policy-Making." *International Feminist Journal of Politics* 17, no. 2: 308–26.

Mies, Maria, and Vandana Shiva. 2014. *Ecofeminism* (2nd ed.). London: Zed Books.

Mohanty, Chandra. 1991. "Introduction Cartographies of Struggle: Third World Women and the Politics of Feminism." In *Third World Women and the Politics of Feminism*, edited by Chandra Mohanty, Ann Russo and Lourdes Torres, 1–47. Bloomington: Indiana University Press.

Mora, Camilo, Daniele Spirandelli, Erik C. Franklin, John Lynham, Michael B. Kantar, Wendy Miles, Charlotte Z. Smith, Kelle Freel, Jade Moy, Leo V. Louis, Evan W. Barba, Keith Bettinger, Abby G. Frazier, John F. Colburn IX, Naota Hanasaki, Ed Hawkins, Yukiko Hirabayashi, Wolfgang Knorr, Christopher M. Little, Kerry Emanuel, Justin Sheffield, Jonathan A. Patz, and Cynthia L. Hunter. 2018. "Broad Threat to Humanity from Cumulative Climate Hazards Intensified by Greenhouse Gas Emissions." *Nature Climate Change* 8, no. 12: 1062–71.

Moraes, Andrea, and Patricia E. Perkins. 2009. "Deliberative Water Management." In *Eco-Sufficiency and Global Justice: Women Write Political Ecology*, edited by Ariel Salleh, 140–54. New York: Pluto Press; Melbourne: Spinifex Press.

Nayak, Nalini. 2009. "Development for Some is Violence for Others: India's Fisherfolk." In *Eco-Sufficiency and Global Justice: Women Write Political Ecology*, edited by Ariel Salleh, 109–20. New York: Pluto Press; Melbourne: Spinifex Press.

Neumayer, Eric, and Thomas Plümper. 2007. "The Gendered Nature of Natural Disasters: The Impact of Catastrophic Events on the Gender Gap in Life Expectancy, 1981–2022." *Annals of the Association of American Geographers* 97, no. 3: 551–66.

Pearse, Rebecca. 2017. "Gender and Climate Change." *WIREs Climate Change* 8: e451. DOI: 10.1002/wcc.451.

Resurrección, Bernadette P. 2013. "Persistent Women and Environment Linkages in Climate Change and Sustainable Development Agendas." *Women's Studies International Forum* 40 (September–October): 33–43.

Robinson, Fiona. 2011. *The Ethics of Care: A Feminist Approach to Human Security*. Philadelphia: Temple University Press.

Smith, Linda Tuhiwai. 2013. *Decolonizing Methodologies. Research and Indigenous Peoples* (2nd ed.). London: Zed Books.

Smyth, Ines, and Caroline Sweetman. 2015. "Introduction: Gender and Resilience." *Gender and Development* 23, no. 3: 405–14.

Tanyag, Maria. 2018. "Resilience, Female Altruism, and Bodily Autonomy: Disaster-Induced Displacement in Post-Haiyan Philippines." *Signs: Journal of Women in Culture and Society* 43, no. 3: 563–85.

True, Jacqui. 2012. *The Political Economy of Violence against Women*. New York: Oxford University Press.

True, Jacqui. 2013. "Gendered Violence in Natural Disasters: Learning from New Orleans, Haiti and Christchurch." *Aotearoa New Zealand Social Work* 25, no. 2: 78–98.

True, Jacqui. 2016. "Women, Peace and Security in Asia Pacific: Emerging Issues in National Action Plans for Women, Peace and Security." Discussion Paper. http://www2.unwomen.org/-/media/field%20office%20eseasia/docs/publications/2016/12/1-nap-jt-for-online-r3.pdf?v=1&d=20161209T065558.

True, Jacqui, and Maria Tanyag. 2018. "Violence Against Women/Violence in the World: Toward a Feminist Conceptualization of Global Violence." In *Routledge Handbook on Gender and Security*, edited by Caron Gentry, Laura Shephered, and Laura Sjoberg, 15–26. London: Routledge.

True, Jacqui, and Yolanda Morales-Riveros. 2018. "Toward Inclusive Peace: Analysing Gender-Sensitive Peace Agreements 2000–2016." *International Political Science Review* 40, no. 1: 23–40.

UN Women. 2015. *Preventing Conflict, Transforming Justice, Securing the Peace: A Global Study on the Implementation of United Nations Security Council Resolution 1325*. New York: United Nations. http://wps.unwomen.org/en.

Vaioleti, Timote. 2014. "Talanoa: Differentiating the Talanoa Research Methodology from Phenomenology, Narrative, Kaupapa Maori and Feminist Methodologies." *Te Reo* 56–57: 191–212.

Van Aelst, Katrien, and Nathalie Holvoet. 2016. "Intersections of Gender and Marital Status in Accessing Climate Change Adaptation: Evidence from Rural Tanzania." *World Development* 79: 40–50.

3 Why gender does not stick

Exploring conceptual logics in global disaster risk reduction policy

Sara Bondesson

Introduction

In the Anthropocene, climate change, globalisation, urbanisation and increasing interconnectedness between physical, human and technological systems are all driving forces in contemporary disasters (Di Baldassarre et al. 2018; Steffen et al. 2011). Disaster vulnerability often follows social and political inequalities (Bankoff, Frerst, and Hilhorst 2004; Bondesson 2017; Enarson and Hearn Morrow 1998; Fothergill, Maestas, and De Rouen Darlington 1999; Fothergill and Peek 2004; Jones and Murphy 2009; Tierney 2014; Wisner 2003) and gendered differences in vulnerability are particularly salient (Arora-Jonson 2011; Bradshaw 2013; Collins et al. 2014; Enarson and Pearson 2016; Neumayer and Plümper 2007). Disaster risk reduction (DRR) is a global field of policy planning and practice that aims to minimise disaster risks and mitigate the adverse impacts of hazards. As such, DRR policymaking must take gendered vulnerabilities to disasters into account, since the lack of gender perspectives yields ineffective and inequitable risk reduction.

In this chapter, I conduct an analysis of the Sendai Framework for action; the central policy document in the global field of DRR. Since this Framework sets the agenda for the wider field of DRR practice across the globe, it is important to scrutinise for anyone interested in problems of gender-based disaster inequality. The Sendai Framework acknowledges issues of gender inequality yet, as discussed in this chapter, does so in a rather limited and somewhat problematic way. To understand the shortcomings I make use of Carol Bacchi's "What's the Problem Represented to Be?" (WPR) approach to policy analysis. With help of this analytical tool, I identify two conceptual logics in the Framework that prevent full incorporation of a gender perspective. Firstly, a relief logic assumes a temporality of acuteness and prescribes male-dominated professional domains as experts. This makes a political analysis of gender inequality unintelligible. The relief logic also renders silent political solutions to alter gender inequalities. Secondly, a techno-managerial logic proposes technical and managerial solutions to problems of disaster risk. This rewrites solutions to structural inequalities as problems that can be solved technologically and managerially – in contrast to the types of political solutions needed to alter gender inequalities.

I am part of the network of research experts who contribute to shaping the DRR agenda. DRR research is very policy-oriented. Practitioners are often integrated in academic conferences and research is regularly made with clear policy recommendations. A feminist activist turned researcher, I found the disaster research field to be alarmingly lacking in social and political inequality analysis. Although there are pockets of research centres around the globe[1] that do interesting work on issues of differentiated vulnerability, much of the research remains focused on subjects such as meteorological modelling, infrastructural vulnerability or financial disaster insurance modelling. No matter their talents, natural and technical scientists are not trained in analysis of social and political dimensions in risk and vulnerability. Therefore, rather than diving head in to the intricacies of including politically marginalised groups in recovery processes (a topic close to my heart), for example, I often find myself simply arguing for the relevance of focusing on social and political dimensions of disasters. In this chapter, I systematically explore these weaknesses of the DRR field by exploring some of the conceptual logics that make it hard for gender sensitive perspectives to cut through the noise of the field's techno-scientific agenda.

Contributing to DRR and WPR policy analysis

With this analysis, I aim to contribute to two main areas: DRR and WPR policy analysis. Firstly, my analysis contributes to a better understanding of why global DRR policy has been so slow to incorporate gender perspectives. Despite ongoing advocacy from civil society-based transnational advocacy networks consisting of researchers, UN personnel and activists integrating gender perspectives into DRR policy, it is a process that has produced both limited and somewhat problematic policy results (as I discuss below) (Bradshaw 2014; DAW/ISDR; Seager 2006; UNISDRb; UNISDR, UNDP, and IUCN 2009). This issue can also be explored through other analytical approaches, ranging from feminist institutional (Krook and Mackay 2011), to norm diffusion approaches (Hollis 2017) or framing (Björnehed and Erikson 2018), to more agentic-oriented approaches (True in Rai and Waylen 2008). However, discursive analysis may offer something particular to an understanding of why it seems so hard to shift the perspectives of the DRR field. Looking at underlying conceptual logics can help explain why a gender perspective does not 'stick' despite multiple advocacy attempts from researchers and activists. This fills an important gap in DRR policy research, since so few discursive readings of the DRR field exist. Simon Hollis (2014) argues that, with a few exceptions, DRR research has a bias toward agency-focused studies. The role of specific organisations, people or communities is highlighted, while there is a lack of studies that focus on the discursive boundaries that could be understood to "demarcate the horizons of possibility" (Hollis 2014, 343). However, as others have argued, investigating discursive boundaries and ideational foundations help in understanding the rules that govern the game (Hollis 2014, 2017; Larsson 2015). Among the exceptions is Sarah Bradshaw, who analyses gender perspectives in the development discourse and compares it with DRR policies (2014), and Reidar Staupe-Delgado and Bjorn Ivar Kruke (2017), who demonstrate that a contingency approach to DRR inhibits proactive preparedness. However, no study sheds light on underlying conceptual logics that may hinder full incorporation of a gender perspective in the Sendai Framework. The second contribution is to the increasing WPR scholarship. To date, WPR applications focus on policy areas, such as education, equity, health, drug and alcohol or development policies in Australia, Sweden, Finland, Canada, the UK and India (Bacchi and Goodwin 2016; Calvo 2013; Carson and Edwards 2011; Hearn and McKie 2010; Payne 2014). Many feminist WPR scholars have used this approach to identify underlying assumptions of existing policies on gender equality. In contrast, I use the tool to analyse how underlying conceptual logics pertaining to other problem areas may prevent integration of a gender perspective into DRR policy.

Gender in global DRR policy

In this section, I establish the context of global DRR policy and introduce the Sendai Framework 2015–2030. I then give an overview of the extent to which the Framework incorporates gender issues and discuss how gender is treated in both limited and problematic ways in the policy.

Global DRR policy and the Sendai Framework 2015–2030

DRR is a global field of policy planning and practice that involves systematic efforts to analyse and reduce the causal factors of disasters. Examples of activities range from vulnerability reduction, land and environmental management, to improvement of preparedness and early-warning systems (UNISDRa). The United Nations International Strategy for Disaster Reduction (UNISDR) has the lead but the larger field of DRR consists of international organisations, national legislatures, NGOs, academia and businesses (Hannigan 2012). The Sendai Framework 2015–2030 is the main policy document of global DRR policy and sets the agenda for DRR practice across the globe. It is a 15-year, voluntary and non-binding agreement for UN member states and came about as a successor to the Hyogo Framework for Action (HFA) 2005–2015. Negotiations and consultations were ongoing 2012–2015, and at the Third UN World Conference in Sendai City, Japan, the Framework was adopted by UN member states. Several thousand stakeholders participated in the conference, including parliaments, civil society representatives, national and local government representatives, private sector actors, researchers and other UN organisations (UNISDRd 2018). In the Sendai Framework, the codified norms of the field are found, ultimately prioritising and ordering themes and areas, and setting the direction. The Framework also functions as a catalyst for the funding and programming of DRR projects across member states and is an important benchmark in terms of programme assessments. The Sendai Framework specifies the expected outcome of the policy as the "substantial reduction of disaster risk and losses in lives, livelihoods and health and in the economic, physical, social, cultural and environmental assets of persons, businesses, communities and countries" (Sendai Framework 2015–2030, 12). To achieve this outcome the goal is to:

> [P]revent new and reduce existing disaster risk through the implementation of integrated and inclusive economic, structural, legal, social, health, cultural, educational, environmental, technological, political and institutional measures that prevent and reduce hazard exposure and vulnerability to disaster, increase preparedness for response and recovery, and thus strengthen resilience.
>
> (Sendai Framework 2015–2030, 12)

The outcome and goal are further operationalised into seven targets and four priorities for action, and these priorities are then thematically ordered by a clarification of the achievements needed at the national, regional and local levels (Sendai Framework 2015–2030).

Limitations of a gender perspective in the Sendai Framework

Regarding gender, there are a few passages in the policy that focus on women's heightened vulnerability (together with vulnerability for other social categories like children, elderly persons and people with disabilities). There are also segments indicating the need for increased participation by women in DRR decision-making. One of the document's guiding principles states that DRR requires "empowerment and inclusive, accessible and non-discriminatory participation, paying special attention to people disproportionately affected by disasters, especially the poorest. A gender, age, disability and cultural perspective should be integrated in all policies and practices, and women and youth leadership should be promoted" (Sendai Framework 2015–2030, 13). Thus, in some of its wording the Framework addresses gender inequality. It also frequently mentions "stakeholder involvement" throughout target areas and priorities for action. Stakeholders are defined as women, children and youth, persons with disabilities, older persons, indigenous peoples, migrants, academia and the media (Sendai Framework 2015–2030, 23).

Promising as these statements may seem, they are both limited and somewhat problematic in three regards, compared to what a gender perspective could entail. First, there are no concrete priorities for action or indicators pertaining to how increased involvement of women should be attained. This lack of indicators is conspicuous. Without specifying what needs to be done, commitments to address problems of gendered vulnerabilities remain empty promises. They will most likely not be translated into concrete national policies, programmes or accountability. Furthermore, this lack of indicators is telling when compared to other problem areas acknowledged in the policy, such as a lack of technical capacity for example, which is followed up by specified and detailed indicators. Second, a focus on women's increased participation in numbers could be said to harbour a simplified notion about descriptive representation. Debates on descriptive and substantive representation have problematised whether more women in decision-making positions lead to political solutions for greater gender equality (Phillips 1995; Pitkin 1967). Although empirical and theoretical developments have been made linking 'critical masses' of women to substantial shifts in agenda-setting (Dahlerup 2006; Kanter 1977), and standpoint feminist theory has emphasised how women's subject positions in social hierarchies shape their outlook on problems and solutions compared to male elites (Harding 1991; Hartsock 1985), many uncertainties remain as to whether more women in politics equals more gender sensitive policies. This leads me to my third and final point, namely that the policy concerns itself with women and women's issues, rather than incorporating a gender perspective. A gender perspective needs to move beyond a mere focus on the number of women in DRR decision-making positions. Instead of focusing on the number of women in DRR decision-making positions, a gender perspective would address the underlying problem of gender inequality that results in skewed distribution of risks and vulnerability. It implies a focus on the social orders that produce gendered vulnerabilities. Disasters are gendered, meaning that the gender order of any given society produces differences in vulnerability (Arora-Jonson 2011; Bradshaw 2013; Collins et al. 2014; Enarson and Pearson 2016; Neumayer and Plümper 2007). The gender order is "the overall structure of gender relations in a particular society, at a particular time in history. It is a patterned system of ideological and material practices, performed by individuals in a society through which power relations between men and women are made, and remade, as meaningful" (Pilcher and Whelehan 2017, 60). The gender order denotes different roles for women, girls, men and boys, or as Eric Neumayer and Thomas Plümper (2007, 551) note, "different socially constructed vulnerabilities that derive from the social roles men and women assume, voluntarily or involuntarily, as well as existing patterns of gender discrimination". A gender perspective highlights how gender roles are produced through social institutions that shape norms of behaviour, as well as access to rights and power. In this sense everyone is gendered since everyone has a social position in the gender order, men and women, as well as genderqueer and non-binary persons, meaning that a mere focus on women is too limited.

Many academic developments in understanding the effects of gender roles, relations and identities in disaster situations have been made during the last decade, most notably Elaine Enarson and Betty Hearn Morrow's *The Gendered Terrain of Disasters* (1998) (see also Cupples 2007; David and Enarson 2012; Eklund and Tellier 2012; Enarson and Pearson 2016; Gaillard et al. 2017; Gault 2005; Mazurana, Benelli, and Walker 2013). In countries marked by gender inequalities, women and girls often carry responsibility for the household and childcare, whereas men take on formal positions in the labour market. Due to social control, women and girls may be restricted in their mobility in comparison to men; they can access medical facilities to a lesser extent, are overrepresented in insecure and informal jobs, have limited opportunities to migrate to find jobs, and are often affected by gender-based and sexual violence increasing after disasters

(Austin 2016; Fothergill, Maestas, and De Rouen Darlington 1999; Sultana 2010; UNFPA). To incorporate a gender perspective in the context of DRR would be to identify and analyse the underlying social orders that produce differentiated vulnerability as well as prescribe political solutions that challenge these social orders. It would demand that DRR policy devises necessary political, economic and social solutions to alter skewed distribution of power, wealth and rights. None of these aspects of a gender perspective are incorporated in the Sendai Framework. It is therefore interesting to dig deeper to understand why the incorporation of a gender perspective is so limited.

Exploring the Sendai Framework through the WPR approach

In this section, I will introduce the theoretical foundations of the WPR approach followed by an outline of the analytical questions that guide my analysis. Global DRR policy is a typical instance of global governance: a global collective action problem exists that needs to be solved through common solutions such as promoting risk and vulnerability reduction (Hollis 2017). Global governance literature has explored the "global language in which epistemic communities influence our political vocabulary and even imagination, as well as their input into framing and legitimizing policy at the global level" (Rai and Waylen 2008, 1). Feminist readings of global governance have often investigated discursive biases favouring the status quo of gender inequality (Peterson and Sisson Runyan 2010; Rai and Waylen 2008; Tickner 1992; Youngs 2004). I inscribe myself in this tradition by making use of a particular analytical tool for policy analysis, namely Bacchi's WPR approach (Bacchi 1999, 2009; Bacchi and Goodwin 2016). Drawing on Foucauldian policy analysis, Bacchi's post-structural[2] WPR approach is a tool for analysing underlying assumptions in policies. Its core premise is that "what one proposes to do about something reveals what one thinks is problematic (needs to change)" (Bacchi in Betslas and Beasley 2012, 21). Therefore, by looking at suggested policy solutions, the approach identifies the problem representations that underpin the policy. WPR analysts refute the idea that policy makers simply react to existing social problems. Instead, they acknowledge how policy solutions produce certain representations of problems. Policy analysts scrutinise the unexamined assumptions that need to be in place for those representations to be intelligible (Bacchi and Goodwin 2016). One example is from my time as a stressed PhD candidate, where the doctoral candidates of the department were offered a stress management course. The solution put forward, cognitive behavioural training to manage stress, implied a certain representation of the problem, which was the individual's inability to manage the workload. However, other possible solutions could have been suggested, for instance structural changes in the type of support doctoral candidates receive from the department, which in turn would represent an alternative understanding of the problem, which is that doctoral candidates are stressed because they lack proper institutional support. As these examples show, problem representations "limit what can be talked about as relevant, shape people's understandings of themselves and the issues, and impact materially on people's lives" (Bacchi in Betslas and Beasley 2012, 22).

Global DRR discourse may harbour implicit representations of reality that highlight certain issues while silencing others (Bacchi and Goodwin 2016). The DRR field could be conceptualised as a 'knowledge regime' (Foucault 1980), in that it may prescribe certain rules for what is considered possible versus impossible social action (Jorgensen and Phillips 1999). It can also be conceptualised as a 'hegemonic discourse' (Bergström and Boréus 2005; Fung 2005; Young 2001), containing assumptions about who is an expert, and widely accepted generalisations about how society operates, which blocks the influence of some people and ideas, while magnifying others (Young 2001). All of these conceptualisations point to similar discursive functions in terms of what is named or unnamed, who is included or excluded, and which power relations are constituted and reproduced in the text. Bacchi's WPR approach neatly captures these various ways of understanding discourses into a useful analytical tool for policy analysis.

By looking at particular policy proposals or solutions, the analyst can draw (interpretive) conclusions about which causes to a problem are presupposed. To do so is less about decoding how agents frame certain issues to their own benefit, but rather about probing for deeper conceptual underpinnings. It is a way of drawing attention to what may be hidden in language that at the outset may seem self-evident. The goal of the approach is to problematise policies by bringing silences into the open (Bacchi 1999).

A note on the concept of 'problem' in relation to the issue of gender inequality is in place here. The WPR approach focuses on implicit problem representations underpinning a policy and puts forward the claim that a representation of the problem is one out of many thinkable others. As Bacchi (2009, xi) states, it is "unwise and inappropriate to think that 'problems' somehow exist in the world". Problems are, in Bacchis writing, shaped and produced by implicit assumptions about reality. In this sense, 'problems' are created within the policy process rather than existing outside of it; in a problematizing process that categorises certain versions of reality as true whereas other versions are downplayed or silenced. Given this, how should one understand the empirically established fact that women in general are more vulnerable than men in relation to disasters, if not as a 'problem' that global DRR policymaking ought to address? Can we even talk about gendered vulnerability as a problem that has an individual existence beyond our (various) understandings of it? However, as Betslas (in Betslas and Beasley 2012, 39), who has conducted WPR analysis of Australian poverty policy, discusses; the focus on problem representations does not "put in question the reality of poverty understood as the experience of extreme disadvantage and deprivation". Neither does an analysis of the implicit problem representations of the Sendai Framework negate the reality of gendered vulnerabilities. Rather, it investigates how this condition is understood (or perhaps how it is neglected or silenced by other competing problem representations).

Questions that guide the analysis

Bacchi suggests several questions that can be used in conducting a WPR analysis (Bacchi 2009). Bacchi's set of analytical questions starts with identifying manifest wordings and phrases of policy proposals. Subsequent questions then continually probe deeper and deeper into the underlying layers of meaning, in order to explore the underpinning ideas that make certain policy proposals intelligible (and render other policy proposals unintelligible). With some slight revisions of Bacchi's suggested questions, I pose the following three questions in my analysis of the Sendai Framework 2015–2030:

* Question 1: What's the problem represented to be?

By asking this question I look at the proposed solutions of the Sendai Framework and by identifying these proposed solutions, try to draw reasonable conclusions about what kinds of problems are assumed to be important, given the way in which the solutions are put forward in the text. If identifying problem representations is a rather straightforward task, the subsequent questions of the WPR approach tease out more interesting analysis. The second question I pose to the policy is:

* Question 2: What conceptual logics underpin the representations of the problem and where do these conceptual logics come from?

The task here is to start identifying conceptual logics that underline the problem representations identified in question one. Such conceptual logics underpin what is explicitly articulated in policy documents. The analysis therefore moves beyond the manifest level of policy text to identify underlying conceptual logics (Braun and Clarke 2006). The function of conceptual logics is to demarcate the type of political analysis that is rendered thinkable or viable within a policy field (Bacchi 2009). This question further begs consideration of the practical field that the conceptual logics stem from. The WPR approach to policy analysis is in a way an anthropological one, as policies and policy processes are seen as cultural projects. Inquiring about policies' sources hence requires a broader look at the larger practical field of which the policy is a part (Bacchi 2009). Therefore, I also ask questions about the broader field of DRR from which the problem representations and conceptual logics emerge. Here I consider previous research about the practices and processes of the DRR field, because these practices and processes lay the foundation for how problems are represented (Bacchi in Betslas and Beasley 2012). In this sense, the analysis moves beyond the Sendai Framework, to include references to the wider field of DRR. The third step in my analysis is a discussion of whether and how the conceptual logics render integration of gender perspectives difficult. In doing so, I address the following questions based on the WPR approach:

* Question 3: What remains unproblematic or silent in these conceptual logics and do the logics close off or limit, certain social interventions?

I pose these questions in order to draw attention to the meaning-making involved in global DRR policymaking. Bacchi states that analysts ought to look at how problem representations close off, or limit, certain social interventions (while promoting others). Focus is on the material impacts that these discursive effects may have, since how problems are represented affect people's lives (Bacchi 2009). Instead of seeing the Sendai Framework as a policy that simply responds to pre-existing problems, by asking these questions I highlight how particular issues dominate the DRR agenda, whereas other issues are downplayed. This in turn affects what will be done, or not done politically to alter or challenge gender inequality in relation to disasters (Bacchi 2009).

Analysis of the Sendai Framework 2015–2030

In the following analysis of the Sendai Framework 2015–2030 I answer the analytical questions stipulated above. In the first section, I highlight two different problem representations that mark the policy text. In the second passage, I discuss which conceptual logics underpin these representations of the problem and discuss from where these conceptual logics stem. In the third section, I reflect on what is left unproblematic or silent in these conceptual logics and if they close off or limit social interventions relating to gender.

Question 1: What's the problem represented to be?

The Sendai Framework is marked by two different problem representations: *lack of technical and managerial capacity* and *lack of capacity to deal with acute management of disasters*. There is frequent promotion in the Sendai Framework of different technical and managerial solutions to deal with disaster risks, such as monitoring and early-warning systems, technical capacity-building, in-situ and space-based earth observations, space technologies, remote sensing, geographical information systems, hazard and climate modelling, forecasting models, as well as a plethora of different cost and benefits risk assessment tools. Emphasis is on prediction, measurement, containment and control (Sendai Framework 2015, 15). For instance, under Priority 1: Understanding Disaster Risk, one of the ways in which this should be done is through the strengthening of "technical and scientific capacity to capitalise on and consolidate existing knowledge and to develop and apply methodologies and models to assess disaster risks, vulnerabilities and exposure to all hazards" (Sendai Framework 2015, 14). In another, "social, economic, educational and environmental challenges and disaster risks" ought to be remedied by the promotion of "investments in innovation and technology development in the long-term, multi-hazard and solution-driven research in disaster risk management" (Sendai Framework 2015, 14). In some segments, the Framework speaks about underlying disaster risk factors, in other words social and economic problems that may come to bear on disaster risks. If interpreted generously, gender inequalities could be an example of such a risk factor, even if this is not explicitly stated. Yet, when solutions are proposed to remedy these underlying risk factors, we find that they should be addressed through "disaster risk-informed public and private investments" (Sendai Framework 2015, 13). It is not immediately evident what is meant by risk-informed public and private investments but suffice is to say, it does not have any clear-cut bearing on gender inequality. Furthermore, taking a look at the solutions promoted, the following quote illustrates how they are often centred on activities that would promote improved acute disaster management:

> [P]romote regular disaster preparedness, response and recovery exercises, including evacuation drills, training and the establishment of area-based support systems, with a view to ensuring rapid and effective response to disasters and related displacement, including access to safe shelter, essential food and non-food relief supplies, as appropriate to local needs.
>
> (Sendai Framework 2015, 20)

Question 2: Which conceptual logics underpin the representations of the problem and from where do these conceptual logics arise?

The two problem representations are in turn underpinned by two conceptual logics: the relief logic and the techno-managerial logic. I discuss these logics in the following sections.

The relief logic

The problem representation of *lack of capacity to deal with acute management of disasters* is underpinned by a conceptual logic of relief. DRR as a field is not primarily about disaster relief per se, but instead about preventing the effects of disasters through various long-term solutions. Yet, there are still tenets within the field that could be said to be characterised by a conceptual logic of relief. This logic is part of a legacy of a shortsighted, problem-solving discourse that characterises the closely related field of humanitarianism. The DRR field first emerged out of the international humanitarian relief field (Hollis 2014). This field was initially institutionalised with the creation of the International Committee of the Red Cross (ICRC) in 1877, followed by the League of Nation's establishment of the International Relief Union (IRU) in 1927, as well as the various international cooperation projects that grew out of them. A report by the UN Secretary General in 1971 and the instigation of the Office of the Disaster Relief Coordinator widened the horizon of humanitarianism to also include prevention, control, preparedness and reconstruction after disasters (Hollis 2014). The last few decades have seen a shift from reactive to long-term proactive disaster risk and vulnerability reduction (Rogers in Chandler and Coaffee 2017; UNISDR, UNDP, and IUCN 2009). With the creation of the International Decade for Natural Disaster Reduction (IDNDR) and a sharp increase in the geographical spread and number of NGOs and international organisations involved in disaster risk management, humanitarianism and disaster risk reduction are now two separate yet interconnected fields (Hollis 2014). This interconnectedness means that the tenets of disaster management (short time spans, focus on relief and recovery rather than prevention, preparedness and societal change) still bleed into the DRR field. Disasters and crises bear with them a certain temporality (Narby 2014) in the sense that managing them is about solving immediate problems rather than underlying problems. Since the disaster management discourse is still an integral part of DRR discourse, DRR analysis still suffers from the somewhat problematic legacy of the more shortsighted body of knowledge that permeates disaster management.

The techno-managerial logic

The second problem representation: *lack of technical and managerial capacity* is underpinned by what I would call a techno-managerial logic. This logic probably stems from the fact that the DRR discourse is generally influenced by technical schemes such as Critical Infrastructure Protection or different rationalistic agendas of, for example, financial risk regulation, as well as organisational management theory (Rogers in Chandler and Coaffee 2017). A techno-managerial logic puts focus on technical and managerial solutions to handle risk and vulnerability. According to this logic, vulnerability can be reduced if, for example, better technical warning systems can be developed, more effective infrastructural protection can be put into place, sensors to measure seismic activity can be improved and organisations can better their ability to deal with logistical challenges. These are all measures that improve technical, organisational and infrastructural disaster preparedness systems. Yet they do less in terms of addressing social and political problems of underlying gender inequality. This is mirrored by tendencies in other instances of global governance, where techno-managerial approaches to global problems have been said to depoliticise global governance, as suggested by Lily Gunngildur Magnusdottiar and Annica Kronsell (2015; see also Kronsell 2013).

Question 3: What remains unproblematic or silent in these conceptual logics and do the logics close off or limit certain social interventions?

Starting with the relief logic, it has a couple of characteristics that function to limit or render difficult the incorporation of a gender perspective. Firstly, it assumes a temporality of acuteness. Long-term analysis of underlying problems is made difficult in light of the temporal aspect. Gender inequality is a historically contingent underlying problem of many societies, and often constitutes the core of social orders in most contexts. However, a relief logic will not be able to capture such endemic problems. The temporal aspect of the relief logic thus precludes political analysis of long-standing historical problems such as gender inequality, as well as makes it hard to promote political solutions that may alter long-standing inequalities. Secondly, the relief logic prescribes certain male-dominated professional domains as experts, most often emergency management organisations. This renders analysis of gender inequality and suggestions for altering the problem less likely. Responsibility for managing disasters has, in many contexts, moved from a military to an emergency service sector (mainly blue light authorities, such as police, fire and ambulance personnel) over the last few decades. Bureaucratic organisations and command-and-control hierarchies have traditionally been the norm of disaster management (Dynes 1994; Gardner 2013; Neal and Phillips 1995; Whittaker, McLennan and Handmer 2015). This in turn means that many activities are "operationalized from the top down through hierarchical and paramilitary organizational structures" (Rogers in Chandler and Coaffee 2017, 18). In light of top-down approaches, disaster relief has traditionally involved little local involvement (McEntire 1997). It is only recently that the importance of integrating local knowledge into disaster preparedness and management has been acknowledged. Although contemporary DRR policymaking increasingly values the knowledge of local communities (Reichel and Fromming 2014), empirical studies still show a lack of realisation of this goal in practical disaster management and preparedness programmes (Allen 2006; Mazurana, Benelli, and Walker 2013; Raymond et al. 2010). Lack of local community involvement often results in differential treatment of social groups. Supraya Akerkar (2007), for example, has shown how certain groups of women were excluded from relief and recovery assistance due to patterns of institutionalised discrimination in countries affected by the 2004 Indian Tsunami. Similar findings are reflected in a study by Subhasis Bhadra (2017). However, when the empowerment of local communities is highlighted in policy texts, what is often referred to is not disaster-affected communities per se but rather first responders in (male-dominated) emergency management organisations such as the police and fire protection services (Rogers in Chandler and Coaffee 2017). Fire fighters, police and ambulance personnel are not usually trained in highlighting gender inequality, or in analysing its root causes or proposing political solutions to alter it. Yet, it is these male-dominated relief-oriented professional domains that are prescribed as experts in the relief logic.

Moving on to the techno-managerial logic, it stresses technical solutions to DRR, which have little room for critical power analysis of gender inequality. Other researchers have come to similar conclusions. Simon Hollis for example claims that DRR policies are characterised by a "pervasive scientific logic of rationality, responsibility and control, assuming an Enlightenment's idea of 'man's control over nature' through technological advances" (Hollis 2014, 352). Kaira Zoe Alburo-Cañete, in a (2014) study of post-disaster intervention in the Philippines, states that the interventions "remained grounded in medico-technical and managerial narratives" and that these narratives rendered issues of gender and sexuality disembodied and apolitical. When gendered and sexual dimensions of disasters are reduced to technical problems with technical solutions, this reproduces vulnerability, Alburo-Cañete argues. In sum, the techno-managerial logic rewrites solutions to structural inequalities as problems that can be technologically and managerially solved. This in contrast to the types of political solutions that are needed to alter gender inequalities.

Conclusions

In this chapter, I have conducted an analysis of the Sendai Framework for action, the central policy document of DRR. Although the Framework acknowledges issues of gender inequality, it does so in a limited and somewhat problematic way. To understand these shortcomings, I made use of Bacchi's WPR approach and identified two conceptual logics that prevent full incorporation of a gender perspective. The relief logic assumes a temporality of acuteness and prescribes male-dominated professional domains as experts, making a political analysis of gender inequality unintelligible. The techno-managerial logic prescribes technical and managerial solutions to problems of disaster risk, which rewrite solutions to structural inequalities as problems that can be solved technologically and managerially.

The analysis contributes to DRR scholarship, mainly because previous studies on gender and DRR policy have not looked at discursive hindrances before. However, research on gender and DRR policy is generally scarce and other types of studies are also needed in order to seek answers as to why incorporating a gender perspective is difficult. Institutional analysis would, for example, be viable, particularly feminist approaches to identify how institutional processes construct and maintain gendered power relations (Krook and Mackay 2011). One could also approach the issue by looking at norm diffusion or norm internalisation (Hollis 2017), or through framing analysis (Björnehed and Erikson 2018) by looking at whether the frames of gender advocacy movements align or contrast with delegates' frames at global DRR conferences. Agentic-oriented theories could also be useful, to highlight how gender activism attempts to shift global governance policy (True in Rai and Waylen 2008). Interesting in this regard is Colin Walch's (2015) notion that the global process of negotiations that preceded the formulations of the Sendai Framework excluded researchers, civil society representatives and NGOs from central decision-making venues (Walch 2015). However, even with a full incorporation of a gender perspective into DRR policy, promises may very well evaporate (Moser and Moser 2005) when translated into implementation processes (Sultana 2010; UNISDRc; UNISDR, UNDP, and IUCN 2009). Global discourses notwithstanding, it is in the implementation processes that real change emerges (or not). Today there is a substantial gap between global DRR goals and outcomes at a local level. The ability of the international community to spearhead local community resilience remains limited (Hollis 2017). Hence, further research on how gender policy at a global level translates into implementation at regional and local levels would be interesting.

The analysis also contributes to other scholarships using the WPR approach. WPR applications have been made in, e.g., Canada, India and Australia and concern areas such as health, drug and alcohol, education, equity or development policy (Bacchi and Goodwin 2016). Feminist WPR scholars have mainly used this approach to analyse different gender-related policies (see, for example, Payne 2014 or Carson and Edwards 2011). Dolores Calvo (2013), for instance, examined the European Union's gender mainstream strategy and found that it conveyed an understanding of gender as a fixed category (Calvo 2013). Jeff Hearn and Linda McKie (2010) showed how policies aimed to secure women's and children's safety from domestic violence fail to problematize the gendered nature of violence. I have contributed with a slightly different type of analysis to the growing field of WPR policy analysis. I demonstrated that the Sendai Framework has not fully integrated a gender perspective due to the dominance of the relief and techno-managerial logics that underpin the policy. Exploring underlying conceptual logics proved to be a fruitful way to understand why a gender perspective does not 'stick'. Because both the relief logic and the techno-managerial logic pertain to a discourse of problem-solving, political analysis of gender inequality is made difficult. As Robert W. Cox (Cox in Keohane 1986, 208) states, problem-solving:

[t]akes the world as it finds it, with the prevailing social and power relationships and the institutions into which they are organized, as the given framework for action. The general aim of problem-solving is to make those relationships and institutions work smoothly by dealing effectively with particular sources of trouble

Incorporating a gender perspective is something that challenges prevailing social and power relationships, which is in contrast to the problem-solving underpinnings of the relief logic and the techno-managerial logic. Due to the pervasiveness of these logics there is not enough fertile soil within global DRR policy for a gender perspective to 'stick', despite multiple attempts from advocates, researchers and activists to change the agenda.

Notes

1 See for example the Natural Hazards Center at Boulder University in Colorado, The Centre for Gender and Disaster at the Institute for Risk and Disaster Reduction in London, or the Superstorm Research Lab at the Institute for Public Knowledge in New York City.
2 Post-structural approaches usually challenge and question Enlightenment assumptions about reason, science and progress, often direct attention to knowledge production practices that may produce inegalitarian forms of rule, emphasise a plurality of practices, and usually insist that realities are contingent and open to challenge and change (Bacchi and Goodwin 2016).

References

Akerkar, Supriya. 2007. "Disaster Mitigation and Furthering Women's Rights: Learning from the Tsunami." *Gender, Technology and Development* 11, no. 3: 357–88.
Alburo-Cañete, and Zoe K. Kaira. 2014. "Bodies at Risk: 'Managing' Sexuality and Reproduction in the Aftermath of Disaster in the Philippines." *Gender, Technology and Development* 18, no. 1: 33–51.
Allen, Katrina M. 2006. "Community-Based Disaster Preparedness and Climate Adaptation: Local Capacity-Building in the Philippines." *Disasters* 30, no. 1: 81–101.

Arora-Jonson, Sema. 2011. "Virtue and Vulnerability: Discourses on Women, Gender and Climate Change." *Global Environmental Change* 21: 744–51.

Austin, Duke, W. 2016. "Hyper-Masculinity and Disaster: The Reconstruction of Hegemonic Masculinity in the Wake of a Calamity." In *Men, Masculinities and Disaster*, edited by Enarson, Elaine, and Bob Pearson, 45–55. New York: Routledge.

Bacchi, Carol. 1999. *Women, Policy and Politics: The Construction of Policy Problems.* London: Sage.

Bacchi, Carol. 2009. *Analysing Policy: What's the Problem Represented to Be?* Melbourne, Australia: Pearson.

Bacchi, Carol, and Susan Goodwin. 2016. *Poststructural Policy Analysis: A Guide to Practice.* New York: Palgrave Macmillan.

Bankoff, Greg, Georg Frerks, and Dorothea Hilhorst. 2004. *Mapping Vulnerability: Disasters, Development and People.* London, UK: Earthscan.

Bergström, Göran, and Kristina Boréus. 2005. *Textens mening och makt: metod i samhällsvetenskaplig text- och diskursanalys.* Lund: Studentlitteratur.

Betslas, Angelique, and Chris Beasley. 2012. *Engaging with Carol Bacchi: Strategic Interventions and Exchanges.* Adelaide: University of Adelaide Press.

Bhadra, Subhasis. 2017. "Women in Disasters and Conflicts in India: Interventions in View of the Millennium Development Goals." *International Journal of Disaster Risk Science* 8: 196–207.

Björnehed, Emma, and Josefina Erikson. 2018. "Making the Most of the Frame: Developing the Analytical Potential of Frame Analysis." *Policy Studies* 39, no. 2: 109–26.

Bondesson, Sara. 2017. *Vulnerability and Power: Social Justice Organizing in Rockaway, New York City, after Hurricane Sandy.* PhD dissertation. Department of Government, Uppsala University.

Bradshaw, Sarah. 2013. *Gender, Development and Disasters.* Northampton: Edward Edgar Publishing.

Bradshaw, Sarah. 2014. "Engendering Development and Disasters." *Disasters Special Issue: Building Resilience to Disasters Post-2015* 39, no. 1: 54–75.

Braun, Virginia, and Victoria Clarke. 2006. "Using Thematic Analysis in Psychology." *Qualitative Research in Psychology* 3, no. 2: 77–101.

Calvo, Dolores. 2013. *What Is the Problem of Gender? Mainstreaming Gender in Migration and Development in the European Union.* PhD dissertation. Department of Sociology and Work Science, University of Gothenburg, Gothenburg.

Carson, Lisa, and Kathy Edwards. 2011. "Prostitution and Sex Trafficking: What Are the Problems Represented to Be? A Discursive Analysis of Law and Policy in Sweden and Victoria, Australia." *The Australian Feminist Law Journal* 34: 63–87.

Collins, Andrew E., Samantha Jones, Bernard Manyena, and Janaka Jayawickrama, eds. 2014. *Hazards, Risks and Disasters in Society.* Amsterdam: Elsevier.

Cox, W. Robert. 1986. "Social Forces, States, and World Orders." In *Neorealism and Its Critics*, edited by Robert O. Keohane, 204–54, New York: Columbia University Press.

Cupples, Julie. 2007. "Gender and Hurricane Mitch: Reconstructing Subjectivities After Disaster." *Disasters: The Journal of Disaster Studies, Policy and Management* 31, no. 2: 155–75.

Dahlerup, Drude. 2006. "The Story of the Theory of Critical Mass." *Politics and Gender* 2, no. 4: 511–22.

David, Emmanuel, and Elaine Enarson, eds. 2012. *The Women of Katrina: How Gender, Race, and Class Matter in an American Disaster.* Nashville: Vanderbilt University Press.

Di Baldassarre, Giuliano, Daniel Nohrstedt, Johanna Mård, Steffi Burchardt, Cecilia Albin, Sara Bondesson, Korbinian Breinl, Frances M. Deegan, Diana Fuentes, Marc Girons Lopez, Mikael Granberg, Lars Nyberg, Monika Rydstedt Nyman, Emma Rhodes, Valentin Troll, Stephanie Young, Colin Walch, and Charles F. Parker. 2018. "An Integrative Research Framework to Unravel the Interplay of Natural Hazards and Vulnerabilities." *Earth's Future* 6: 305–10.

Dynes, Russell R. 1994. *Situational Altruism: Toward an Explanation of Pathologies in Disaster Assistance*. University of Delaware, Disaster Research Center (DRC). Preliminary Paper, no. 201.

Eklund, Lisa, and Siri Tellier. 2012. "Gender and the International Crisis Response: Do We Have the Data and Does it Matter?" *Disasters* 36, no. 4: 589–608.

Enarson, Elaine, and Betty Hearn Morrow. 1998. *The Gendered Terrain of Disaster: Through Women's Eyes*. New York: Praeger Publishers.

Enarson, Elaine, and Bob Pearson, eds. 2016. *Men, Masculinities and Disasters*. New York: Routledge.

Fothergill, Alice, and Lori A. Peek. 2004. "Poverty and Disasters in the United States: A Review of Recent Sociological Findings." *Natural Hazards* 32: 89–110.

Fothergill, Alice, Enrique G. M. Maestas, and JoAnne DeRouen Darlington. 1999. "Race, Ethnicity and Disasters in the United States: A Review of the Literature." *Disasters* 23, no. 2: 156–73.

Foucualt, Michael. 1980. *Power/Knowledge: Selected Interviews and Other Writings, 1972–1977*. New York: Pantheon Books.

Fung, Archon. 2005. "Deliberation Before the Revolution: Toward an Ethics of Deliberative Democracy in an Unjust World." *Political Theory* 33, no. 2: 397–419.

Gaillard, Jean-Christophe, Kristinne Sanz, Benigno C. Balgos, Soledad Natalia M. Dalisay, Andrew Gorman-Murray, Fagalua Smith, and Vaito'a Toelupe. 2017. "Beyond Men and Women: A Critical Perspective on Gender and Disaster." *Disasters* 41, no. 3: 429–47.

Gardner, Robert O. 2013. "The Emergent Organization: Improvisation and Order in Gulf Coast Disaster Relief." *Symbolic Interaction* 36, no. 3: 237–60.

Gault, Barbara. 2005. *The Women of New Orleans and the Gulf Coast: Multiple Disadvantages and Key Assets for Recovery Part I: Poverty, Race, Gender and Class*. Washington, DC: Institute for Women's Policy Research.

Hannigan, John. 2012. *Disasters Without Borders: The International Politics of Natural Disasters*.Cambridge, UK: Polity Press.

Harding, Sandra. 1991. *Whose Science? Whose Knowledge? Thinking from Women's Lives*. Buckingham: Open University Press.

Hartsock, Nancy C. M. 1985. *Money, Sex, and Power: Toward a Feminist Historical Materialism*. Boston: Northeastern University Press.

Hearn, Jeff, and McKie, Linda. 2010. "Gendered and Social Hierarchies in Problem Representation and Policy Processes: 'Domestic Violence' in Finland and Scotland." *Violence Against Women* 16, no. 2: 136–58.

Hollis, Simon. 2014. "Competing and Complimentary Discourses in Global Disaster Risk Management." *Risk, Hazards & Crisis in Public Policy* 5, no. 3: 342–63.

Hollis, Simon. 2017. "Localized Development Gaps in Global Governance: The Case of Disaster Risk Reduction in Oceania." *Global Governance* 23: 121–39.

Jones, Eric C., and Arthur D. Murphy, eds. 2009. *The Political Economy of Hazards and Disasters*. New York: Altamira Press, Rowman & Littlefield Publishers, Inc.

Jorgensen, Marianne W., and Louise Phillips. 1999. *Diskursanalys som teori och metod*. Lund: Studentlitteratur.

Kanter, Rosabeth Moss. 1977. "Some Effects of Proportions on Group Life: Skewed Sex Ratios and Responses to Token Women." *American Journal of Sociology* 82, no. 5: 965–90.

Keohane, Robert O., ed. 1986. *Neorealism and Its Critics*. New York: Columbia University Press.

Kronsell, Annica. 2013. "Gender and Transition in Climate Governance." *Environmental Innovation and Societal Transitions* 7: 1–15.

Kronsell, Annica, and Gunngildur Lily Magnusdottir. 2015. "The (In)Visibility of Gender in Scandinavian Climate Policy-Making." *International Feminist Journal of Politics* 17, no. 2: 208–326.

Krook, Mona Lena, and Fiona Mackay. 2011. *Gender, Politics and Institutions*. New York: Palgrave Macmillan.

Larsson, Oscar. 2015. *The Governmentality of Meta-governance: Identifying Theoretical and Empirical Challenges of Network Governance in the Political Field of Security and Beyond*. PhD dissertation. Department of Government, Uppsala University.

Mazurana, Dyan, Prisca Benelli, and Peter Walker. 2013. "How Sex- and Age-disaggregated Data and Gender and Generational Analyses can Improve Humanitarian Response." *Disasters* 37, no. 1: 68–82.

McEntire, David A. 1997. "Reflecting on the Weaknesses of the International Community During the IDNDR: Some Implications for Research and its Application." *Disaster Prevention and Management* 6, no. 4: 221–33.Moser, Caroline, and Annalise Moser. 2005. "Gender Mainstreaming since Beijing: A Review of Success and Limitations in International Institutions." *Gender & Development* 13, no. 2: 11–22.

Narby, Petter. 2014. *Time of Crisis: Order, Politics, Temporality*. PhD dissertation. Department of Political Science, Lund University, Lund.

Neal, David M., and Brenda D. Phillips. 1995. "Effective Emergency Management: Reconsidering the Bureaucratic Approach." *Disasters* 19, no. 4: 327–37.

Neumayer, Eric, and Thomas Plümper. 2007. "The Gendered Nature of Natural Disasters: The Impact of Catastrophic Events on the Gender Gap in Life Expectancy, 1981–2002." *Annals of the Association of American Geographers* 97, no. 3: 551–66.

Payne, Sarah. 2014. "Constructing the Gendered Body? A Critical Discourse Analysis of Gender Equality Schemes in the Health Sector in England." *Current Sociology* 62, no. 7: 956–74.

Peterson, V. Spike, and Anne Sisson Runyan. 2010. *Global Gender Issues in the New Millennium* (3rd ed.). Boulder: Westview Press.

Phillips, Anne. 1995. *The Politics of Presence*. Oxford and New York: Clarendon Press, Oxford University Press.

Pilcher, Jane, and Imelda Whelehan. 2017. *Key Concepts in Gender Studies*. London: Sage Publications.

Pitkin, Hanna F. 1967. *The Concept of Representation*. Berkeley, CA: University of California Press.

Rai, Shirin M., and Georgina Waylen, eds. 2008. *Global Governance: Feminist Perspectives*. New York: Palgrave Macmillan.

Raymond, Christopher M., Ioan Fazey, Mark S. Reed, Lindsay C. Stringer, Guy Robinson, and Anna C. Evely. 2010. "Integrating Local and Scientific Knowledge of Environmental Management." *Journal of Environmental Management* 91: 1766–77.

Reichel, Christian, and Fromming, Urte Undine. 2014. "Participatory Mapping of Local Disaster Risk Reduction Knowledge: An Example from Switzerland." *International Journal of Disaster Risk Science* 5: 41–54.

Rogers, Peter. 2017. "The Etymology and Genealogy of a Contested Concept." In *The Routledge Handbook of International Resilience*, edited by David Chandler and Jon Coaffee, 13–25. New York: Routledge.

Seager, Joni. 2006. "Noticing Gender (or not) in Disasters." *Disasters Geoforum* 37: 2–3.

Staupe-Delgado, Reidar, and Bjorn Ivar Kruke. 2017. "El Niño-induced Droughts in the Columbian Andes: Towards a Critique of Contingency Thinking." *Disaster Prevention and Management* 26, no. 4: 382–95.

Sultana, Farhana. 2010. "Living in Hazardous Waterscapes: Gendered Vulnerabilities and Experiences of Floods and Disasters." *Environmental Hazards* 9, no. 1: 43–53.

Tickner, J. Ann. 1992. *Gender in International Relations: Feminist Perspectives on Achieving Global Security*. New York: Columbia University Press.

Tierney, Kathleen. 2014. *The Social Roots of Risk: Producing Disasters, Promoting Resilience*. Stanford: Stanford University Press.

True, Jacqui. 2008. "Gender Mainstreaming and Regional Trade Governance in Asia-Pacific Economic Cooperation (APEC)." In *Global Governance: Feminist Perspectives*, edited by Shirin M. Rai and Georgina Waylen, 129–59. New York: Palgrave Macmillan.

UNISDRa. "What is Disaster Risk Reduction?" www.unisdr.org/who-we-are/what-is-drr.

UNISDRb. "Gender and a Leading Role for Women in Disaster Risk Reduction." www.unisdr.org/we/advocate/gender.

UNISDRc. 2014. "Towards the Post-2105 Framework for Disaster Risk Reduction (HFA2): Women as a Force in Resilience Building, Gender Equality in Disaster Risk Reduction." Background Paper. www.unisdr.org/2015/docs/gender/BackgroundPaper_Gender InclusionHFA2.pdf.

UNISDRd. "United Nations Office for Disaster Risk Reduction." www.unisdr.org/.

United Nations. 2015. "Sendai Framework on Disaster Risk Reduction 2015–2030." www.preventionweb.net/files/43291_sendaiframeworkfordrren.pdf.

United Nations Office for Disaster Risk Reduction (UNISDR), United Nations Development Programme (UNDP), and International Union for the Conservation of Nature (IUCN). 2009. *Making Disaster Risk Reduction Gender-Sensitive: Policy and Practical Guidelines*. Geneva, Switzerland. www.unisdr.org/we/inform/publications/9922.

United Nations Population Fund (UNFPA). "Humanitarian Emergencies." www.unfpa.org/emergencies.

Walch, Colin. 2015. "Expertise and Policy-Making in Disaster Risk Reduction." *Nature Climate Change* 5: 706–07.

Whittaker, Joshua, Blythe McLennan, and John Handmer. 2015. "A Review of Informal Volunteerism in Emergencies and Disasters: Definition, Opportunities and Challenges." *International Journal of Disaster Risk Reduction* 13: 358–68.

Will, Steffen, Åsa Persson, Lisa Deutsch, Jan Zalasiewicz, Mark Williams, Katherine Richardson, Carole Crumley, Paul Crutzen, Carl Folke, Line Gordon, Mario Molina, Veerabhadran Ramanathan, Johan Rockström, Marten Sheffer, Hans Joachim Schellnhuber, and Uno Svedin. 2011. "The Anthropocene: From Global Change to Planetary Stewardship." *Ambio: A Journal of the Human Environment: Research and Management* 40: 739.

Wisner, Ben. 2003. "Sustainable Suffering? Reflections on Development and Disaster Vulnerability in the Post-Johannesburg World." *Regional Development Dialogue* 24, no. 1: 135–48.

Young, Iris Marion. 2001. "Activist Challenges to Deliberative Democracy." *Political Theory* 29, no. 5: 670–90.

Youngs, Gillian. 2004. "Feminist International Relations: A Contradiction in Terms? Or: Why Women and Gender are Essential to Understanding the World 'We' Live In." *International Affairs* 80, no. 1: 75–87.

4 Women as agents of change?

Reflections on women in climate adaptation and mitigation in the Global North and the Global South

Misse Wester and Phu Doma Lama

Introduction

Climate change is undoubtedly the most serious challenge facing our planet today. The consequences of extreme weather events, loss of natural resources and threatened livelihoods are presently being felt across the globe. Today, it is clear that we live in the era of the Anthropocene and that this is a result of human activities. These activities have given rise to a number of unintended and perhaps unforeseeable effects and consequences, direct and indirect, which together have defied national borders (Oppenheimer et al. 2014). As these consequences are not evenly distributed across regions, the Global South is seen as being more exposed to the negative effects of climate change. The Global South therefore appears to be more vulnerable than the Global North to the ramifications of climate change. Vulnerability is often perceived as a combination of exposure to risk, poverty and lack of resources; factors that influence the possibilities to adjust or adapt to climate change. Women in particular are often identified as the most vulnerable group and even more so when referring to women living in the so-called 'developing world' (Mertz et al. 2009; Reid and Huq 2007). The United Nations' sustainable development goals as set out in Agenda 2030 have gender equality and the empowerment of women and girls as their fifth goal. Undoubtedly, women experience the effects of climate change differently in cases where they are less equipped to navigate negative consequences due to inequalities and a lack of access to resources. At the same time it is important to point out that discourses portraying women as vulnerable and poor have encountered significant criticism, both theoretically and empirically (Nightingale 2006; Rocheleau, Thomas-Slayter, and Wangari 1996; Van Aelst and Holvoet 2016).

Despite such critique, the image of vulnerable women persists in the present work on climate change (Arora-Jonsson 2014). If women in the Global South are seen as vulnerable, women in more affluent societies, such as the Global North, are often seen as responsible for making sustainable decisions and caring for the environment to a much greater extent than men (Hawkins 2012; Wester 2012). These images of women as either a homogenous group or as differentiated but vulnerable and disempowered groups, are often contrasted with the portrayal of women as having altruistic intentions and responsibilities. These different sets of images are often contradictory and fail to address the social context in which gender is created, sustained and experienced. In this chapter, we argue that most efforts aimed at using women's positions relative to climate change will be insufficient if not conducted with great care and analytical sensitivity.

In this contribution, we show that mitigation and adaptation measures of climate change often target women, either as untapped resources for combating climate change or as active and powerful participants in adaptation or mitigation efforts. We demonstrate that even if discussions of climate change attempt to include gender, actual efforts rarely challenge the status quo but rather reinforce stereotypical perceptions of gender or may even result in greater vulnerability. The chapter is structured as follows: first, we review relevant literature on gender and climate change, focusing in particular on gender mainstreaming; a tool used in many different contexts. After this, we discuss the perceived roles of women, first in the Global South, then in the Global North. We exemplify with our research conducted in these two areas to illustrate the analytically challenges we have identified. Finally, we conclude by discussing the implications of this for the larger field of gendered ramifications of climate change.

Gender, climate and disasters: the role of women

Ever since a gender perspective on climate change was introduced, women have been identified as an important target group when it comes to climate change adaptation and mitigation efforts, something explicitly stated in the IPCCs fourth assessment report (Parry et al. 2007). In this, women are recognised as suffering more than men from the negative consequences of climate change, while the effects of climate change more generally are seen to cause a number of changes in situations and livelihoods for both men and women, but in different ways. Disaster studies have similarly pointed to how pre-existing vulnerabilities play a large role in the gendered consequences of climate change (Blaikie et al. 1994). Examples include how gendered roles, such as caring for children or the elderly, affect women to a larger extent than men and how they tend to increase women's afflictions in times of crises. In addition, social norms limit access to resources, resulting in a disproportionate death toll among women compared to men during disasters (Enarson and Fordham 2001). This is also the case after the acute phase of a disaster has passed. One example highlighting vulnerability in post-disaster situations is when men migrate in search of alternative means of livelihood, leaving women to fend for their families without a male head of the household. Impoverishment of women due to loss of productive assets in a post-disaster situation can also force women into low-wage labour, thus adding to this vulnerability discourse (Nelson et al. 2002).

There has been increased acknowledgement of the fact that vulnerability to climate change (or indeed to any hazard) is rooted in social and economic factors rather than in biophysical ones (Mikulewicz 2018). This inclusion of social dimensions involves an understanding of the fact that contextual factors are socio-economic, not physical, and that they are responsible for shaping community and individual adaptation to climate change (Kelly and Adger 2000; Mearns and Norton 2009; O'Brien et al. 2006). It is within this conversation of social vulnerability that gender roles and gender relations have come to be recognised as factors shaping vulnerability to climate change (Terry 2009). However, despite multiple research efforts to understand biophysical vulnerabilities and the interplay with social structures, climate policies continue to locate vulnerabilities in physical events (Nightingale 2017). Of course, the negative effects of climate change are felt across a variety of groups, as risk exposure interacts with gender, age and socio-economic status, meaning that climate change has major implications for marginalised groups, such as women, children and transgendered, elderly and disabled people (Costello et al. 2009; Denton 2002). However, for clarity of argument, we focus only on women in this contribution.

Gender mainstreaming

Efforts to include gender in not just areas of climate change and disasters can, in a variety of contexts, be traced back to a frustration over a lack of significant development in improving women's position at the policy level after the UN Women's Conference in Mexico in 1975. This frustration culminated in the Beijing Conference in 1995, which called on governments to recommit to gender equality. This had global implications and led many governments to address gender in policymaking (Alston 2014). One tool used to integrate a gender perspective on more strategic levels is gender mainstreaming, which has been defined as:

> [T]he process of assessing the implications for women and men of any planned action, including legislation, policies or programmes, in all areas and at all levels. It is a strategy for making women's as well as men's concerns and experiences an integral dimension of the design, implementation, monitoring and evaluation of policies and programmes in all political, economic and societal spheres so that women and men benefit equally and inequality is not perpetuated.
>
> (ECOSOC 1997/2, 27)

This focus on policymaking is particularly explicit in the Council of Europe's definition of gender mainstreaming as: "The (re)organisation, improvement, development and evaluation of policy processes, so that a gender equality perspective is incorporated in all policies at all levels and at all stages, by the actors normally involved in policy-making" (Council of Europe 1998, 15).

Although gender mainstreaming is a well-recognised strategy to include and integrate gender, or rather women, in formal policies; it has nonetheless been criticised for becoming a mere metaphor (Connelly et al. 2000). A major criticism against gender mainstreaming is its focus on women's inclusion, by increasing the number of women in decision-making positions for instance, rather than on approaches for achieving the goal of gender equality as originally proposed (Zalewski 2010, 25). Part of the problem is that gender mainstreaming is surrounded by a conceptual ambiguity concerning its goal. Such lack of clarity may persist even when the goal, such as equity, is explicitly stated (Rubery 2002, 502). This ambiguity could be related to whether the mainstreaming policy has one of three aims: first, to increase the number of women in areas dominated by men, thereby creating a 'sameness' between men and women; or second, to seek to explore the differences between the genders so that women's unique contribution is recognised because men and women are inherently different; or third, to transform the very relations of power (Alston 2014, 290). We would argue that the first two aims should be explored in greater detail in the area of gender and climate if the third is to be achieved. The conflation of gender mainstreaming (strategy) and gender equality (goal) has further added to the confusion. In many instances gender mainstreaming becomes a goal in itself, leading to differences in implementation (Andersson 2017). Mainstreaming policies fail to challenge power relations and may in a number of circumstances even reinforce them (Woodward 2008). In this regard, Derbyshire (2012) emphasises how women-centric projects targeting women as beneficiaries can overlook gender equity concerns and thus fail to bring fundamental changes to such concerns.

Despite these critical voices, which are also mirrored in the gender and development literature (Chant 2016), there is a continued tendency to treat women as a homogenous category and to implicitly conceive of gender differences as something inherent. When it comes to climate change and mitigation efforts, feminist research first criticised the field of climate change for being gender-blind, leading to calls for gender mainstreaming in climate work (Denton 2002; Masika 2002). This original blindness could be attributed in part to the dominance of masculine discourses in climate change research with its focus on the natural sciences and on technological solutions (Terry 2009). However, even if gender is now being included in the mainstream climate debate, there are apparent contradictions in the portrayal of the role of women as both victims and saviours (Arora-Jonsson 2011; MacGregor 2010). Such assumptions have come to justify the inclusion of poor rural women in community-based adaptation actions in the Global South and a reliance on virtuous women in the Global North who are supposed to increase the efficiency and effectiveness of policies (Hemmati and Röhr 2009; Terry 2009). Despite their relative differences, women in such portrayals are therefore often viewed as the bearers of change.

It is critical to focus on the more vulnerable position of women and other marginalised groups in climate change. Including women in decision-making processes on issues that concern them, should be taken for granted. However, how this inclusion should occur and on what basis women ought to be included deserve further discussion. Should women be included because they contribute something unique due to their gender? Should women be included because they have a right to be included? Such questions raise concerns about the possibility to think of women as an overall encompassing group and the role of contextual factors. Based on the criticism levelled by feminist scholars, we can safely assume that context is indistinguishable from gender (see especially Part III of this volume). Gender is part of local contexts and social fabrics, yet such insights are not sufficiently reflected in current climate change adaptation and mitigation efforts.

In what follows, we demonstrate the significance of contextualising climate change adaptation and mitigation strategies that include a gender perspective, as such strategies may otherwise fail to bring about any substantial change.

Perspectives from the Global South

Women in the Global South are seen as both vulnerable and as an untapped resource (Arora Jonsson 2011). The livelihoods of women in the Global South, particularly in rural areas where women collect firewood, fetch water and depend on biomass and other such natural resources, become demonstrative of how close women's livelihoods are to environments that are climate sensitive (Denton 2002). This makes women in the Global South close to nature, as they depend on raw natural resources for their everyday needs. They are seen as an untapped resource because of their assumed altruism, an altruism that is to be used to benefit society as a whole. In other words, if given a chance, women's inclusion in climate adaptation would result in contributions that could save their families, their communities and ultimately the planet (Brickell and Chant 2010).

Such framings bolster the justification for women's participation based on their pro-environmental behaviour. This gives rise to notions that women will be more climate efficient and hence indispensable for the success of any project aiming at efficient climate adaptation. The numerical underrepresentation serves as a starting point to accomplish the goal of the project. Linking back to the aforementioned criticism directed at gender mainstreaming, our fieldwork in the Maldives illustrates that even if such efforts are well-intended, they are limited and might fail to address the local context in which women live.

The untapped women in the Maldives?

In the Maldives, there is an acute lack of women in decision-making roles at the national level, with only five females among 85 parliamentarians (ADB 2014). There have been efforts made to increase women's participation in the electoral process and to increase the number of women candidates. In 2018, the first female secretary general was appointed and in 2008, the ban on women holding presidential office was removed. However, there are many social norms at the grassroots level that dictate how women and men should behave that are not reflected in the national policies. This is true not only for politics, but also in climate change adaptation. The findings below come from our research in the Global South and describe climate adaptation efforts that address women, but fail to take the social context into account. Our examples are drawn from fieldwork conducted in the Maldives in 2016 (see Lama 2018). The aim of the study was to understand adaptation to changes on two islands in the Maldives, one where tourism was an important sector and one where it was not. These were selected in order to provide a rich picture of adaptation processes. Initially the study did not have an explicit focus on gender but as the interviews progressed, the gendered nature of adaptation efforts was exposed. This became particularly visible when the mobility of women in comparison to men became apparent. Further research included investigating how the national programme for adaptation to climate change in the Maldives affects women at the local level. The programme document outlines gender equality and empowerment of women as one of its main goals, to be achieved by means of poverty alleviation and creation of job opportunities. This programme has identified target areas that have to be strengthened, including the areas of tourism and fishing along with an emphasis on integration of climate change adaptation with disaster management (NAPA 2006).

The adaptation programmes that specifically target the tourism and the fishing industry can be accused of being insensitive in the analysis of their impacts. These programmes are highly political. There is a contestation of needs and interests, and the exercising of power is decisive in determining whose needs and interests count, albeit in subtle and implicit ways. As an island nation, the Maldives is highly reliant on tourism (even if the contribution of air travel to climate change is not acknowledged) for job opportunities and for adding costumers to local markets. Also, commercial fishery is seen as an important activity. Fish is a significant resource in the Maldives and as this sector is sensitive to climate change, projects aimed at the fishery sector have been introduced.

Our fieldwork in the Maldives (Lama 2018), particularly on the two islands of Maafushi and Kudafari, indicates that climate adaptation measures aimed at these two sectors (i.e., tourism and fishing) have resulted in a decrease in the mobility for women and an increase in the mobility of men over time. For example, our study shows that with the increase of guesthouse tourism activities and harbour development, local women are discouraged from walking along the beaches of Maafushi since visitors usually wear light and revealing clothing or swim wear, something deemed inappropriate in the area and to which local women should not be exposed. The results of our study thus indicate that local women can no longer move freely on their island, resulting in a loss of social and informal trading networks. In terms of occupation, local women's participation in the livelihood activities of the island was found to be low and mostly associated with low-income jobs. For example, in the tourism sector, a common job occupied by local women was cooking for guesthouse managers or cleaning the clothes of temporary workers, jobs associated with the lowest social status.

During interviews, women identified these activities as home based and well suited to the traditional gender roles of women to cook, clean and take care of children, and therefore encouraged. This can be contrasted to social norms that link men with occupations that allow them to be more physically mobile, while restricting women to the confines of their homes. Even if women want to go out and work, a lack of childcare and social norms militate against the mobility of women. Our study concludes that the increase in infrastructural activities to promote the sustainability of tourism decreases native women's mobility and opportunity to trade with their neighbouring islands. In combination with the increased mobility for men who can migrate and find work on other islands, this particular strategy aimed at strengthening the Maldives' economy has resulted in increased social isolation and decreased mobility of local women.

In addition, both islands are seeing a decrease in the number of women engaged in fishing-related livelihoods. This must be seen in relation to the increased mechanisation of the fishing sector, with the introduction of fish-aggregating devices to facilitate fish finding. This aligns well with the adaptation goal mentioned in the National Adaptation Programme of Action Strategy (2006) to introduce technologies in order to increase local food production. The mechanisation of fish processing has failed to account for women's role in the fishing sector, leading to women being phased out of the process (El-Horr and Pande 2016). The very means of production used in fishing are owned by men (ADB 2014), yet again decreasing women's access to resources. Even if the fishing industry is supported by national efforts, the circumstances of local women who face the challenges of climate change are not likely to improve. Thus, any efforts to increase women's participation in the fishing sector, as intended during the post-tsunami recovery phase, without considering the issue of ownership of means of production, will have limited or no value for achieving gender equality.

These empirical observations stand in sharp contrast to the projected effects of climate adaptation strategies. Strengthening the two industries that would supposedly make Maldivians more resilient in response to climate change has instead increased the vulnerability of certain groups of women on the local islands.

The official policy of including more women at a national level has become the basis for increasing the participation of women in adaptation projects by creating livelihood opportunities that promise empowerment and independence. Examples of alternative livelihoods include baking, sewing and handicraft making, activities that often take place through women's development committees. Other efforts aim to involve women in projects related to water, soil and energy conservation and waste management on the local islands (see UNDP 2016 for details). However, during a conversation with one of the international organisation's representative involved with environmental conservation projects at the community level in the Maldives, the participation of women was described only at the implementation level. That is, to ensure attendance without considering the local context. The representative pointed out that the alternative livelihood projects place an extra burden on women's domestic chores while claiming to provide alternative livelihoods, a problem that remained unaddressed in official policies. Creating an extra burden included making women solely responsible for the implementation and success of a project, but without assigning them ownership and control over the benefits (as also observed in other programmes; e.g. Leach 2007). This means that women are given a double burden of performing their traditional duties while also participating in adaptation projects, which does little to address the increased vulnerability women experience. Here, women's participation is crucial for the success of the project, but as project designers do not consider social context or social realities they fail to see such involvement as an additional burden.

Perspective from the Global North

We now move our attention from the Global South to the Global North to continue our investigation of how women are involved and included in climate change efforts. In our previous example, women in the Global South were seen as instrumental to climate change programmes by being involved in projects that would strengthen communities in the face of climate change. However, by not taking context into consideration, these programmes have left groups of women more vulnerable because of decreased mobility and increased labour. Turning to the Global North, women in this part of the world are also seen as crucial for addressing climate change. The increase of women in contexts that are relevant to climate change is believed to bring about change, as it is known that women perceive climate change as more urgent than men do and as they contribute less to certain climate emissions (Magnusdottir and Kronsell 2015). So, is there a link between climate mitigation acts and gender?

In our project on climate mitigation and female representation, we investigated how women who serve on the boards of companies that contribute greatly to greenhouse gas (GHG) emission in Sweden and Norway perceive climate change, and if their perceptions influence their behaviour. The background to this study is that the Nordic countries have largely headed the call for greater inclusion of women in decision-making processes. Norway was the first country in the world to introduce legislation in 2006 that prescribed quotas for the boards of companies to increase female representation. This legislation meant that the boards have to consist of at least 40 percent of women in order to comply.

In recent years, similar legislation has been introduced in countries like Spain, France, Iceland, Italy, the Netherlands and Belgium, whereas Sweden has opted for working with recommendations and voluntary measures (Machold et al. 2013). Evaluations indicate that if boards are more gender-balanced, the overall climate discussion is better, meetings are longer and include more questions from both men and women (Matsa and Miller 2013). An increase in the number of women on boards also improves the company's work on strategic issues (Nielsen and Huse 2010). So, in theory, this would mean that if women have a greater sense of urgency when it comes to climate change, and act on these convictions, and if they have more influence on strategic decisions, an increase in the number of women would heighten the likelihood for companies to work towards a reduction of greenhouse gases.

We conducted interviews and did a survey study of women and men who serve on the boards of those companies that make the largest contributions to GHG in Sweden: production and transportation. The survey and interviews were conducted in both Sweden and Norway, but the results discussed here are restricted to Sweden. In the survey, we included questions on how the respondents perceived climate change, if and how they believed the company worked with climate mitigation, and if climate was something that was discussed during the board meetings. Our results are based on some 400 responses and follow the same pattern of risk perception found in other studies. Women in our study regarded the consequences of climate change as more serious than the men. This does not mean that the men in our study denied climate change, as the overall concern about climate change was high across the sample. It merely means that women reported significantly higher levels of concern. Both men and women believed that the consequences of climate change would affect the business of their relevant company.

This led us to conclude that companies tend to regard the issue of climate change as highly relevant and as a question which should be addressed by the board. In other words, if climate change is likely to affect the company, the board is obliged to do something about it since it is their responsibility to work towards the survival of the company. We attempted to investigate the working climate on the boards to see if the atmosphere was such that their members felt they could talk freely about issues that concerned them. We did this by using items that focus on possibilities to influence the process, as well as on levels of cooperation and conflict (Mathisen, Ogaard, and Marnburg 2013). Our results indicate that both men and women on the boards felt that they had the ability to influence the decisions made, in addition to having a good level of cooperation and low levels of conflict.

In other words, we identified a situation where all conditions were favourable for those actors who make a substantial contribution to the emission of greenhouse gases to generate actions or make decisions that would address climate change. Here we have a group of powerful women, all dedicated to the environment, seeing climate change as an urgent and relevant issue and with access to decision-making processes that could make a difference (not forgetting the altruistic character of women). Surely, this must be reflected in the decisions made. Alas, this was not the case. In the vast majority of these companies, climate change or actions taken to mitigate climate change (for example, by reducing energy use, using renewable sources or funding innovation), were not topics addressed by the boards at all.

This example illustrates that a mere inclusion of women is not enough. Even if the situation for women in the Global North is different from women in the Global South with regard to access to decision-making and participatory processes, the impact of this involvement is not certain. It would seem that a critical mass does not equal critical acts (Childs and Krook 2009).

Are all women the same?

While women in the Global North are portrayed as being more removed from nature, they are at the same time regarded as having similar obligations as women in the South (see Hawkins 2012). As mothers and caregivers, the stereotypical view is that women's efforts should be devoted to improving the lives of others (Hawkins 2012). Our examples indicate that this holds true regardless of where women live and what their circumstances are.

Getting the numbers right

Increasing women's participation in programmes designed to identify alternative livelihoods or aimed at empowering women, whether in the Global North or South, fails to account for power relations and the heterogeneity of women (Arora-Jonsson 2011). The global average for the number of women involved at the highest decision-making level, measured by the percentage of women in parliament, was 23.4 percent in 2017 (UN Women 2019). In the Nordic countries, this number was higher; i.e., 41.7 percent, while in Asia or the Pacific, the numbers decrease to 19.6 percent and 15 percent. The most common ministerial posts held by women were positions related to the environment, natural resources and energy (UN Women 2019). This distribution could suggest that there is still a lack of women's involvement at the highest levels of decision-making, and where women are included, they follow the pattern of being responsible for the environment rather than other areas. Women are not a homogenous group, neither within nor between countries, making a single focus on gender unfortunate (Nagel 2016). Using community forestry in Nepal as an example, Nightingale (2006) shows how women are viewed as subordinate to men when it comes to decision-making in the management of forests. At the same time, however, positions based on caste give higher caste women more access to resources and participation than lower caste women (see also Oven et al. this volume). This example clearly indicates that women are not a homogenous pro-environmental category and that gender roles are not fixed but can be reconfigured depending on the context. Despite this, there are few examples of projects or mindsets that aim to change the status quo.

Efforts to include women in the Global South, even when acknowledging the unequal material impact of climate change on men and women, often fail to critically assess "the values and ideological constructs that shape the material realities of gender equality and ultimately lead to planetary crises" (MacGregor 2010, 235). Most projects, in our experience, recycle a certain kind of participation. It frequently takes the form of adding women to climate change adaptation projects that might support women to improve livelihoods, but it does so without providing them with access to positions of power or changing the context that discourages them from participating in the first place. At the heart of this lies the failure to take context, such as power, roles and expectations, into account. In Nepal for instance, where access to livelihood and decision-making is shaped by gender along class, caste and ethnic lines, addressing the power relations of participation becomes pertinent in the context of climate change adaptation (Nightingale 2006, 2017; see also Bista 1991; Eriksen, Nightingale, and Eakin 2015). Our results from the Global North indicate the same pattern in the sense that women are restrained in and by their contexts. If women are included because they are perceived to represent an alternative perspective, which certainly can be questioned, they cannot be expected to act in accordance with this 'alternative' in any given situation.

The gendered division of responsibilities

We observe that women are usually involved as receivers of aid or beneficiaries of well-intended programmes, but they seldom drive projects which they have themselves identified as being effective and necessary. Women are rarely the developers of technological solutions, but rather the users of clean energy technological innovations. Of particular interest regarding energy use in the Global South is that women are the ones that collect and use firewood for household needs. Here, some efforts have been aimed at reducing the climate impact by offering alternative stoves for women to cook on (Denton 2002; Muttarak and Chankrajang 2015), but addressing the gendered division of labour is much more difficult. Introducing and providing women with cleaner options for household firewood does nothing to address the division of labour in the households and does not include women in the decision-making process to a greater extent. In the North, women are encouraged to enter the field of energy under the commonly cited pretext that female representation is still low in many areas of technology and in the private sector (Rosen 1994). Important to note here is that in the North, there is still a division of labour and gendered roles that extend to the family. Men are more likely to make decisions on matters that have an effect (or no effect) on climate mitigation (Wester and Eklund 2011). For example, more men than women own cars, but women have been portrayed as villains for contributing to the congestion (and safety violations) when driving their children to school (Terry 2009). Also, when it comes to energy saving behaviour in the North, many of the 'smart' solutions introduced in households are perceived as high-tech solutions, where for example real-time feedback on energy behaviour is presented on interactive platforms, but this is rarely used by the households themselves (Nilsson et al. 2018). Men also use more energy than women (Magnusdottir and Kronsell 2015), but few campaigns address men. These innovations are technical in nature, developed sometimes with little regard for the users. These technological innovations are part of a male discourse and, from our perspective, since the main responsibilities of the home fall on women, the users are part of the female discourse.

In some cases, efforts have been made to highlight the division of gender roles in the household rather than to hide it. In her research, Hawkins (2012) points out that several cases of ethical consumption, such as buying products where the producer donates part of the profit to charitable projects, are targeted at women and sometimes framed as 'mothers helping mothers'. These tasks are then used as a unifying factor to bring together women across borders. However, noble as this might be, using gender roles to create unification might be problematic in its own way. It creates the idea that women can be brought together into a homogenous category that is notoriously portrayed as altruistic in terms of serving family and/ or community needs. Such assumptions are profoundly engrained in the discourse on climate change adaptation (Brickell and Chant 2010). This is problematic, we suggest, as it implies that women have a pure morality thanks to which they will virtually always put the well-being of others before their own in terms of shouldering the main responsibility for children, extended family and possibly also for their communities (Brickell and Chant 2010).

Failing to incorporate these insights into climate change mitigation or adaptation strategies might reproduce the same patterns over and over again. It means that women continue to be portrayed as agents of change at the same time as they are prevented from fully participating as agents of change.

From inclusion and integration to deconstruction and recreation

Climate change is happening now, making implementation of mitigation or adaptation strategies urgent. Extreme weather events and rising sea levels are consequences of climate changes that currently affect many regions of the world and it is anticipated that these effects will be more widely experienced. This implies that if policy- and decision-makers are serious in their efforts to reduce different and unequal impacts, a more careful analysis of gendered power relations should be included.

Analysing examples of gender invisibility in vulnerability assessments, MacGregor (2010) argues that vulnerability assessment becomes a procedural exercise of measuring and calculating. There is hardly any room left for the voices of the vulnerable and/or women, as their vulnerabilities become simplified through indicators (MacGregor 2010, 228). This echoes our main point that policies encouraging participation as representation alone are not enough. Policies with a limited and uncritical conceptualisation of gender end up including women numerically. At a local level, this may involve women in ways that burden them and reinforce unpaid labour by making women responsible for projects without remuneration.

Increasing participation for underrepresented women is fundamental. However, as we have argued, participation for representation alone must be questioned. Even at higher levels of decision-making, participation is not a guarantee that climate issues will be raised and that gender equality will be achieved. In the case of climate change adaptation, there is still limited knowledge of how a gender perspective can be useful for adaptation and mitigation (Dymén, Langlais, and Cars 2014).

Participatory approaches that fail to account for power relations, however well-intended, could result in a 'tyranny of the group' and a 'tyranny of decision making' in the sense that powerful interests co-opt the interest of others (Cooke and Kothari 2001, 7–8). For gender mainstreaming in climate change mitigation and adaptation policies, participation may end up making women rubber stamps in decision-making and fail to bring about necessary changes. At the same time, the transformative potential of participation to bring about social change cannot be ignored (Mohan 2001). Alternatives to electoral representation include women bureaus within the state machinery and within civil society and social movements, as well as alliances between them (Walby 2005).

Taking these points of criticism into consideration, does this imply that participation for gender mainstreaming must be abandoned all together or can it be resuscitated and revitalised to transform and challenge power relations that structurally and discursively make women vulnerable?

Any attempt at mainstreaming requires what Rees (1998) calls 'tinkering', 'tailoring' and 'transforming'. This involves appreciating the context by identifying and understanding the situations where women find themselves unable to voice their opinions (even when they have a voice) – situations which restrain them and make them ultimately vulnerable to climate and its effects. If, for example, adaptation policies can be formulated at rather abstract levels, then a 'tinkering' would entail a gender mainstreaming component. Tailoring this policy to fit local contexts would require an in-depth understanding of what this policy would mean for local communities or specific groups. Done properly, it would create fertile soil to bring about transformation to the context itself. This is of course not the only way to proceed, and the path is far from being linear. The intention is rather to draw attention to the importance of understanding and transforming the constraining contexts in which women find themselves. There is a need to go beyond inclusion and integration (tinkering and tailoring) in the spaces where hegemonic masculinity exerts its powers and influence (Carrigan, Connell, and Lee 1985). The aim should be a deconstruction of these spaces and their recreation to realise gender equity (transforming). Critical inquiries into the inclusion of women as agents of change in climate change adaptation and mitigation policies in both the Global North and South, could serve as starting points.

In addition, any projects aimed at transforming the status quo should acknowledge that gender goes deep. We internalise what we think we can contribute and what we cannot. If women do not see that they have qualities as leaders or decision makers, they will not act as if they do. Further, if women are seen as primary caretakers of their families, what can they do to change this image and the behaviours connected to this role? We run the risk of playing the numbers game where greater inclusion of women is achieved, but where no substantial changes in climate change mitigation occur, such as decreasing GHG emissions. In the Maldives, greater inclusion led to greater vulnerability even though women were included at the highest levels of politics. Also, few women in top-level management means them being alone in a male context. Is it reasonable to assume that these women will argue for the importance of gender relevant issues such as climate change even when not acknowledged by other members of a group with which they collaborate?

Arora-Jonsson (2011, 296) notes that the inclusion of women in the area of environmental policy has been beneficial since it has contributed to a "decentering of the male subject of environmental policy". Many calls have been made for a move from 'women in development' to 'gender in development', but such calls are not taken seriously in many climate change policies targeting women. There is also a lack of inclusion of the male perspectives on climate change and on the developed mitigation and adaptation strategies. It is striking that despite particular aspects of the male context, ranging from lower risk perceptions of climate change to increased mobility (voluntary or involuntary) due to climate change, few climate change mitigation or adaptation strategies have targeted them. Focusing on women and their role in climate adaptation and mitigation remove male practices from the climate debate altogether. In addition, policies that focus on women only could end up considering men and masculinity as a non-issue or as a group who enters the climate narrative as polluters and patriarchs. Although disaster studies have made headway in this regard by pointing out concerns of hegemonic masculinities (e.g., Cornwall 2000; Enarson and Pease 2016; Gaillard et al. 2017), climate change studies and development policies have paid little attention to men and masculinities (Alston and Kent 2008; Mearns and Norton 2009). There are examples of how men are affected by climate change, as in our own fieldwork in the Maldives, as well as in other places (Davidson 2012), but the wider inclusion of men in climate change adaptation policies is still lacking.

Such inclusion would involve not only understanding the role expectations of men, but also the behaviours that put men in a position of privilege in some circumstances while also keeping them from seeking help in times of stress (Alston and Kent 2008). Inquiry into hegemonic masculinities is a matter of urgency, particularly with the steady rise of gender discourses being part of the climate agenda, where the role of men is increasingly being recognised (Vincent et al. 2014). Such an enquiry would emphasise that power relations and hegemonic masculine behaviour during and after climate stress, such as being altruistic or violent, are not assumed and normalised. Climate change does not by itself cause or worsen gender inequalities. The two are linked in terms of effective mitigation and adaptation strategies for combating the negative effects of climate change. As such they necessitate the consideration of power relations and context likely to benefit the position of certain subjectivities, such as class, caste, race, nationality etc.

Linked to the issue of female representation is the absence of a critical discussion of gendered divisions of labour in climate adaptation and mitigation strategies. It seems impossible to expect women to participate and contribute to climate change mitigation and adaptation if we fail to address the context in which this occurs. Even if women are being included in participatory or decision-making processes, they cannot make decisions or act in ways that are outside their contextual reach. It is unrealistic to expect women who work and live in one cultural context, to let their 'altruistic selves run free' and start caring for their community in a context where this is not possible. Such an approach runs the risk of introducing gender policy or gender mainstreaming in ways that keep women in their present positions and ignore the context in which these strategies are to be enacted. From a global perspective, it would seem that women as a group have limited access to power. Of course there are great variations within this group with extremes at both ends of the scale, with some women having access to significant resources and good positions within a power structure, and some not. This is also true for men, as class, race and caste are also factors that affect an individual's position in society. Still, access to power is relative in cases where women have less than men, regardless of context.

Women in the South may have had the benefit of numerous education programmes funded by benign international donor organisations (such as UNDP and Asian Development Bank), but this makes little difference without a platform from which to incorporate this knowledge. In other words, gender must be placed in a particular context (Arora-Jonsson 2011). For gender mainstreaming and participation in climate change policies, this means understanding gender as a process enmeshed in politics and power where gender concerns are a product of history and shaped by a particular context. Such an increased focus on context might help align gender concerns, not from the vantage point of climate, but rather from societal vulnerabilities.

Thus, what is required is an approach that captures different experiences of climate change shaped by relations of power, including gender, class, caste, nationality and race (Taylor 2014). An intersectional analysis, armed with a critical feminist approach that probes power relations and account for different experiences of climate change, could be a way forward (Kaijser and Kronsell 2014). Such an approach could serve as a starting point to move from tinkering and tailoring for inclusion and integration towards transformation through deconstruction and recreation of spaces for co-learning and collective action.

Conclusions

With gender discourse gaining importance as part of the climate change agenda, understanding the ways in which gender is addressed and integrated in mitigation and adaptation policies forms an integral part of a feminist inquiry in climate change scholarship. Among these are inclusionary actions such as gender mainstreaming in development policy, which have gained currency in climate change policies. In this chapter, we have outlined ways in which gender is addressed and included in the climate change agenda of the Global North and South. One of the common threads that link the inquiry on gender in climate change in both contexts is the problematic inclusion of the image of sameness, either as one homogenous group, or differentiated into one vulnerable group and one altruistic group. Our examples show the danger of including women in ways that fail to consider the context and power relations that shape the division of labour, making inclusion a futile exercise that does nothing to reduce gender inequality or address climate change.

We have outlined some ways, in no way exhaustive, to caution against the reuse of such images in mainstreaming and in participation, as these may be limited to getting the numbers right. What we propose is a feminist inquiry that seeks to critically understand the bases of inclusion. Such an inquiry should focus on the contexts in which women find it difficult to exercise their agency. A central part of this inquiry is to go beyond a focus on women alone towards a focus on power relations and contexts in which women, men and other gender minorities find themselves powerless to deal with climate change and where the exercise of power becomes privileged. Such an inquiry forms a starting point to deconstruct structures and discourses that include women in problematic ways. Only then can we move towards recreating spaces for participation to address hegemonic power structures.

References

Alston, Margaret. 2014. "Gender Mainstreaming and Climate Change." *Women's Studies International Forum* 47: 287–94.

Alston, Margaret, and Jenny Kent. 2008. "The Big Dry: The Link Between Rural Masculinities and Poor Health Outcomes for Farming Men." *Journal of Sociology* 44, no. 2: 133–47.Andersson, Renée. 2017. "The Myth of Sweden's Success: A Deconstructive

Reading of the Discourses in Gender Mainstreaming Texts." *European Journal of Women's Studies* 25, no. 4: 455–69.

Arora-Jonsson, Seema. 2011. "Virtue and Vulnerability: Discourses on Women, Gender and Climate Change." *Global Environmental Change* 21, no. 2: 744–51.

Arora-Jonsson, Seema. 2014. "Forty Years of Gender Research and Environmental Policy: Where do We Stand?" *Women's Studies International Forum* 47: 295–308.

Asian Development Bank (ADB). 2014. *Maldives Gender Equality Diagnostic of Selected Sectors*. Report. ADB. www.adb.org/documents/maldives-gender-equality-diagnostic-selected-sectors.

Bista, Dor Bahadur. 1991. *Fatalism and Development: Nepal's Struggle for Modernization*. Kolkatta and Patna, India: Orient Blackswan.

Blaikie, Piers, Terry Cannon, Ian Davis, and Ben Wisner. 1994. *At Risk: Natural Hazards, People's Vulnerability and Disasters*. London and New York: Routledge.

Brickell, Katherine, and Sylvia Chant. 2010. "The Unbearable Heaviness of Being' Reflections on Female Altruism in Cambodia, Philippines, The Gambia and Costa Rica." *Progress in Development Studies* 10, no. 2: 145–59.

Carrigan, Tim, Bob Connell, and John Lee. 1985. "Toward a New Sociology of Masculinity." *Theory and Society* 14, no. 5: 551–604.

Chant, Sylvia. 2016. "Women, Girls and World Poverty: Empowerment, Equality or Essentialism?" *International Development Planning Review* 38, no. 1: 1–24.

Childs, Sarah, and Mona Lena Krook. 2009. "Analysing Women's Substantive Representation: From Critical Mass to Critical Actors." *Government and Opposition* 44, no. 2: 125–45.

Connelly, M. Patricia, Tania Murray Li, Martha MacDonald, and Jane L. Parpart. 2000. "Feminism and Development: Theoretical Perspectives." In *Theoretical Perspectives on Gender and Development*, edited by Jane L. Parpart, Patricia Connelly, M. Patricia Connelly, and V. Eudine Barriteau, 51–159. Ottawa: International Development Research Centre (IDRC).

Cooke, Bill, and Uma Kothari, eds. 2001. *Participation: The New Tyranny?* London and New York: Zed books.

Cornwall, Andrea. 2000. "Missing Men? Reflections on Men, Masculinities and Gender in GAD." *IDS Bulletin* 31, no. 2: 18–27.

Costello, Anthony, Mustafa Abbas, Adriana Allen, Sarah Ball, Sarah Bell, Richard Bellamy, Sharon Friel, Nora Groce, Anne Johnson, Maria Kett, Maria Lee, Caren Levy, Prof Mark Maslin, David McCoy, Bill McGuire, Hugh Montgomery, David Napier, Christina Pagel, Jinesh Patel, Jose Antonio Puppim de Oliveira, Nanneke Redclift, Hannah Rees, Daniel Rogger, Joanne Scott, Judith Stephenson, John Twigg, Jonathan Wolff, and Craig Patterson. 2009. "Managing the Health Effects of Climate Change: Lancet and University College London Institute for Global Health Commission." *The Lancet* 373, no. 9676: 1693–733.

Council of Europe, Steering Committee for Equality between Women, Men. Group of Specialists on Mainstreaming, European Committee for Equality between Women, and Men, Group of Specialists on Mainstreaming. 1998. "Gender Mainstreaming: Conceptual Framework, Methodology and Presentation of Good Practice: Final Report of Activities of the Group of Specialists on Mainstreaming (EG-S-MS)." Policy Document. www.unhcr.org/3c160b06a.pdf.

Davidson, Joanna. 2012. "Of Rice and Men: Climate Change, Religion, and Personhood Among the Diola of Guinea-Bissau." *Journal for the Study of Religion, Nature & Culture* 6, no. 3: 363–81.

Denton, Fatma. 2002. "Climate Change Vulnerability, Impacts, and Adaptation: Why Does Gender Matter?" *Gender & Development* 10, no. 2: 10–20.

Derbyshire, Helen. 2012. "Gender Mainstreaming: Recognising and Building on Progress. Views from the UK Gender and Development Network." *Gender & Development* 20, no. 3: 405–22.

Dymén, Christian, Richard Langlais, and Göran Cars. 2014. "Engendering Climate Change: The Swedish Experience of a Global Citizens Consultation." *Journal of Environmental Policy & Planning* 16, no. 2: 161–81.

El-Horr, Jana, and Rohini Prabha Pande. 2016. "Understanding Gender in Maldives: Toward Inclusive Development." *Policy Document. Directions in Development, Countries and Regions (World Bank Series)*. Washington, DC: World Bank. https://openknowledge.worldbank.org/handle/10986/24118.

Enarson, Elaine, and Maureen Fordham. 2001. "From Women's Needs to Women's Rights in Disasters." *Global Environmental Change Part B: Environmental Hazards* 3, no. 3: 133–36.

Enarson, Elaine, and Bob Pease, eds. 2016. *Men, Masculinities and Disaster*. London and New York: Routledge.

Eriksen, Siri H., Andrea J. Nightingale, and Hallie Eakin. 2015. "Reframing Adaptation: The Political Nature of Climate Change Adaptation." *Global Environmental Change* 35 (November): 523–33.

Gaillard, Jean-Christophe, Kristinne Sanz, Benigno C. Balgos, Soledad Natalia M. Dalisay, Andrew Gorman-Murray, Fagalua Smith, and Vaito Toelupe. 2017. "Beyond Men and Women: A Critical Perspective on Gender and Disaster." *Disasters* 41, no. 3: 429–47.

Hawkins, Roberta. 2012. "Shopping to Save Lives: Gender and Environment Theories Meet Ethical Consumption." *Geoforum* 43, no. 4: 750–59.

Hemmati, Minu, and Ulrike Röhr. 2009. "Engendering the Climate-Change Negotiations: Experiences, Challenges, and Steps Forward." *Gender & Development* 17, no. 1: 19–32.

Kaijser, Anna, and Annica Kronsell. 2014. "Climate Change Through the Lens of Intersectionality." *Environmental Politics* 23, no. 3: 417–33.

Kelly, Mick P., and Neil W. Adger. 2000. "Theory and Practice in Assessing Vulnerability to Climate Change and Facilitating Adaptation." *Climatic Change* 47, no. 4: 325–52.

Lama, Phu Doma. 2018. "Gendered Consequences of Mobility for Adaptation in Small Island Developing States: Case Studies from Maafushi and Kudafari in the Maldives." *Island Studies Journal* 13, no. 2: 111–28.

Leach, Melissa. 2007. "Earth Mother Myths and Other Ecofeminist Fables: How a Strategic Notion Rose and Fell." *Development and Change* 38, no. 1: 67–85.

MacGregor, Sherilyn. 2010. " 'Gender and Climate Change': From Impacts to Discourses." *Journal of the Indian Ocean Region* 6, no. 2: 223–38.

Machold, Silke, Morten Huse, Katrin Hansen, and Marina Brogi, eds. 2013. *Getting Women on to Corporate Boards: A Snowball Starting in Norway*. Cheltenham and Northampton, MA: Edward Elgar Publishing.

Magnusdottir, Gunnhildur, and Annica Kronsell. 2015. "The (in)Visibility of Gender in Scandinavian Climate Policy-Making." *International Feminist Journal of Politics* 17, no. 2: 308–26.

Masika, Rachel, ed. 2002. *Gender, Development and Climate Change*. Oxford: Oxfam International.

Mathisen, Gro Ellen, Torvald Ogaard, and Einar Marnburg. 2013. "Women in the Boardroom: How Do Female Directors of Corporate Boards Perceive Boardroom Dynamics?" *Journal of Business Ethics* 116, no. 1: 87–97.

Matsa, David A., and Amalia R. Miller. 2013. "A Female Style in Corporate Leadership? Evidence from Quotas." *American Economic Journal: Applied Economics* 5, no. 3: 136–69.

Mearns, Robin, and Andrew Norton, eds. 2009. *Social Dimensions of Climate Change: Equity and Vulnerability in a Warming World.* Washington DC: World Bank Publications.

Mertz, Ole, Kirsten Halsnæs, Jørgen E. Olesen, and Kjeld Rasmussen. 2009. "Adaptation to Climate Change in Developing Countries." *Environmental Management* 43, no. 5: 743–52.

Mikulewicz, Michael. 2018. "Politicizing Vulnerability and Adaptation: On the Need to Democratize Local Responses to Climate Impacts in Developing Countries." *Climate and Development* 10, no. 1: 18–34.

Ministry of Environment, Energy and Water. 2006. "National Adaptation Program of Action (NAPA)." Policy Document. Republic of Maldives.

Mohan, Giles. 2001. "Participatory Development." In *The Companion to Development Studies,* edited by Desai Vandana and Robert B. Potter, 49–53. London, UK Routledge.

Muttarak, Raya, and Thanyaporn Chankrajang. 2015. "Who Is Concerned About and Takes Action on Climate Change? Gender and Education Divides among Thais." *Vienna Yearbook of Population Research* 13: 193–220.

Nagel, Joane. 2016. *Gender and Climate Change.* New York: Routledge, Taylor and Francis Group.

Nelson, Valerie, Kate Meadows, Terry Cannon, John Morton, and Adrienne Martin. 2002. "Uncertain Predictions, Invisible Impacts, and the Need to Mainstream Gender in Climate Change Adaptations." *Gender & Development* 10, no. 2: 51–59.

Nielsen, Sabina, and Morten Huse. 2010. "The Contribution of Women on Boards of Directors: Going beyond the Surface." *Corporate Governance: An International Review* 18, no. 2: 136–48.

Nightingale, Andrea. 2006. "The Nature of Gender: Work, Gender, and Environment." *Environment and Planning D: Society and Space* 24, no. 2. 165–85.

Nightingale, Andrea J. 2017. "Power and Politics in Climate Change Adaptation Efforts: Struggles over Authority and Recognition in the Context of Political Instability." *Geoforum* 84(August): 11–20.

Nilsson, Anders, Misse Wester, David Lazarevic, and Nils Brandt. 2018. "Smart Homes, Home Energy Management Systems and Real-Time Feedback: Lessons for Changing Energy Consumption Behavior from a Swedish Field Study." *Energy and Buildings* 179: 15–25.

O'Brien, Karen, Siri Eriksen, Linda Sygna, and Lars Otto Naess. 2006. "Questioning Complacency: Climate Change Impacts, Vulnerability, and Adaptation in Norway." *AMBIO: A Journal of the Human Environment* 35, no. 2: 50–56.

Oppenheimer, Michael, Maximiliano Campos, Rachel Warren, Joern Birkmann, George Luber, Brian O'Neill, and Kiyoshi Takahashi. 2014. "Emergent Risks and Key Vulnerabilities." In *Climate Change 2014: Impacts, Adaptation, and Vulnerability. Part A: Global and Sectoral Aspects. Contribution of Working Group II to the Fifth Assessment Report of the Intergovernmental Panel on Climate Change,* edited by Christopher B. Field et al., 1039–100. Intergovernmental Panel on Climate Change (IPCC). Cambridge: Cambridge University Press.

Parry, Martin, Martin L. Parry, Osvaldo Canziani, Jean Palutikof, Paul Van der Linden, and Clair Hanson, eds. 2007. "Climate Change 2007: Impacts, Adaptation and Vulnerability." In *Working Group II Contribution to the Fourth Assessment Report of the IPCC.* Cambridge and New York: Cambridge University Press.

Rees, Teresa. 1998. *Mainstreaming Equality in the European Union: Education, Training and Labour Market Policies.* London: Routledge.

Reid, Hannah, and Saleemul Huq. 2007. "Community-Based Adaptation: A Vital Approach to the Threat Climate Change Poses to the Poor." Briefing Paper. International Institute for Environment and Development (IIED). London: IIED.

Rocheleau, Dianne, Barbara Thomas-Slayter, and Esther Wangari. 1996. *Feminist Political Ecology: Global Issues and Local Experiences. International Studies of Women and Place.* New York: Routledge.

Rosen, A. 1994. "Adam, Eve and the Controversial Rib: Gender, Technology, Conflict and Universalism." *Dialogue and Humanism* 4, nos. 2–3: 23–30.

Rubery, Jill. 2002. "Gender Mainstreaming and Gender Equality in the EU: The Impact of the EU Employment Strategy." *Industrial Relations Journal* 33, no. 5: 500–22.

Taylor, Marcus. 2014. *The Political Ecology of Climate Change Adaptation: Livelihoods, Agrarian Change and the Conflicts of Development.* London: Routledge.

Terry, Geraldine. 2009. "No Climate Justice Without Gender Justice: An Overview of the Issues." *Gender & Development* 17, no. 1: 5–18.

UN Women. 2019. "Women in Politics 2017 Map." www.unwomen.org/en/digital-library/publications/2017/4/women-in-politics-2017-map.

United Nations Development Programme (UNDP). 2016. "Country Programme Document for Maldives (2016–2020)." Country Report. DP/DCP/MDV/3. http://repository.un.org/handle/11176/312230.

United Nations Economic and Social Council (ECOSOC). 1997. "Mainstreaming the Gender Perspective into All Policies and Programmes in the United Nations System." Policy Report. New York: UN ECOSOC.

Van Aelst, Katrien, and Nathalie Holvoet. 2016. "Intersections of Gender and Marital Status in Accessing Climate Change Adaptation: Evidence from Rural Tanzania." *World Development* 79: 40–50.

Vincent, Katharine, Petra Tschaker, Jon Barnett, Marta G. Rivera-Ferr, and Alistair Woodward. 2014. "Cross-Chapter Box on Gender and Climate Change." In *Climate Change 2014: Impacts, Adaptation, and Vulnerability*, edited by Christopher B. Field et al., 105–07. Intergovernmental Panel on Climate Change (IPCC). Cambridge and New York: Cambridge University Press.

Walby, Sylvia. 2005. "Gender Mainstreaming: Productive Tensions in Theory and Practice." *Social Politics: International Studies in Gender, State & Society* 12, no. 3: 321–43.

Wester, Misse. 2012. "Risk and Gender: Daredevils and Eco-Angels." In *Handbook of Risk Theory*, edited by Sabine Roesner, Rafaela Hillerbrand, Sandin Per Sandin, and Martin Peterson, 1029–48. Springer: Netherlands.

Wester, Misse, and Britta Eklund. 2011. "'My Husband Usually Makes those Decisions': Gender, Behavior, and Attitudes toward the Marine Environment." *Environmental Management* 48, no. 1: 70–80.

Woodward, Alison E. 2008. "Too Late for Gender Mainstreaming? Taking Stock in Brussels." *Journal of European Social Policy* 18, no. 3: 289–302.

Zalewski, Marysia. 2010. "'I Don't Even Know What Gender Is': A Discussion of the Connections Between Gender, Gender Mainstreaming and Feminist Theory." *Review of International Studies* 36, no. 1: 3–27.

5 Industrial/breadwinner masculinities

Understanding the complexities of climate change denial

Paul Pulé and Martin Hultman

Introduction

Modern Western men and masculinities are shaped by socialised performances that are conditioned rather than predetermined (Butler 1990). In this chapter, we consider the lives of those men who occupy the most privileged positions in society in the Global North and the masculine socialisations that define them. These performances are assertive, self-serving, entitled, aggressive/violent, and myopically caring, have broadened beyond heteronormative, suburban, Protestant, educated, sporty, white 'gentlemen' to include characteristics that are relatable to much wider groups of men such as: poor, working-class, aggressive, overtly xenophobic and racist, patriotic, un- or under-educated white men, coupled with notions of entitled paternalism, exclusivity, authority and the economic primacy that accompanies wealth. Claims of marginalisation and the degrees to which Western white men across this wider spectrum claim being 'left behind' have become acute throughout the Global North, and have instigated fresh backlashes that are affronting generations of hard fought gender equity gains (Kimmel 2017 [2013]). Drawing from examples in Western Europe and the United States, we focus on those whose primacy blinds them to their impacts on society and environment – individual men and masculinist constituencies who are enmeshed with fossil fuel addicted industrialisation and corporatisation, are commonly aligned with climate change denial and whose allegiances are emboldened by traditional socialisations of masculine identities that we refer to here and elsewhere as 'industrial/breadwinner masculinities' (Hultman and Pulé 2018).

The 2015 Conference of Parties (COP21) in Paris heralded broad international agreement to mitigate global climate change. Subsequent conventions in Marrakech (COP22), Bonn (COP23) and Katowice (COP24) have seen progressive refinements as well as increasing parochial contention about compliance to those agreements. To date, international cooperation has persisted in the form of regulatory debate resulting in some reform, keeping hope for effective responses alive. However, and despite the consensus reached by the vast majority of researchers about the social and ecological peril that is upon us, achieving and actioning effective and binding responses to anthropogenic climate change has been far from the smooth start to a new beginning that was the great promise of the Paris Accord (Plumer 2018). So much so that more than 20,000 scientists have drafted a 'warning to humanity' (a reworking of an original letter to the Union of Concerned Scientists from some 25 years ago) unequivocally calling our attention to the sobering fact that the threats of a climate crisis have worsened and more rapidly than the most pessimistic of internationally agreed predictions (IPCC 2018). We are in the midst of comprehensive changes to every biotic system on the planet due to anthropocentrically induced carbon pollution coupled with unfettered materialism and consumption rates, burgeoning population growth, catastrophic biodiversity loss, resulting in untold levels of stress upon all of life on Earth. These concerns collectively indicate that we are transitioning beyond 'climate change' and are now entering an era of 'climate breakdown' (IPCC 2018). Despite this, Saudi Arabia, Kuwait, Russia and the United States (with tacit support from Australia), as member states with heavy investments in fossil fuel industries, recently combined forces to obstruct international proceedings that are designed to commit nations to strict carbon pollution controls (Doherty 2018).

The significance of these steps in interrupting global cooperation on climate change is increasingly acute. Corporate attempts to expand and extend fossil fuel development are proving victorious, and are spurred on by lobby groups and researchers whose cumulative efforts assure humanity's collision course with apocalyptic scale concerns. The persistence of vocal deniers to snub climate data has been accompanied by rises in xenophobic isolationism in response to refugee crises and in the most overt cases, outright white supremacy; these concurrent trends singing from near identical song sheets, seeking recruits for their causes from very similar discontented (esp. male) white, working and middle classes constituencies, particularly throughout the Global North (Lockwood 2018). The correlations despite class disparities are blaringly evident; they share in common an addictive allegiance to the hegemonic allegiances of hyper-masculinities or the hierarchicalisation of wealth distribution generated by natural resource exploitation (Connell 1995).

In our recently published monograph, we interrogated hyper-masculinities through an exploration of industrial/breadwinner typologies in order to expose an alliance between the masculine identities of industrial elites and working-class workers (as well as their middle-class managers) at the 'coal face' of industrial productivity and corporatisation throughout the Global North (Hultman and Pulé 2018). We consider these typologies of masculinist industrialised addiction to be concerning not only on account of their impacts on society and environment but also because their impulses are being met by tepid government regulations seeking systemic compromise and reform at-best, reflective of what we refer to as 'ecomodern masculinities' (Hultman 2017; Hultman and Pulé 2018). While acknowledging that critical analyses of traditional notions of hyper-masculinities do not provide us with the whole story, we suggest that both typologies (industrial/breadwinner in particular but also regulatory and reformist or ecomodern masculinities) share in common a tendency to yield to a 'white male effect' (or a dauntless approach to global through to personal risk) coupled with climate change denial or weak/non-existent climate concerns (Finucane et al. 2000, 160; Slovic et al. 2005; McCright and Dunlap 2011; Dunlap and McCright 2015). Both typologies represent formidable bulwarks against transformative change towards a truly sustainable future. Here, as elsewhere, we argue that the most effective path towards a deep green future requires 'ecological masculinities' as a third way forward for the benefit of all life on Earth (Pulé 2013; Hultman and Pulé 2018). While further explication of ecomodern and ecological masculinities is deferred to those previously published works with others joining the fray (Gaard 2015; Pease 2019), we keep our focus in this chapter on a critical analysis of industrial/breadwinner masculinities, reflective of this typology's most acute intersections with white male effect and its compounding impacts of climate change denial. Of course, these phenomena are intertwined and lay at the very heart of our global challenges, calling us to not simply mitigate carbon pollution, but to also transition beyond the toxic forms of masculinities that we believe to be their primary cause.

The challenge of global climate change

Climate change denial is spreading throughout Europe, North America and Australasia (which – to no great surprise – represent the same regions that have the largest per capita human carbon footprints on the planet). This corresponds with growing populist movements that have adopted climate change denial as one of their imperatives, making it critical to investigate this denial within broader social, political and ideological contexts (Hultman and Anshelm 2017). Such trends forth the need to expose climate change denial for what it is: a tactic of wealthy – mostly white Western – men (supported by working- and middle-class white men in particular) to re-assert social, economic and political power and control over natural resource extraction and wealth distribution while wantonly disregarding the deleterious global, regional and local impacts of anthropogenic climate change on the current and future fecundity of society and Earth (Brulle 2014). Aligned with Greta Gaard (2015, 24), we concur that "climate change may be described as white industrial capitalist hetero-male supremacy on steroids, boosted by widespread injustices of gender and race, sexuality and species", implicating climate change denial as obtuse expressions of the hyper-masculine socialisations that Mary O'Brien (1981) referred to as Western 'malestream' norms. Our examination of the intersection between climate change denial and white male effect considers the intersection of power and resource inequalities based on gender, class and race (along with ableness, sexual orientation, ethnicity and age). These variables have reasserted Global Northern white men's primacy not only through recent waves of anti-feminist backlashes, persistently shaping men's values and actions, but also to further obfuscate the intrinsic value of non-human nature as collectively 'otherised' by a male-dominated world (Warren 2000). Recent considerations of violent white supremacist inspired extremism in the US corroborate these concerns (Kimmel 2017 [2013]; Kimmel 2018).

It is important to reiterate that the complexities of human-induced climate change are existential performances with severe social and ecological implications. The biotic, political and personal consequences of climate change highlight a pressing need to transform energy supplies, infrastructural development, mobility, consumption patterns and the very ways we conceptualise ourselves and our relationships within and beyond human communities. Such comprehensive challenges disrupt the very fabric of our social, economic and political machinations. For example, consider the spreading consumer habits of global human populations along with unprecedented increases in extreme weather events (such as rogue storms, droughts, floods and heat waves of greater intensity over longer periods of time) (ECIU 2018), coupled with increased, average annual atmospheric temperatures, ocean acidification and warming, sea-level rise accompanied by accelerated ice sheet shrinking, glacial retreats and loss of snow cover the world over (IPCC 2014). Additionally, consider the IPCC (2014, 40) research that noted the period between 1983 and 2012 was the warmest 30-year period in the last 800 years, with the last 10 years being successively warmer than any earlier decades since 1850 with a 0.65–1.06°C temperature increase over the entire planet between 1880 and 2012. The National Aeronautics and Space Administration (NASA) concurred, concluding that 2016 was the warmest year on record since 1880 (and 2017 the third and 2018 the fourth warmest), with a mean global temperature increasing by 0.8°C (NASA 2018; NASA 2019). Effectively, thermal expansion on a planetary scale has contributed to an 81 millimetre rise in global ocean levels since 1993; past global climatic changes of these proportions that occurred during the Pleistocene (approx. 12,000 years ago) resulted in complete transformations of surface vegetation, regional mass extinctions of plants and animals and sea-level changes near 130 metres (Pittock 2009, 2–4; Spratt and Lisiecki 2016). The most recent climate science predicts biotic risks that are as or even more severe (IPCC 2018).

We have also been alerted to the increased geological risks associated with climate change (earthquakes, tsunamis, landslides, avalanches, glacial outburst floods and volcanic eruptions) representing other sources of likely widespread devastation to humans and non-humans alike (McGuire 2012, 9). Despite the historical record and contemporary exacerbation of these global symptoms of climate change, a dearth of political will from international leaders to respond with conviction and haste to protect our common future persists. Sadly, even processes of environmental regulation and industrial reform reflective of the ecological modernisation movement have deferred to human economic interests ahead of comprehensive social and environmental care. This raises the question: why has there been so much lethargy towards effectively tackling these planetary-scale problems? Even more concerning is the growing vocalisation of climate change denial, particularly, and ironically, from men (of the Global North and their prime beneficiaries) who stand to loose the most from these alarming concerns. Notably, the intersections between climate crises and gender has been thin, and while gaining increased attention, requires more explicit exposure (Enarson and Pease 2016; Gaard 2015; Hultman and Pulé 2018). The root causes of climate change can be found in 'business as usual' approaches to international machinations, locating our notion of industrial/breadwinner masculinities, in particular, at the very core of global social and environmental problems.

Industrial/breadwinner masculinities

For our purposes, the term industrial/breadwinner masculinities is used here inter-changeably with Western hyper-masculinised 'patriarchal' and 'normative' mas-culinities (which we apply primarily to men, but also the masculinities adopted by some women as well). Their combined efforts ensure that men – Western white men in particular (and the hegemonic systems that support their primacy) – are socially, economically and politically advantaged over all others.

Notably, we use the term 'industrial' to emphasise the ways that the harmful social and environmental implications of industrialisation are backgrounded for the sake of capital growth. In the modern context, the beneficiaries of extrac-tive dependent industrialisation are not only the owners of the means of produc-tion, but also include fossil fuel and mining executives, financial managers and bankers, corporate middle and senior level managers and administrators (along with the families and communities that benefit from their surpluses) – the vast majority of direct beneficiaries being Western, white and male. We also include shareholders in this typology given they reap the profits of the companies that they have invested in, recognising that the demographics of this group can be quite variable, and include women investors who also benefit financially from hyper-masculinised systems as do those women (and others), granted substan-tially fewer in number, who are corporate leaders or heads of state or closely bonded to the benefits accorded Western white men through familial associations (Connell and Wood 2005). These prime beneficiaries of hyper-masculinities col-lectively represent those individuals who claim pride of place as the principal controllers of corporate capitalism and laud that primacy at the expense of those human and other-than-human others who are marginalised by a male-dominated world (Anshelm and Hultman 2014).

Clearly, we cannot simply attribute our social and ecological problems to (West-ern white) men alone. However, the dominance of men and male-dominated cul-ture stood against otherised people (specifically: women and LGBTIQ+ persons) and other-than-human life has been centuries in the making. Natural resource extraction, surplus production, wealth creation and the capacities to acquire and protect surpluses from others has long resulted in consolidated benefits for men ahead of all others. Building on Carolyn Merchant's (1990) defining text *The Death of Nature: Women, Ecology and the Scientific Revolution*, we have noted that men have been historically rewarded for pursuing exploitative prac-tices, despite far-reaching social and environmental costs. For centuries, mascu-line hegemonisation has implemented systematic levels of organised oppression against any challenges to the hubris that accompanies male domination. We have seen further refinements and an acceleration of that consolidation of wealth into the hands of ruling industrial elites, almost all of whom have been and continue to be men, gaining renewed traction in the 21st century as disparities between rich and poor have widened.

Accompanying those who own or directly benefit from ownership of the means of production are those working for them to generate surplus wealth. Judith

Stacey's (1998, 267) *Brave New Families: Stories of Domestic Upheaval in Late-Twentieth-Century America* introduced us to the term 'breadwinner', which refers to those working-class individuals who are commonly found at the 'coal-face' of extractive practices. Throughout human history, those individuals have also largely been men and in the context of Western social constructs they have also been historically white. Typically, breadwinner masculinities represent those individuals who toil in mines, work on manufacturing assembly lines, swing hammers, move goods and grow crops (roles that are increasingly performed by people of colour in some Global Northern contexts such as the US) – practices that variably have deleterious environmental consequences. These workers are closely related to industrial masculinities discussed above, but represent a distinct and economically, politically and socially constrained group that serves as the 'foot soldiers' rather than the 'lieutenants' and 'generals' of Global Northern industrial means of production.

Like their industrial counterparts, breadwinner constituents are dependent upon resource extraction, representing individuals on the opposite end of a class hierarchy who are similarly deeply invested in the continued success of a corporatisation and the commoditisation of Earth's resources. That the mechanisms of global capitalism are demonstrating increasing fragility (think here of: the Global Financial Crisis of 2007–2008; the European economic disaster of Brexit; the rise of populism; increasing rates of social and environmental refugees; growing xenophobia and nationalistic isolationism) is telling.

The dismantling of democracy in the pursuit of preserved profitability that advantage the few, with supposed 'trickle-down' benefits to the many is proving to be fundamentally flawed as a mechanism for wealth distribution, further to consequential social and environmental costs. Unfortunately, reactions to this state of affairs heavily influenced by industrial/breadwinner typologies have been misdirected towards scapegoating those who are 'otherised' rather than holding to account the harmful characteristics of these typologies themselves and the very systems that have been designed to privilege them at the expense of so many others. Such an analysis sheds some light on self-professed billionaire Donald Trump's shock success in the US 2016 presidential election, his extreme wealth accompanied by an impetuous bravado that offered a sense of security and paternalism to working- and middle-class white men in particular in the wake of growing frustration for those who continue to find the promises of corporate capitalism alluring even if largely out of their reach. Viewed through this lens, we begin to understand why a hyper-masculinised industrialist such as Trump would offer the most surety to so many working white individuals as the fracturing of global capitalism becomes ever more socially and environmentally evident. In order to 'Make America Great Again', industrialists and working men necessarily joined forces to reassert the privileges of masculine hegemonisation in a desperate attempt to protect and preserve Global Northern, white, male primacy. Accordingly, industrial/breadwinner typologies wed owners of the means of production and their workers to the pursuits of industrial growth through Earth exploitation, noting that each requires the other to thrive, at increasingly alarming cost to all of life (Anshelm and Hultman 2014).

We note the influence of fossil fuel companies and their associated infrastructures (such as: automobile manufacturing and use, energy production along with cooling and heating, military technologies and the waging of war, including industries associated with civil aviation and other forms of transportation), provides us with acute examples of the links between industrial/breadwinner masculinities and climate change denial.

Industrial/breadwinner masculinities and climate change denial

Intersections between industrial/breadwinner masculinities and climate change denial reveal a disconnection between modern Western malestreams and Earthcare. As global social and environmental concerns gain momentum, climate change denial has ramped up to cast climate science as oppositional to the implied securities and assumed entitlements of masculine primacy (Anshelm and Hultman 2014). Aaron McCright and Riley Dunlap (2011) noted that the 1997 Kyoto Protocol triggered an initial reflex of conservative political activity (particularly in the US) buoyed by a small group of dissident and contrarian scientists who have lent their credentials to think tanks that have championed economic rationalism over climate change concerns. This trend has since intensified. It is well recognised that to maintain an illusion of controversy, industrialists, special interest groups and public relations firms have increased their sophistication and capacities to manipulate climate data in order to promote pro-industrial agendas. Consequently, the contemporary intersections between climate change denial and industrial/breadwinner typologies expose the reliance of owners of the means of production and their workers on industrial growth and corporate capitalism (Anshelm and Hultman 2014).

A central strategy has been to pit emotive views, reflective of socio-political and economic biases aligned with populist agendas, against overwhelming field data and analyses by global experts. This has created a false impression of an even debate. Many of these (mostly white male) contrarian voices have participated in generating climate controversy as industry-funded researchers who also hold strong beliefs in global market forces and a general mistrust of regulatory and/or centralised government policies (McCright and Dunlap 2011; Anshelm and Hultman 2014). As is the case with industrial/breadwinner typologies, climate change deniers represent a well-resourced cadre of industrial researchers, corporate leaders, big-business owners and special interest groups who continue to throw massive resources at attempting (with varying degrees of success) to convince us that global warming is a 'normal' geological cycle and concerns ought to be considered nothing more than hysteria (Farrell 2016).They claim that climate science is drummed up by a politically correct left to the detriment of the supposed 'good life' that has long been the great promise of centuries of uninterrupted male domination (Oreskes and Conway 2010). Their willingness to misrepresent and subjectively interpret climate science is difficult for many to understand in the wake of overwhelming evidence validating concerns and is, for some, considered a crime against humanity and the planet (Savransky 2018). A telling example of this is the use of the public relations firm APCO by ExxonMobil to confuse popular disquisition on climate change that, unsurprisingly, was the same firm engaged by Philip Morris to confuse the health risks of tobacco smoking in the early 1990s (Monbiot 2006). Also consider veiled acknowledgement by ExxonMobil of the severe consequences of emissions from coal, oil and gas that has been noted but intentionally downplayed by company management in order to extend consumer markets and preserve profits (Supran and Oreskes 2017).

A foundational explanation for the protestations of climate change denial is obvious. The mere suggestion that we live on a finite planet that is being rapidly transformed by anthropogenic factors such as carbon pollution directly challenges the primacy of those who stand to benefit the most from unfettered industrialisation (Anshelm and Hultman 2014; Supran and Oreskes 2017). Effectively, to ignore or contest climate science is a reflex of prioritising people before planet and a select few in particular who gain the most through societal and natural resource exploitation. As an additional example, consider some of Sweden's vocal climate deniers, who have organisational affiliations in sectors where business research as well as science and technology studies meet. Per-Olof Eriksson, a former board member of Volvo and CEO of SECO Tools and Sandvik, wrote an influential article in the leading Swedish business paper *Dagens Industri* declaring his doubts that carbon emissions affect the global climate. Ingemar Nordin, Professor of Philosophy of Science, joined the fray by stating that the IPCC's "selection and review of scientific evidence are consistent with what politicians wanted", which he considered to be just cause for treating such reports with suspicion (Hultman and Anshelm 2017). Economy Professors Marian Radetzki and Nils Lundgren claimed that in 2009, the IPCC deliberately constructed their models 'in an alarmist direction', alleging that climate science was being dramatised by those with special interests and hidden agendas to slow economic growth (Hultman and Anshelm 2017). While extremely well funded and as a result disproportionately visible, such contrarian views represent but a small proportion of the Swedish climate debate. Further to viewing these climate change deniers as anti-science, we argue that it is important to also understand how their very identities (even if some contrarian individuals are women or queer) have been shaped by industrial modernisation and how this configuration biases their interrogation of climate science precisely because the data affronts these individuals at the level of personal and professional identity, while also interrupting their literal or presumed acquisition of resources, power, privilege and domination that an industrial/breadwinner allegiance affords.

Granted, industrial/breadwinner typologies have the most to lose from a complete redefinition of global systematics towards a truly sustainable future that places all life on equal footing. However, they are also and ironically, likely to be the biggest losers of collapsing global social, economic, political and environmental systems as well (Anshelm and Hultman 2014; McCright and Dunlap 2011). Adding race to these complexities reveals another important consideration.

White, male and in denial

With recent successes of populist governments, attention has shifted towards the plights of working- and middle-class white men in the wake of an increasingly heterogeneous world. These groups have risen up in many Global Northern nations to vocally (and at times violently) support nationalist/white suprema-cist/neofascist male-dominated leadership and hyper-masculine agendas that prioritise xenophobic and isolationist responses to global problems (con-sider recent electoral outcomes of the populist net gains in the 2019 EU elec-tions, 17%+ in Andalucía in Spain, Austria, Finland, Denmark, Hungary, Italy, Sweden, Switzerland and the United States – as well as similar gains of right-wing movements in Brazil, India, Thailand and Turkey beyond the Global North) (Hochschild 2016; Youngs 2018). This is not a new analysis. Other researchers have also critiqued hyper-masculine hegemonies, highlighting "[d]elusions, of hyper-separation, transcendence, and dominance [that] . . . engender denial of the many global [social and] environmental crises" (Alaimo 2009, 28). Clearly, studies like this expose the limitations of socialisations associated with being Western, white and male, bringing with them a heavy reliance on anthro-pocentric notions that Earth's natural resources are humanity's for the taking and that the wealth they generate is then distributed in accordance with presumed orders of entitlement that follow racial divisions of privilege.

In a world that favours white people ahead of people of colour, conundrums that confront white working- and middle-class persons are concealed behind the nor-malisation of white primacy. For white men in particular, conditioning encourages creativity, initiative, motivation, drive – a freedom to move forward with inten-tion and an expectation to then gain the greatest rewards along the way, despite (arguably in spite of) the risks – to be economically, politically and socially suc-cessful protector/providers. In doing so they are promised some of the spoils of profiteering and the (often illusionary) assurance of achieving the most revered of heights in society as modelled by the wealthy. Goals in a growth addicted society, of becoming rich, gaining power and with that having the capacity to shape the world in their own image that leaves them feeling safe, create powerful incentives to support demagoguery, as we have seen in many Global Northern nations in recent years. Hyper-masculinities necessitates constraints on success within the confines of such a system, leaving, many men feeling angry and hurt about the struggle to achieve their version of a promised dividend through the accoutre-ments of male (white) domination so much so that the ranks of populist and white supremacy groups have experienced increased support at public gatherings and organised demonstrations throughout the Global North that is second only to the rise of fascism in the 1930s, sharing concerns about the impact of climate science on growth in alignment with industrial/breadwinner trends (BBC 2017; Hultman and Anshelm 2017). Like climate change deniers, populist support for veiled or overt white supremacy is, at its root, a fear-based response to impending global social and environmental changes, which ignore the realities and resiliences that we gain as people and a planet through celebrations of intersectionalities.

Conclusions

Clearly, those masculinities most closely aligned with resource extractivism and associated corporate services are not only straining human societies and Earth's living systems by being complicit in populist support for climate change denial.

Industrial/breadwinner masculinities also dominate global machinations and in doing so, seek to control narratives that have great bearing on the ways that we shape current and future international cooperation agreements. These trends also dictate socialisations of future generations of men and masculinities in the direction of revisionist social and ecological injustices (Fleming 2010). As we have demonstrated, it is no coincidence that the characteristic features of white male effect are entangled with industrial/breadwinner typologies and the climate change denial since they align (indeed are dependent upon) the presumed entitlements of race, socio-economics, corporitisation and anthropocentrism (Anshelm and Hultman 2014; Hultman and Pulé 2018; McCright and Dunlap 2011). In exposing this nexus, it is our intention to work towards a more relational future that encourages broader, deeper and wider care for the global commons can emerge.

References

Alaimo, Stacy. 2009. "Insurgent Vulnerability and the Carbon Footprint of Gender." *Women, Gender & Research* nos. 3–4: 22–35.

Anshelm, Jonas, and Martin Hultman. 2014. "A Green Fatwā? Climate Change as a Threat to the Masculinity of Industrial Modernity." *NORMA: International Journal for Masculinity Studies* 9, no. 2: 84–96.

Begely, Patrick, and Jacqueline Maley. 2017. "White Supremacist Leader Mike Enoch to Visit Australia." *Sydney Morning Herald*, May 13. www.smh.com.au/politics/federal/white-supremacist-leader-mike-enoch-to-visit-australia-20170513-gw46fn.html.

BBC (British Broadcasting Commission). 2017. "White Supremacy: Are US Right-Wing Groups on the Rise?" www.bbc.com/news/world-us-canada-40915356.

Brulle, Robert. 2014. "Institutionalizing Delay: Foundation Funding and the Creation of US Climate Change Counter-Movement Organizations." *Climatic Change* 122, no. 4: 681–94.

Butler, Judith. 1990. *Gender Trouble: Feminism and the Subversion of Identity*. New York: Routledge.

Connell, Raewyn. 1995. *Masculinities*. Berkeley: University of California Press.

Connell, Raewyn, and Julian Wood. 2005. "Globalization and Business Masculinities." *Men and Masculinities* 7, no. 4: 347–64.

Doherty, Ben. 2018. "Australia's Silence During Climate Change Debate Shocks COP24 Delegates." *The Guardian*, December 9. www.theguardian.com/environment/2018/dec/10/australias-silence-during-climate-change-debate-shocks-cop24-delegates?CMP=Share_iOSApp_Other.

Dunlap, Riley, and Aaron McCright. 2015. "Challenging Climate Change: The Denial Countermovement." In *Climate Change and Society: Sociological Perspectives,* edited by Riley Dunlap and Robert Brulle, 300–32. New York: Oxford University Press.

Enarson, Elaine, and Bob Pease. 2016. *Men, Masculinities and Disaster*. Oxon: Routledge.

ECIU (Energy Climate and Intelligence Unit). 2018. " Heavy Weather: Tracking the Fingerprints of Climate Change, Three Years after the Paris Agreement." https://eciu.net/reports/2017/heavy-weather.

Farrell, Justin. 2016. "Corporate Funding and Ideological Polarization about Climate Change." *Proceedings of the National Academy of Sciences* 113, no. 1: 92–97.

Finucane, Melissa, Paul Slovic, C. K. Mertz, James Flynn, and Theresa Satterfield. 2000. "Gender, Race, and Perceived Risk: The 'White Male' Effect." *Health, Risk & Society* 2, no. 2: 159–72.

Fleming, James. 2010. *Fixing the Sky: The Checkered History of Weather and Climate Control*. New York and Chichester: Columbia University Press.

Gaard, Greta. 2015. "Ecofeminism and Climate Change." *Women's Studies International Forum* 49(March–April): 20–33.

Hochschild, Arlie. 2016. *Strangers in Their Own Land: Anger and Mourning on the American Right*. New York: The New Press.

Hultman, Martin. 2017. "Natures of Masculinities: Conceptualising Industrial, Ecomodern and Ecological Masculinities." In *Understanding Climate Change through Gender Relations*, edited by Susan Buckingham and Virginie le Masson, 87–103. Oxon: Routledge.

Hultman, Martin, and Jonas Anshelm. 2017. "Masculinities of Global Climate Change: Exploring Examples of Industrial, Ecomodern, Industrial and Ecological Masculinity." In *Climate Change and Gender in Rich Countries: Work, Public Policy and Action*, edited by Marjorie Cohen, 19–34. Abington: Routledge.

Hultman, Martin, and Paul Pulé. 2018. *Ecological Masculinities: Theoretical Foundations and Practical Guidance*. Oxon: Routledge.

IPCC (Intergovernmental Panel on Climate Change). 2014. "Climate Change 2014: Synthesis Report." https://www.ipcc.ch/site/assets/uploads/2018/02/SYR_AR5_FINAL_full.pdf.

IPCC (Intergovernmental Panel on Climate Change). 2018. "Global Warming of 1.5°C." https://www.ipcc.ch/site/assets/uploads/sites/2/2018/07/SR15_SPM_version_stand_alone_LR.pdf.

Kimmel, Michael. 2017 [2013]. *Angry White Men: American Masculinity at the End of an Era*. New York: Nation Books.

Kimmel, Michael. 2018. *Healing From Hate: How Young Men Get Into – and Out of – Violent Extremism*. Oakland: University of California Press.

Lockwood, Matthew. 2018. "Right-Wing Populism and the Climate Change Agenda: Exploring the Linkages." *Environmental Politics* 27, no. 4: 712–32.

McCright, Aaron, and Riley Dunlap. 2011. "Cool Dudes: The Denial of Climate Change among Conservative White Males in the United States." *Global Environmental Change* 21, no. 4: 1163–72.

McGuire, Bill. 2012. *Waking the Giant: How a Changing Climate Triggers Earthquakes, Tsunamis, and Volcanoes*. Oxford: Oxford University Press.

Merchant, Carolyn. 1990. *The Death of Nature: Women, Ecology and the Scientific Revolution*. New York: HarperCollins.

Monbiot, George. 2006. *Heat: How to Stop the Planet from Burning*. Toronto: Doubleday Canada.

NASA (National Aeronautics and Space Administration). 2018. "Long-term Warming Trend Continued in 2017: NASA, NOAA." https://climate.nasa.gov/news/2671/long-term-warming-trend-continued-in-2017-nasa-noaa/

NASA (National Aeronautics and Space Administration). 2019. "Global Temperature." https://climate.nasa.gov/vital-signs/global-temperature/

O'Brien, Mary. 1981. "Feminist Theory and Dialectical Logic." *Signs: Journal of Women in Culture and Society* 7, no. 1(Autumn): 144–57.

Oreskes, Naomi, and Erik Conway. 2010. "Defeating the Merchants of Doubt." *Nature* 465, no. 7299: 686–87.

Pease, Bob. 2019. "Recreating Men's Relationship with Nature: Toward a Profeminist Environmentalism." *Men and Masculinities* 22, no. 1: 113–23.

Pittock, A. 2009. *Climate Change: Science, Impacts and Solutions* (2nd ed.). Collingwood: CSIRO Publishing.

Plumer, Brad. 2018. "Climate Negotiators Reach an Overtime Deal to Keep Paris Pact Alive." *The New York Times*, December 15. https://nyti.ms/2QTOSqY?smid=nytcore-ios-share.

Pulé, Paul. 2013. *A Declaration of Caring: Towards Ecological Masculinism*. PhD dissertation, Murdoch University.

Radetzki, Marian, and Nils Lundgren. 2009. "En Grön Fatwa har Utfärdats." *Ekonomisk Debatt* 37, no. 5: 57–65.

Savransky, Rebecca. 2018. "Schwarzenegger Planning to Sue Oil Companies for 'Knowingly Killing People All Over the World'." *The Hill*, March 12. http://thehill.com/blogs/blog-briefing-room/news/377894-schwarzenegger-targeting-oil-companies-for-knowingly-killing.

Slovic, Paul, Ellen Peters, Melissa Finucane, and Donald Macgregor. 2005. "Affect, Risk, and Decision Making." *Health Psychology* 24, no. 4S: S35–S40.

Spratt, Rachel, and Lorraine Lisiecki. 2016. "A Late Pleistocene Sea Level Stack." https://www.clim-past.net/12/1079/2016/cp-12-1079-2016.pdf

Stacey, Judith. 1998. *Brave New Families: Stories of Domestic Upheaval in Late-Twentieth-Century America*. Berkeley: University of California Press.

Supran, Geoffrey, and Naomi Oreskes. 2017. "Assessing ExxonMobil's Climate Change Communications (1977–2014)." *Environmental Research Letters* 12, no. 8: 1–18.

Warren, Karen. 2000. *Ecofeminist Philosophy: A Western Perspective on What It Is and Why It Matters*. Lanham: Rowman & Littlefield Publishers, Inc.

Youngs, Richard. 2018. "The Ordinary People Making the World More Right-Wing." https://carnegieeurope.eu/2018/11/05/ordinary-people-making-world-more-right-wing-pub-77647.

Part II

6 Climate change and 'architectures of entitlement'

Beyond gendered virtue and vulnerability in the Pacific Islands?

Nicole George

Introduction

Pacific Island peoples are often described as living at the 'front line' of global climate change and the first communities to be negatively impacted as we enter an era known as the Anthropocene. Their vulnerability to a future where human activity contributes to the intensification of environmental dangers in the form of extreme weather events, sea-level rise and ocean acidification has seen some Pacific communities likened to 'canaries in a coal mine', or the inhabitants of a 'fragile paradise' doomed to extinction (Boom and Lederswach 2011; Deloughrey 2013; Nunn 2016). In this chapter, I examine how women from the Pacific Islands are featured in, and are able to shape debate on their experiences of, and responses to, the impacts of climate change. As the following discussion will explain, powerful gendered tropes shape where and how women participate in this discussion. To uncover the gendered politics of climate change in the Pacific Islands, I adopt a 'feminist critical perspective' as it is explained by Annica Kronsell, an approach that I develop to both illuminate "the material and normative dimensions of the gender power order" (2015, 74) and to reflect critically on how this order shapes women's and men's differing capacities to participate in, and influence responses to, environmental security in the Pacific region.

As for many other parts of the globe, debate on how women are impacted by climate change in the Pacific tend to be framed by, and reinforce, well defined narratives of gendered virtue and vulnerability (Arora-Jonsson 2011). To sharpen the feminist critical perspective I develop in this chapter, I describe a 'gendered architecture of entitlement', showing how the prevailing narratives of virtue and vulnerability that are evident in the debate on sea-level rise in the Pacific influence where and how women may participate in responses to the environmental insecurity that afflicts their communities. Although transnational political influence also shapes this advocacy, there is a strong tendency for Pacific women to embrace political postures that emphasise feminised and indigenous capacities for resistance to climate change as part of global environmental sustainability campaigns. My discussion shows that strongly idealised representations of contemporary Oceanic femininity are often expected, and indeed, central to these political projects, and that Pacific women may face censure and obstruction when they choose not to embrace this ideal. Thus, I show that on the issue of environmental security in the Pacific region, women enter a terrain of debate that is already strongly gendered, and have, at present, little capacity to transform that terrain in ways that challenge expectations about the roles and standing of women more generally in society.

I begin this discussion with a review of some of the ways gendered perspectives of climate change have been developed in the academic literature, focusing on the broad themes that emerge from this discussion. From here I introduce the architecture of entitlement concept, and show that this helps me to extend my study of the prevailing gendered order that shapes climate governance in the Pacific. I contend that the idea of a gendered architecture of entitlement illuminates where and how understandings of 'rightful' claims to authority operate to both enable and constrain women's agency in response to environmental insecurity challenges. In the third section of the chapter, I discuss three responses to climate vulnerability in the Pacific Islands that have involved or been led by women; the first is the 'fragile paradise' photography that has been used to document community responses to climate change in Kiribati, the second is the 'climate warrior' resistance movements involving young Pacific islanders as part of the Pacific 350 campaign (McNamara and Farbotko 2017), and the third, the Tulele Peisa resettlement project, initiated and coordinated by Carteret Islander Ursula Rakova, who I interviewed in 2014 (Connell 2017; George 2016a). I develop the discussion to consider how each example sits within, is guided by, or attempts (with mixed results), to disrupt a broader gendered architecture of entitlement that is both enabling and potentially constraining for women participants.

Gender and climate change

While the study of climate change has been dominated by scientific and economistic modes of analysis, a growing academic literature, particularly from political geography, has argued for more attention paid to the social impacts of this global challenge (Adger et al. 2011; Adger and Kelly 1999; Barnett 2003; Barnett and Adger 2007; Campbell and Barnett 2010; Farbotko and Lazrus 2012). Some of this work has included passing commentary on gender, but it is only in the last decade that academic attention has been more directly concentrated on the gendered impacts of climate change (Alston 2013; Arora-Jonsson 2011; Denton 2002; Kronsell 2013; Mcleod et al. 2018; Mcnamara and Farbotko 2017; Terry 2009, 6). For the most part, the topic has drawn the interest of researchers within the field of feminist development studies, and is focused on understanding how gender shapes individual and community capacities to absorb, adapt to, and make decisions about the impacts of climate change. While enquiry which places an explicit focus on gender, climate change and security is less common, these questions are now also being taken up with more concentration in the gender and climate change research and have lately become a more prominent focus of some studies (Alam, Bhatia, and Mawby 2015; Detraz 2009).

Three key themes emerge from this body of work as a whole. The first is informed by the increasingly powerful view that climate change will constitute a disruptive challenge to the present international order and generate security risks, tensions and violence that will make increased human contestation over scarce resources become more likely. Hence there has been some discussion of the security risks that are, or will be posed to women by 'environmental conflict' (Detraz 2009, 347). This work tends to echo well-rehearsed feminist approaches to peace and conflict studies, and examines the particular dynamics of environmental conflict, how these may make women vulnerable to violence and other types of hardship, or conversely how they may see women included or excluded from conflict-transition processes.

A second theme emphasises the gendered impacts of environmental disaster as distinct events (floods, hurricanes and cyclones, earthquakes or bushfires) – what might be called a disaster relief and recovery perspective. This work offers a gendered analysis of how women and men become differentially vulnerable to disaster-related physical harm and death, are differentially exposed to risk, disadvantage or gendered discrimination and violence, as displaced populations or in longer-term processes of disaster 'recovery' (Alston 2013, 289; Terry 2009, 7–8). This work has offered important insights into the gendered nature of environmental disasters, the fact that women are more likely to perish in disasters due to restrictions on their movements, or be exposed to violence in the disaster's aftermath, as households deal with the pressures of lost and damaged property, as well as the economic impacts of these types of crises (Barkha 2018). This literature is complimented by the growing literature on 'disaster' management, which has begun to reflect both on the gendered character of disaster events themselves but noted the important, but usually unpaid, and under-recognised community welfare and pastoral care roles that are often shouldered by women in the disaster recovery process (Enarson and Fordham 2001; Yonder, Akcar, and Gopalan 2005).

A third theme in the climate change literature considers the gendered insecurities that result from more attenuated processes of environmental degradation (Detraz 2009, 351). Some of this work has attempted to assign 'responsibility' for greenhouse gas emissions that contribute to climate change on the basis of gender. This work contends that male ownership of vehicles and general levels of economic consumption, particularly in western and developed contexts, is usually more elevated than women's and can be interpreted as suggesting that they are also contributing to a high proportion of polluting emissions (Jonsson-Latham 2007, cited in Terry 2009). By contrast, women's vulnerabilities within climate-change-affected communities as potentially disadvantaged economic earners or marginalised decision-makers are illustrated, prompting questions about how these impact upon women's capacities for resilience and adaptability to climatic change (Alston 2013; Kronsell 2013; Terry 2009).

As Kronsell has demonstrated, these similar gendered dynamics are also evident at the highest levels of climate change governance, where economistic, scientific and technical knowledge has predominated over study of the human and social impacts of climate change. These former approaches do not privilege gender, but that does not mean they are not gendered (Kronsell 2013). Rather, as a critical feminist enquiry reveals, these spheres of authority and realms of practice are constructed in ways that privilege male interests or capacities as the norm. Within this masculinised sphere, alternative forms of knowledge or sites of action that might be more beneficial to women are frequently subordinated. So, for example, as Terry has shown, the marketisation of carbon, and carbon trading, as a means by which to incentivise a reduction in carbon emissions has become institutionalised as part of the global climate governance regime, but this mechanism puts many women at a disadvantage because they generally have more restricted access to land, credit or information and thus lack the same capacities as men for equal participation. While it is important not to overgeneralise this argument and recognise that intersectional differences of ethnicity, class, age and physical ability may also be important factors which endow some women with more access to market-based resources than others (c.f. Crenshaw 1994; Yuval-Davis 2006), it is also true that in a more general sense, markets tend to function in ways that privilege masculine ambition and privilege and hence are beneficial to the standing of men in greater numbers than to women.

The contention that the subordinated standing of women and the masculinised nature of contemporary climate change governance are coterminous challenges which shape where and how women are able to respond to the ongoing impacts of gradual environmental degradation is most relevant to the empirical discussion of gender and climate change in the Pacific Islands that I develop in this chapter. Dramatic instances of environmental disaster are not a novelty for the Pacific Islands and do elicit strong attention and deep sympathy from international actors (bilateral aid agencies and international non-government organisations, for example) eager to provide assistance. Likewise, the idea that conflict may occur in this part of the world as island populations are required to contend with scarcity and food insecurity, or the depletion of the resources that are necessary to sustain life is not fanciful for the peoples of this region. Even the most rudimentary study of contemporary Pacific history will demonstrate that the movement of peoples in this region, has often been prompted by environmental resource depletion (as a result of human and climactic pressures) and that community re-settlement initiatives have not always been negotiated harmoniously (Connell 2017). Yet, given the sheer scale of the regional impact, it is perhaps more important to examine how women and men in many Pacific Island communities are positioned to manage the attenuated, 'slow violence' of climate change (Nixon 2011, 2) manifest principally through the phenomenon of sea-level rise.

In drawing out how environmental disaster can be experienced as a form of climate-change related 'slow violence', I am inspired by Rob Nixon's characterisation of the graduated and compounding impacts of environmental degradation. Nixon's work draws attention to the environmental hazards or 'violence' that take the form of "delayed destruction, that is dispersed across time and space, an attritional violence that is typically not viewed as violence at all" (Nixon 2011, 2). In the Pacific region, the slow violence of rising sea levels is manifest in phenomena such as coastal erosion and the loss of property and arable land, water and soil salinisation, ocean acidification and disrupted rainfall patterns. These attenuated or creeping forms of environmental hazards have already contributed to the displacement of some communities in Pacific Island countries, and will threaten the displacement of many more into the future (Connell and Lutekhaus 2017). These phenomena are experienced as forms of violence for those Pacific Island communities that are affected because of the powerful physical and spiritual connections to place that are formed by indigenous communities, and the expectations that the region's indigenous people, and in matrilineal cultures, particularly women, have a duty to care for and protect their land (c.f. Sirivi and Havini 2004).

To draw out the ways in which relations of gendered power shape how different groups of women and men are able to contribute to, and shape the direction of, governance in response to these phenomena, I argue that it is helpful to consider how these contributions function to uphold or disrupt a broader 'architecture of entitlement'. References to an 'architecture of entitlement' have been most common within the field of political geography where this concept has been developed to examine how social groups' differential access to resources impacts upon their security and makes them differentially vulnerable to climate change phenomena because of their varying capacities to access limited resources (Adger and Kelly 1999). The term has been invoked to illustrate how individuals, groups and institutions are engaged in contested processes of resource distribution (political and material) to manage the impacts of climate change and how this contestation is shaped by socially constructed ideas about legitimate 'entitlements'.

Adger and Kelly's (1999, 257) reflections on the "institutional determinants of entitlement distribution" are most pertinent to the analysis presented here, and extend beyond an interest in formalised political structures to also include the more diffuse "rules of the game". Adger and Kelly's contention that "social constructs effectively constrain individual action and thought" is significant here, as is their observation that "social hierarchies and resource entitlement inequalities are rarely overturned in the course of adaptation [to climate change]". Hence "external changes such as climatic extremes and other natural hazards" work to 'reinforce' the social and institutional hierarchies that shape how resources, material and political (power), are distributed in conditions of environmental insecurity (Adger and Kelly 1999, 257).

The critical point in all of this is that phenomena such as climate change do not occur in isolation from the broader socio-cultural, economic or political hierarchies that are already present in Pacific islands contexts and may do much to reinforce or even exacerbate existing inequalities. This means, on the one hand, that the same gendered barriers that have long been recognised as obstructive to women's political and economic ambitions, also operate in ways which prevent women from playing critical roles in the response to climate change phenomena. That is to say that, the difficulties that Pacific women have faced, in almost all of the region's states, to be accepted as equally credentialed, meritorious and effective decision-makers in formal political spheres, will shape how everyday communities receive women's efforts to speak with authority on how they might best meet the challenges of climate change. These challenges have been elaborated upon expansively by women writers of the region (Chattier 2015; Huffer 2006; Liki and Slatter 2015; Meredith 2014; Sepoe 2002; Soaki 2017) as well as by outside observers (Baker 2015; Corbett and Liki 2016; Spark and Corbett 2018; Wood 2015).

On the other hand, the same features of the prevailing gendered architecture that entitle women to speak in a particular fashion, on particular sorts of subject matters, may also legitimate certain forms of 'feminised' gender activism on climate change. Notwithstanding the points made previously, this means, that women's claims to political leadership, where they have been successful, are often carefully framed in ways that emphasise their dutiful obedience to feminised gendered norms of care-giving, social reproduction and welfare. Yet while these are socio-political realms where women may have an uncontested authority as 'entitled' leaders, the reach of that authority is necessarily restricted. Women may draw on their customary and maternal standing to legitimate their demands to be part of certain forms of debate, but this legitimation depends on women behaving in ways that show strict adherence to protocols of 'rightful' and virtuous gendered comportment (George 2016a, 2016c). To sum up, this means that a gendered architecture of entitlement, tends to operate in ways that circumscribe the types of agendas that many women, although of course not all, are able to successfully pursue in the public domain.

In the pages that follow, I show that women may legitimately discuss climate change in ways that replicate tropes of passive, gendered vulnerability; the insecurities that sea-level rise and consequent food insecurity pose for themselves and their dependents. Women may also legitimately position themselves as bearers of cultural identity, cultural knowledge and cultural connections to place. These cultural claims can be mobilised in ways that legitimise women's leadership in the area of climate change resilience. The women who are at the helm of these projects are often celebrated because they work within a trope that supports the prevailing gendered architecture of entitlement. But women who challenge this architecture by making claims to different sorts of authority, or who voice aspirations of leadership not as mothers, not in ways that reflect women's caring or nurturing obligations, but in ways that may be in more direct competition with established spheres of male authority, are subject to a very different and sometimes hostile response because they are perceived to flout established norms of gendered entitlement. In the following section, I now apply this analytical lens in a more grounded fashion to assist my analysis of women's contributions to debate on climate change in the Pacific region, by referring to the three examples introduced earlier.

Gendering the 'fragile paradise'

Discussions of Pacific Islanders' vulnerability to climate change, is frequently framed by the idea that these populations are at the frontline of this global challenge. Their plight, depicted in recent films such as the *Hungry Tide* (Zubrycki 2011), *Paradise Drowned: Tuvalu: The Disappearing Nation* (Tourell 2001) and *There Once Was an Island: Te Henua E Nnoho* (March 2010) adhere to a well-rehearsed narrative of the 'fragile paradise' that may soon disappear. Features of this narrative include images of white sandy beaches, coconut palms waving in a dappled sunlight. Island peoples perform traditional dances. Their children run playfully along the water's edge. Aerial shots emphasise the terrestrial insecurity of atoll dwelling peoples who seek to build their livelihood on strips of land wedged by a vast ocean that, although once was nurturing, now seems menacing too.

These films regularly feature women, but they are often put into the frame in ways that emphasise the broader narrative of vulnerability. A striking image taken by Justin McManus and appearing in the *Sydney Morning Herald* in 2013 to accompany a story about climate change refugees in Kiribati is stereotypical in this regard (Figure 6.1).

It features the tiny frame of a woman in the middle of a wide angle shot. She seems to stand precariously, on a thin sliver of land with her back to the camera. Her vulnerability is accentuated by the immensity of the surrounding sea, the expansive sky above and her separation from the settlement that is located more distantly, across the water, on the horizon. Tellingly, this figure is swathed in black, protecting herself from the sun, but perhaps this black is also a symbol of future mourning, indicative of this tiny figure's precarity in contrast to the immense body of water that threatens her connections to 'place'.

Another set of images appearing as part of the *Land is Life* short film series, which was shot in Kiribati and Tuvalu in 2011 by Rodney Dekker presents a similarly gendered portrait of women's vulnerabilities but also their capabilities in response to climate change (Dekker 2011; *Land Is Life* 2011). While these films

Figure 6.1 Justin McManus for *Sydney Morning Herald*, 17 October 2013

Source: http://www.smh.com.au/environment/climate-change/climate-change-refugee-seeks-asylum-from-rising-seas-20131017-2vnv1.html

put a stronger focus on resilience to climate change than the image discussed above, women's responses to sea-level rise tend, similarly, to uphold the broader gendered architecture of entitlement. The result of this is that the standing and authority of women and men in relation to challenges of climate change, is represented in strongly contrastive ways. Local male politicians and bureaucrats ruminate on the risks posed to their populations behind large desks, in front of banks of computer screens. They are presented as the bearers of political and technical authority on climate change. By contrast, women's authority is framed in terms which emphasise their dutiful, caring and reproductive roles as nurturers of family and the environment.

They are shown as cultivators, working in coastal regeneration projects such as mangrove nurseries or involved in the development of new fishing techniques to improve food security. Their vulnerabilities to sea-level rise are illustrated in reference to their carer responsibilities. Hence many images depict women with children, but also women caring for the graves of ancestors. Images that show women performing cultural songs and dances, remind us also that women are the nurturers of culture, and connections to place, in these communities.

When considered together, these images reinforce a strongly gendered narrative of women's agency and resilience in response to the slow violence of sea-level rise and its related hazards. Sometimes this point is reinforced graphically when women are shot standing knee-deep in water or staring out at the watery horizon. But the contrastive ways in which women's and men's roles are represented in these films, also reinforces the prevailing gendered architecture that establishes relations of power between women and men in climate governance generally. When read from a critical feminist perspective, these images naturalise a gendered architecture of power. Men are shown in the film as leaders with technical and political acumen. They are endowed with an authority that is global as well as local. They describe the challenges that face their community but they couple this with strong criticism of the region's near neighbours, Australia and New Zealand, who are described as ignoring their responsibilities in promoting a stronger prevention focus in international negotiations on environmental protection (c.f. Fry 2015). By contrast, women as bearers of responsibility for the care and nurturing of land and life in this setting, offer views of climate change phenomena that are contained by the focus on family, community and place. They are entitled to speak about the localised and familial domain but their capacities to have authority beyond this realm are not promoted in these films. This is of course in keeping with broader regional characterisations of the formal political domain as a masculinised terrain (see above). But the films also operate visually to suggest that women are disinterested in capacities for critical political action in the face of this impending challenge. So while men are depicted as engaged political actors with perspectives to share on the international stage, the film's focus on women's resilience and adaptability seems to suggest an acceptance of climate change as a "sealed fate" for Pacific Islanders and position women as only able to appreciate its local and immediate impacts on family and community life (McNamara and Farbotko 2017, 18).

Gender and the pacific climate warriors

An alternative response to the environmental challenges that face the Pacific Islands region, has become evident in regional advocacy that concentrates more centrally on the goal of climate change prevention rather than adaptation. The work of the Pacific Climate Warriors is a particularly noteworthy example of this genre of regional response. This network of regional actors was initially mobilised as part of the global 350.org campaign initiated by Bill McKibben in 2008. The Pacific branch of this network, which is known both as Pacific.350 and the Pacific Climate Warriors, describe themselves as a grassroots, youth-led network, working with communities across 15 Pacific Island countries to 'fight' climate change (About Pacific 350 n.d.). In contrast to the characterisations of Pacific Islands populations facing a 'doomed future' (McNamara and Farbotko 2017, 18), the activities of this group take a more strongly activist posture. On the one hand, they work to build awareness of the political dimensions of the climate change challenge within Pacific Island communities. At the same time, they also engage in political campaigns staged internationally to draw attention to the issue of greenhouse gas emissions and the culpability of states that continue to abstain from global agreements to control these emissions.

To support their claim that "we are not drowning, we are fighting", the Pacific Climate Warriors present themselves visually, and rhetorically, as engaged in a battle to defend their islands.

Figure 6.2 Navneet Narayan for 350.org Pacific, 21 January 2014.

Source: https://globalvoices.org/2014/01/21/pacific-climate-warriors-we-are-not-drowning-we-are-fighting/

Figure 6.3 Navneet Narayan for 350.org Pacific, 21 January 2014.

Source: https://globalvoices.org/2014/01/21/pacific-climate-warriors-we-are-not-drowning-we-are-fighting/

Their advocacy materials incorporate visual images of young Pacific Islanders in powerful warrior poses, draped in tapa cloth, or pandanus skirts (Navneet
Narayan 2014; Figures 6.2; 6.3), their bodies made to shine with coconut oil,
in the tradition of many cultures of the region where this is read as a signifier
of 'strength' and power (Hermkens and Lepani 2017, 8). Unlike the images
described in the previous section, the women depicted in these campaigns are not
diminished or dwarfed by the natural environment. Here too, the photographed
figures appear 'knee deep' in water, or at the water's edge. But these representations allow the figures a centrality, they are not dwarfed by the ocean, they are the
ocean's people, a theme that resonates powerfully with Epeli Hau'ofa's emblem-
atic rendering of Pacific Islanders as united and drawing strength through expressions of an Oceanic identity (Hau'ofa 1998).

In contrast to the ways gender roles are distinguished in the films described in
the earlier section, the Pacific Climate Warrior campaign features many images
of women and men together, supporting each other in a fashion that suggests a
mutual strength. As McNamara and Farbotko note, this visualisation of male and
female collaboration challenges the long-standing patriarchal trope of the male
warrior which has had a powerful resonance in Pacific Island societies (2017).
Since the period of first European contact, constructions of Pacific Island masculinity have frequently reinforced the idea that men are 'martial warriors' (Jolly

2016, 305) and that violence, in the form of tribal fighting and war-making is, as a result, something typically inherent to the region's indigenous population. In contrast to the gendered construction of men as war-makers, women are often assigned cultural and religious roles as peacemakers in their families and communities (cf. George 2014, 2016c; Sirvi and Havini 2004). As I have shown elsewhere, these distinctions can also be invoked in ways that lead to a questioning of women's merit or ability to participate in political deliberation of a more formal variety. That toughness and gritty determination expected of male warriors, is also seen as an attribute required of political leaders (George 2016c). It is therefore noteworthy that the Pacific Climate Warrior campaign includes representations of Pacific women who embody power in ways similar to men. This gender collaboration poses a novel and perhaps provocative challenge to the architecture of entitlement (depicted above) that more generally structures participation in deliberations on climate governance.

Nonetheless, there is something of an orientalist gaze informing the construction of these images too, and it is difficult not to view them in ways that recall the objectifying tropes of the 'noble savage' that have long inflected external depictions of the Pacific region (c.f. Smith 1989). While immensely beautiful, these images also recall older colonial-era portraits that required participants to 'perform' their indigeneity for foreign, read metropolitan and coloniser, audiences. Margaret Jolly has investigated the gendered implications of European representations of Pacific peoples in the colonial era, and argued that these helped to perpetuate an eroticised perspective, particularly of Polynesian women, who were frequently represented in 'sexually provocative' and 'tantalizing' ways (Jolly 1997, 102). As Linda Tuhiwai Smith's work on indigenous epistemologies makes clear, this body of work has also helped to sustain that much larger framing of indigenous peoples as the metropolitan sphere's Other, legitimating their subordination to external authority on the grounds that they were 'lesser' and required 'civilisation' to Western norms.

It seems important then, to ask about the ways in which the Pacific Climate Warrior campaign constitutes a 'performance' of indigeneity, and how this sits in dialogue with earlier colonial-era images that contributed to the exoticisation of Pacific Island peoples. It might be productively argued that this campaign is 'reclaiming' and subverting that tradition, particularly through its reworking of gendered roles. Similar themes of this sort can also be found in the provocative work of Samoan-Japanese, and fa'a fafine,[1] artist Shigayuki Kihara. Kihara's 2007 photographic triptych titled *In the Fashion of a Woman* which replicates colonial photographic representations of Oceanic peoples and particularly the trope of the 'dusky maiden' (Tamaira 2010, 1), but in ways that subvert gendered binaries, reminding the viewer, unexpectedly, of the place of transgender people within Pacific communities.

In their own way, the Pacific Climate Warriors campaign may express a similar visual challenge to the colonial tropes described above. The campaign is certainly designed to remind the major polluting nations of the world that their actions contribute to Pacific peoples' ongoing subordination and vulnerability. Nonetheless the heavily stylised and staged performance of Pacific indigeneity also seem to re-appropriate those colonial tropes in ways that idealise and homogenise the regions cultures and its peoples, and thus, in their own ways may also have a subordinating or marginalising impact for local communities. What I mean here is that these images can function as a benchmark to assess the 'authenticity' of Pacific Islanders' claims for recognition in the global debate on climate change. Those who perhaps wish to be part of this debate, but are unable to identify with these 'indigenised' renderings of agency, may find this kind of campaign intimidating or alienating. They may also find that the broader global impact of their message may be compromised if they are unable or unwilling to articulate their arguments in ways that are anchored by these powerful cultural references.

Gender and the warrior poet: Kathy Jetnil-Kijiner

This more critical lens is useful for examining the much celebrated address made by Marshallese poet, Kathy Jetnil-Kijiner who spoke at the UN Climate summit in New York in 2014 (Jetnil-Kijiner 2014). Selected from more than 500 civil society candidates vying for the opportunity to address the gathering, Jetnil-Kijiner's appearance was a source of great pride for Pacific Island activist communities who felt empowered by what they viewed as global recognition of their situation. Her global audience inside the United Nations Assembly, and beyond the walls of that building was captivated by the evocative words of her poem *Dear Matefele Peinam*, directed to her seven-month-old daughter; a performance which made clear Jetnil-Kijiner's determination to ensure her Marshall Islands home would not drown and would remain always there for her child.

But like the Pacific Climate Warriors, Jetnil-Kijiner's address was also a cultural performance for a global audience. She was carefully dressed in the traditional clothing of her place, to remind her audience of her indigenous background. The framing of the address reminded her audience likewise, that she was a mother, and so, like the women described in the first section of this chapter, it was her maternal and caring responsibilities, her gendered sense of virtue and duty to family, as well as to her sinking islands, that seemed to lie at the heart of her activism.

Jetnil-Kijiner's adherence to the gendered architecture of entitlement that structures climate change advocacy was something of a shift from the tone of her previous body of work. Prior to her appearance before the United Nations, Jetnil-Kijiner's poetry tended to convey a much more confrontational tone. Her poem entitled *History Project*[2] gives voice to her outrage over the objectified way Marshallese people were treated during the United States programmes of nuclear testing in the northern Pacific Islands region, in the years following the Second World War. Likewise, her poem *Tell Them*,[3] performed at the London Poetry Parnassus staged in conjunction with the 2012 Olympics, and also focused on the issue of sea-level rise and climate change, conveys angry frustration and defiance over the predicament faced by the Marshal Islands. Notably, neither confirms to an architecture which entitles Pacific women to speak or be heard politically, neither references norms of maternal duty, nor motherhood. While there are references made in *Tell Them* to the cultural practices of basket weaving and jewellery production that are frequently done by women, the poem also allows us to understand Jetnil-Kijiner as someone who moves between the local and the international, someone who travels, who has friends from elsewhere, who shares gifts from her home with those not of that place. That is to say, her response to, and understanding of climate change are not framed only by her identity as a Marshallese woman (in contrast to the *Land is Life* depictions of women), but as a Marshallese woman with knowledge and experience that has come from beyond her islands. It was therefore surprising to see the gendered architecture of entitlement shape, so strongly, her 2014 recital before the United Nations. Here she stood, in Marshallese dress, like the Pacific Warriors, 'performing her culture', and like the virtuous climate change resilient women, performing her gender, talking to her child as part of the address. Indeed, as it concluded she was joined by her child and her husband at the podium, as if to reinforce the gendered rightfulness of her claim to be heard.

Like the women from the *Land is Life* films discussed earlier, the choreography of this address made Jetnil-Kijiner appear dwarfed by her surroundings, and diminutive within the 'masculinised' order of the United Nations General Assembly chamber. Jetnil-Kijiner as the indigenous mother/poet was positioned to the right of an elevated central podium occupied by the United Nations hierarchy which allowed UN General Secretary Ban Ki Moon and his male deputies to gaze down from on high at the performance. Certainly Jetnil-Kijiner's address elicited a strong and emotional response from General Assembly representatives in the room who gave her a standing ovation. Nonetheless the spatial configuration of this moment, showed very clearly the architecture of entitlement that shaped the encounter; where the hierarchy of power was located, and where it was more marginal. Jetnil-Kijiner as a Pacific women was located physically below, and to one side of those for whom the power of the stage accrues more naturally. The spatial organisation of this event was a clear reflection of the prevailing gendered architecture of entitlement shaping Pacific women's responses to climate change more generally.

Women intruding into the 'men's' world

The last example I would like to discuss is a more localised one, demonstrating where and how a prevailing architecture of entitlement also shapes women's capacities for leadership when communities are forced to respond to climate change related displacement. In this context too, the scope of women's political agency is shaped by gendered protocols, or, a gendered 'architecture' that establishes appropriate and rightful, read 'entitled', forms of gendered activity and authority.

The gendered challenges navigated by women in communities that are displaced by sea-level rise were made evident to me personally during an interview I conducted in June 2014 with Ursula Rakova from the Carteret islands, a group of low-lying atoll islands governed by the Autonomous Region of Bougainville (ARoB).

This interview was one of a series of discussions that I undertook with women leaders in Bougainville in a two-week period, to learn about their capacities to progress debate on gendered security in formal and informal decision-making forums. Rakova's insights were particularly interesting given her leadership role, and her efforts to build security for her community, displaced in recent years due to seawater inundation of their atoll islands.

With the assistance of church and kinship networks, and frustrated by a lack of government attentiveness to the growing urgency of environmental insecurity in the Carteret Islands, Rakova independently established a resettlement project for a number of families in Bougainville's north-western Tinputz province in the preceding years. The resettlement project is known as Tulele Peisa which translates as "sailing the waves on our own" (Connell 2017, 88). As this name suggests, Rakova's political position on climate change stridently resists the 'wishful sinking' rhetoric that emphasises the innate vulnerability of 'environmental refugees' (Farbotko 2010, 47; Farbotko and Lazrus 2012, 383). Her efforts to see her re-settled community remain "strong and self-reliant" (Rakova et al. 2009) have therefore involved efforts to establish a cocoa trading cooperative, and the development of reciprocal exchange relations with neighbouring communities to avoid resentments growing in a context where post-conflict relations between communities remain fragile.[4] This enterprise has been pursued as a money earning venture for the community, but it is also one of the few options that the community has to earn income given the fact that land is tightly held across most of Bougainville (Connell 2017). This has meant that the Tulele Peisa settlement has had little opportunity to earn cash income from agricultural produce or from fishing.

Rakova may be internationally renowned for her resourceful response to displacement but on her new home soil she also spoke critically about the gendered challenges this entails. As she reflected on her role as a community leader, and more generally on the place of women in her community, she noted that the Carterets women in the new Tinputz setting were continually required to renegotiate their roles and authority relative to that of men and that prior adherence to norms of matrilineal authority could no longer be assumed; perhaps a reflection that Carterets women are now living a great distance from the traditional lands that once gave them status (Boege 2011). Rakova also spoke critically of the patriarchal cultural practices that existed alongside matrilineal values in Carterets society and the burdens this placed on women. In the new settlement she described her desire to change customary meeting practices which required women to 'crawl' while men sat and the onerous food preparation obligations borne by women during weeks-long male initiation celebrations.

Rakova also described how her efforts to build a cocoa trading collective were frustrated by male economic and political leaders, beyond the settlement, who she said, repeatedly questioned the appropriateness of her intrusion into what they described as 'the man's world'[5] (Rakova, personal testimony, June 2014). Her testimony on this theme, points to the challenges that women generally face in many Pacific societies when they pursue employment or business opportunities.

For example, research conducted in Bougainville and PNG on this theme tends to show that women are often punished for taking up opportunities that will mean that they have to work in close proximity to men. They may face accusations from their spouses or other family members that they are neglecting family duties by working outside the home (Eves et al. 2018). Gina Koczberski's research on the way cash income is managed in family managed palm oil leaseholds, in the East New Britain jurisdiction of Papua New Guinea, is also revealing, and shows that men tend to view participation in the cash domain as their entitlement, and likewise, assert their authority over familial spending in line with "indigenous moral norms and ideologies" that equate women's access to cash as something that "undermines power hierarchies in the home" (Koczberski 2007, 1175, 1182). When understood in this light, the claim that Ravoka's efforts to develop a cocoa trading perspective is a venture that trespasses into the male domain, reflects ideas that she, as a woman, is 'intruding' inappropriately into the male realm of business. She is showing an inappropriate interest in cash itself, an interest, according to prevailing gendered norms, she is not necessarily considered 'entitled' to pursue. Her business and leadership aspirations reflect neither gendered tropes of maternal and caring virtue, nor passive vulnerability, and for this her aspirations draw criticism.

Assessments of the viability of the Tulele Peisa resettlement have often reflected upon the difficulties that the project continues to face, and identified the challenges of accessing land for agricultural purposes as their cause (Boege 2011; Connell 2017). These challenges are not inconsiderable. But the gendered aspect of this scenario is rarely accounted for. The fact that this response to climate change-related insecurity is headed by a woman who looks to circumnavigate the obstacles that prevent her community producing cocoa, by instead creating a cocoa trading cooperative, a domain more usually considered the preserve of men, is not insignificant. Rakova's refusal to conform to the expectations that shape assessments about the appropriate responses women have to conditions of environmental security also compound the problems that have beset this resettlement initiative.

Conclusions

The slow violence of sea-level rise in the Pacific Islands will not capture the international headlines in the same way that more dramatic incidents of environmental disaster such as cyclones, earthquakes and floods, or conflict over environmental resources, regularly do. Environmental disaster and environmental conflict have their own gendered impacts that have been well-researched and analysed by others. My concern in this chapter is alternatively focused, aiming to draw out how gender shapes responses to the attenuated impacts of climate change, and particularly the slow-violence sea-level rise. Through a gendered analysis of the experiences and images I have discussed in this chapter, my aim is to demonstrate how an existing architecture of entitlements, shapes and attributes 'rightfulness' to women's and men's responses to the phenomena of sea-level rise in the

Pacific region. In particular, I show how this architecture is gendered and encourages activity that emphasises women's caring virtue, and docile vulnerability in response to climate change. I also argue that it generates hostile and unsympathetic conditions for women who reject these postures and instead pursue aspirations to lead their communities in ways that are understood to inappropriately challenge men's standing.

This is not to say that women are not entitled to authority, or are completely absent from regional deliberations on this question. As I argue, women have played key leadership roles in their communities and on the international stage to build awareness of and respond to the damaging impacts of climate change phenomena. But, as this chapter makes clear, women's credibility, their legitimacy and their right to be heard in debates on sea-level rise rely, in many cases, upon their adherence to firmly established norms of gendered virtue and vulnerability. Indeed these tropes make women's responses to climate change legible for domestic and international audiences. As my discussion of the *Land is Life* series of short films demonstrates, references to women's maternal duty, their roles as bearers of customary identity, and of ties to place, all provide the basis of women's authority. When women speak, it is in ways that reflect upon the localised familial and communal challenges that are already falling on their shoulders as sea-level rise continues.

It is certainly true that the Pacific Climate Warriors campaigns of resistance to sea-level rise in the Pacific challenge this gendered architecture of entitlement by emphasising, visually at least, the idea of men and women mutually empowered to meet the challenges that climate change will put before them. But representations of gendered virtue are not entirely absent from this campaign. References to indigenised cultural 'performance' pervade the stylised artwork of the campaigns in ways that are strongly reminiscent of colonial era indigenous portraits that contributed to the exoticisation and eroticisation of Pacific peoples in the Occidental gaze. Kathy Jetnil-Kijiner's performance before the United Nations in 2014, while electric, defiant and powerful at one level, was at another, also ultimately one that adhered closely to the prevailing gendered architecture that shapes so much of women's advocacy on this issue. Jetnil-Kijiner thus 'performed' her indigenous and maternal identity before her august UN General Assembly audience. Her message was understood and resonated perhaps, simply because it adhered so strongly to a theme of maternal gendered virtue, underscored by a tone of vulnerability.

Ursula Rakova's response to climate change in the region, subverts rather than accommodates the gendered architecture of entitlement, and she is treated as a kind of dissident, as result. As a would-be cocoa trader, and resettlement leader, Rakova does not embrace the narrative of gendered vulnerability this is a more common feature of regional debates on climate change. Indeed she actively rejects it. But at the same time, her efforts to engage in business, to participate in the cash economy on terms equal with men, and to lead her community in ways that assert her authority over men, subverts the narrative of gendered virtue. Put simply, she faces resistance to her work because she asks for entry into domains that are not considered her entitlement. The gendered perspective of this story is important to bring to light, and to contrast with the previous examples, because it demonstrates in stark terms the discriminatory and constraining aspect of the architecture of entitlement that structures Pacific debate on climate change more generally. When women's virtue and vulnerability are foregrounded as part of their response to climate change, international and local sympathies are easily won. When women's climate change activism strays from this well-worn path, the message seems far more difficult to digest.

Notes

1 Polynesian term for transgender individual.
2 www.youtube.com/watch?v=DIIrrPyK0eU
3 https://jJetnil-Kijiner.wordpress.com/2011/04/13/tell-them/
4 Bougainville experienced a brutal secessionist conflict in the 1990s that accounted for the lives of roughly 10 percent of the population. A peace agreement was signed in 2001, but the tensions that generated conflict remain unresolved in some parts of the community.
5 By this, I understood Rakova to be suggesting that men felt cocoa trading in Bougainville was a domain that they 'rightfully' controlled.

References

"About 350 Pacific." n.d. https://350pacific.org/about/.

Adger, W. Neil, Jon Barnett, F. Stuart Chapin III, and Heidi Ellemor. 2011. "This Must be the Place: Underrepresentation of Identity and Meaning in Climate Change Decision-Making." *Global Environmental Politics* 11, no. 2: 1–25.

Adger, W. Neil, and Mick P. Kelly. 1999. "Social Vulnerability to Climate Change and the Architecture of Entitlement." *Mitigation and Adaption Strategies for Global Change* 4, nos. 3–4: 253–66.

Alam, Mayesha, Rukamani Bhatia, and Briana Mawby. 2015. "Women and Climate Change: Impact and Agency in Human Rights, Security and Economic Development." Report. The Georgetown Institute for Women, Peace and Security. https://giwps.georgetown.edu/sites/giwps/files/Women%20and%20Climate%20Change.pdf.

Alston, Margaret. 2013. "Gender and Climate Change in Australia and the Pacific." In *Research, Action and Policy: Addressing the Gendered Impacts of Climate Change*, edited by Margaret Alston and Kerri Whittenbury, 175–88. Springer, Netherlands.

Arora-Jonsson, Seema. 2011. "Virtue and Vulnerability: Discourses on Women, Gender and Climate Change." *Global Environmental Change* 21, no. 2: 744–51.

Baker, Kerryn. 2015. "Pawa Blong Meri: Women Candidates in the 2015 Bougainville Election." SSGM Discussion Paper 2015/14. Canberra: Australian National University. http://ssgm.bellschool.anu.edu.au/sites/default/files/publications/attachments/2016-07/dp-2015-14-baker-online.pdf.

Barkha, Betty. 2018. "Natural Disasters and Threats for Women in Pacific Small Island States: A Scoping Study in Progress 2006–2016." Policy Brief 2/2018. Monash Gender, Peace and Security. Melbourne: Monash University. http://docs.wixstatic.com/ugd/b4ae f1_2ee19e0e20894a94b55277fdaf854be3.pdf.

Barnett, Jon. 2003. "Security and Climate Change." *Global Environmental Change* 13, no. 1: 7–17.

Barnett, Jon, and Neil Adger. 2007. "Climate Change, Human Security and Violent Conflict." *Political Geography* 26, no. 6: 639–55.

Boege, Volker. 2011. "Challenges and Pitfalls for Resettlement Measures: Experiences in the Pacific Region." Working Paper No. 102. Centre on Migration, Citizenship and Development. Bremen, Germany. www.uni-bielefeld.de/tdrc/ag_comcad/downloads/workingpaper_102_boege.pdf.

Boom, Keely, and Aleta Lederswach. 2011. "Human Rights or Climate Wrongs: Is Tuvalu the Canary in the Coal Mine?" *The Conversation*. https://theconversation.com/human-rights-or-climate-wrongs-is-tuvalu-the-canary-in-the-coal-mine-3830.

Campbell, John and Jon Barnett. 2010. *Climate Change and Small Island States Power, Knowledge and the South Pacific*. London: Routledge.

Chattier, Priya. 2015. "Women in the House (of Parliament) in Fiji: What's Gender Got to Do with It?" *The Round Table* 104, no. 2: 177–88.

Connell, John, and Nancy Lutkehaus. 2017. "Environmental Refugees? A Tale of Two Resettlement Project in Coastal Papua New Guinea." *Australian Geographer* 48, no. 1: 79–95.

Corbett, Jack, and Asenati Liki. 2016. "Intersecting Identities, Divergent Views: Interpreting the Experiences of Women Politicians in the Pacific Islands." *Politics & Gender* 11, no. 2: 320–44.

Crenshaw, Kimberlé. 1994. "Mapping the Margins: Intersectionality, Identity Politics and Violence Against Women of Colour." *Stanford Law Review* 43, no. 6: 1241–99.

Dekker, Rodney. 2011. "Land is Life Pacific." *Oxfam Australia*. www.flickr.com/photos/oxfamaustralia/sets/72157626001415932/.

DeLoughrey, Elizabeth. 2013, The Sea Is Rising: Narrating Climate Change in the Pacific *International Institute, Fall Symposium:"What Is the Future for Islands*. 9 November located at https://www.youtube.com/watch?v=ZkcJZKRmM6w [accessed 15 June 2018].

Denton, Fatma. 2002. "Climate Change Vulnerability, Impacts, and Adaptation: Why Does Gender Matter?" *Gender & Development* 10, no. 2: 10–20.

Detraz, Nicole. 2009. "Environmental Security and Gender: Necessary Shifts in an Evolving Debate." *Security Studies* 18, no. 2: 345–69.

Enarson, Elaine, and Maureen Fordham. 2001. "From Women's Needs to Women's Rights in Disasters." *Environmental Hazards* 3, nos. 3–4: 133–36.

Eves, Richard, Genevieve Kouro, Steven Simiha, and Irene Subalik. 2018. "Do no harm research: Bougainville." Report. Department of Pacific Affairs, Australian National University (ANU), International Women's Development Agency (IWDA) and Australian Aid. http://dpa.bellschool.anu.edu.au/sites/default/files/publications/attachments/2018-03/do_no_harm_bougainville_low_res_0.pdf.

Farbotko, Carol. 2010. "Wishful Sinking: Disappearing Islands, Climate Refugees and Cosmopolitain Experimentation." *Asia Pacific Viewpoint* 51, no. 1: 47–60.

Farbotko, Carol, and Heather Lazrus. 2012. "The First Climate Refugees? Contesting Global Narratives of Climate Change in Tuvalu." *Global Environmental Change* 22, no. 2: 382–90.

Fry, Greg. 2015. "Recapturing the Spirit of 1971: Towards a New Regional Political Settlement in the Pacific." State, Society & Government in Melanesia (SSGM) Discussion Paper 2015/3. Australian National University. http://dpa.bellschool.anu.edu.au/experts-publications/publications/1267/recapturing-spirit-1971-towards-new-regional-political.

George, Nicole. 2016a. "Institutionalising the Women Peace and Security Agenda in the Pacific Islands: Gendering the 'Architecture of Entitlements'?" *International Political Science Review* 37, no. 3: 375–89.

George, Nicole. 2016b. "Lost in Translation: Human Rights, Gender Violence and Women's Capabilities in Fiji." In *Gender Violence and Human Rights in the Western Pacific*, edited by Aletta Biersack, Margaret Jolly, and Martha Macintyre, 81–125. Canberra: Australian National University Press.

George, Nicole. 2016c. "Light, Heat and Shadows: Women's Reflections on Peacebuilding in Bougainville." *Peacebuilding* 4, no. 2: 166–79.

Hau'ofa, Epeli. 1998. "The Ocean in Us." *The Contemporary Pacific* 10, no. 2: 391–410.

Hermkens, Anna-Karina, and Katherine Lepani. 2017. "Introduction: Revaluing Women's Wealth in the Contemporary Pacific." In *Sinuous Objects: Revaluing Women's Wealth in the Contemporary Pacific*, edited by Anna-Karina Hermkens and Katherine Lepani, 1–36. Acton: Australian University Press.

Huffer, Elise. 2006. "A Desk Review of the Factors Which Enable and Constrain Women's Political Representation in Forum Island Countries." In *A Women's Place is in the House: The House of Parliament: Research to Advance Women's Political Representation in Forum Island Countries: A Regional Study Presented in Five Reports*, edited by Elise Huffer and Pacific Islands Forum, 1–56. Suva: Pacific Islands Forum Secretariat.

Jetnil-Kijner, Kathy. 2014. "Statement and Poem by Kathy Jetnil Kijiner, Climate Summit, Opening Ceremony. United Nations. 23 September. Located at https://www.youtube.com/watch?v=mc_IgE7TBSY [accessed 26 April 2019].

Johnsson-Latham, Gerd. 2007. "A Study on Gender Equality as a Prerequisite for Sustainable Development: Report to the Environment Advisory Council, Sweden. Located at https://cdn.atria.nl/epublications/2007/study_on_gender_equality_as_a_prerequisite_for_sustainable_development.pdf [accessed 26 April 2019].

Jolly, M. (2016). Men of War, Men of Peace: Changing Masculinities in Vanuatu. *The Asia Pacific Journal of Anthropology*, 17, nos 3–4: 305–23.

Jolly, Margaret. 1997. "From Point Venus to Bali H'ai: Eroticism and Exoticism in Representations of the Pacific." In *Sites of Desire? Economies of Pleasure: Sexualities in Asia and the Pacific*, edited by Lenore Manderson and Margaret Jolly, 99–122. Chicago: Chicago University Press.

Koczberski, Gina. 2007. "Loose Fruit Mamas: Creating Incentives for Smallholder Women in Oil Palm Production in Papua New Guinea." *World Development* 35, no. 7: 1172–85.

Kronsell, Annica. 2013. "Gender and Transition in Climate." *Governance Environmental Innovation and Societal Transitions* 7 (June): 1–5.

Kronsell, Annica. 2015. "Feminism." In *Research Handbook in Climate Governance*, edited by Karin Bäckstrand and Eva Lövbrand, 73–83. Cheltenham: Edward Elgar Publishing Limited.

"Land Is Life." YouTube video, (n.t.), posted by "Rodney Decker." February 10, 2011. https://vimeo.com/24465420.

Liki, Asenati, and Claire Slatter. 2015. "Control, Alt, Delete: How Fiji's New PR Electoral System and Media Representation Affected Election Results for Women Candidates in the 2014 Election." *Journal of Pacific Studies* 35, no. 2: 71–88.

March, Briar. 2010. *There Once Was an Island: Te henua e nnoho*. Documentary Film. Newburgh: New Day Films. Available at: www.thereoncewasanisland.com/.

Mcleod, Elizabeth, Seema Arora-Jonsson, Yuta J. Masuda, Mae Bruton-Adams, Carol O. Emaurois, Berna Gorong, C. J. Hudlow, Robyn James, Heather Kuhlken, Barbara Masike-Liri, Emeliana Musrasrik-Carl, Agnes Otzelberger, Kathryn Relang, Bertha M. Reyuw, Betty Sigrah, Christina Stinnett, Julita Tellei, and Laura Whitford. 2018. "Raising the Voices of Pacific Island Women to Inform Climate Adaption Policies." *Marine Policy* 93 (July): 178–85.

McManus, Justin. 2013. "Rising Sea Levels Have Already Forced the Relocation of Some Villagers in the Kiribati Islands." *The Sydney Morning Herald.* www.smh.com.au/environment/climate-change/climate-change-refugee-seeks-asylum-from-rising-seas-20131017-2vnv1.html.

McNamara, Karen E., and Carol Farbotko. 2017. "Resisting a 'Doomed' Fate: Aan Analysis of the Pacific Climate Warriors." *Australian Geographer* 48, no. 1: 17–26.

Meredith, Measina. 2014. "Factors Preventing Women Entering Electoral Politics in Samoa." In *Presentation to the Pacific Islands Political Science Association Conference*, June 6, University of French Polynesia.

Narayan, Navneet. 2014. "We Must Draw on Our Heritage and Ancestral Strength to Defend Our Homes." *Global Voices.* https://globalvoices.org/2014/01/21/pacific-climate-warriors-we-are-not-drowning-we-are-fighting/.

Nixon, Rob. 2011. *Slow Violence and the Environmentalism of the Poor.* Cambridge, MA, London: Harvard University Press.

Nunn, Patrick. 2016. "Rise and Fall: Social Collapse Linked to Sea Level in the Pacific." *The Conversation.* https://theconversation.com/rise-and-fall-social-collapse-linked-to-sea-level-in-the-pacific-56268.

Rakova, Ursula, Luis Patron, and Citt Williams. 2009. "How-to Guide for Environmental Refugees." *Our World.* http://ourworld.unu.edu/en/how-to-guide-for-environmental-refugees/.

Sepoe, Orovu. 2002. "To Make a Difference: Realities of Women's Participation in Papua New Guinea Politics." Report. Development Studies Network. https://crawford.anu.edu.au/rmap/devnet/devnet/gen/gen_civil.pdf.

Sirivi, Josephine Tankunani, and Marylin TaleoHavini. 2004. *As Mothers of the Land: The Birth of the Bougainville Women for Peace and Freedom.* Canberra: Pandanus Books.

Smith, Bernard. 1989. *European Vision and the South Pacific.* Melbourne: Oxford University Press.

Soaki, Pauline. 2017. "Casting Her Vote: Women's Political Participation in Solomon Islands." In *Transformations of Gender in Melanesia*, edited by Martha Macintyre and Ceridwen Spark, 95–116. Canberra: Australian National University Press.

Spark, Ceridwen, and Jack Corbett. 2018. "Emerging Women Leaders' Views on Political Participation in Melanesia." *International Feminist Journal of Politics* 20, no. 2: 221–35.

"Statement and Poem by Kathy Jetnil-Kijiner, Climate Summit 2014 – Opening Ceremony." YouTube video, 6:51, posted by "United Nations." 23 September 2014. www.youtube.com/watch?v=mc_IgE7TBSY.

Tamaira, A. Marata. 2010. "From Full Dusk to Full Tusk: Reimagining the 'Dusky Maiden' Through the Visual Arts.'" *The Contemporary Pacific* 22, no. 1: 1–35.

Terry, Geraldine. 2009. "No Climate Justice Without Gender Justice: An Overview of the Issues." *Gender and Development* 17, no. 1: 5–18.

Tourell, Wayne. 2001. *Paradise Drowned: Tuvalu: The Disappearing Nation.* Documentary Film. Dunedin, New Zealand: Natural History New Zealand. www.nzgeo.com/video/paradise-drowned/.

Wood, Terence. 2015. "Aiding Women Candidates in Solomon Islands: Suggestions for Development Policy." *Asia & the Pacific Policy Studies* 2, no. 3: 531 43.

Yonder, Ayse, Sengul Akcar, and Prema Gopalan. 2005. *Women's Participation in Disaster Relief and Recovery.* Report. New York: SEEDS 22, The Population Council. https://www.preventionweb.net/publications/view/2731

Yuval Davis, N. 2006. "Intersectionality and Feminist Politics." *European Journal of Women's Studies* 13, no. 3: 193–209.

Zubrycki, Tom. 2011. *The Hungry Tide*. Documentary Film. Canberra: Ronin Films. www.roninfilms.com.au/feature/5989/hungry-tide.html.

7 Gender as fundamental to climate change adaptation and disaster risk reduction

Experiences from South Asia

Emmanuel Raju

Introduction

I was speaking to my mother about disasters. She grew up in rural Tamil Nadu in a village called Christianpet outside Vellore and was reminded of her young days when drought hit their homeland in the mid 1970s. Predominantly an agrarian village that experienced a prolonged drought at this time, my mother explained that water was mostly the responsibility of women in the household. The men would mostly take care of agriculture. "I used to fetch water, cook for the family and then go to college", she explained. My mother would visit the farms on Sundays and holidays while the other women in the family would visit the farms daily to deliver a cooked lunch to the men and use the opportunity to pick up some vegetables that could not be sold commercially. She added: "My grandmother would then sell some of those vegetables in the town market, as a source of personal income". The drought not only affected the agricultural produce of the village but also the small incomes and livelihoods of women. When I asked more about these conditions, my mother's answer was more or less 'who cared?'. I am thus wondering what has changed with regard to gender roles and responsibilities since my mother's childhood in the context of my own recent fieldwork. This led me to reconsider disaster risk reduction, a subject I have been working on for the past decade, and how gender permeates disasters in terms of hard-hitting unanswered questions.

Gender and the Anthropocene

> A generalized belief in women's vulnerability silences contextual differences. Gender gets treated not as a set of complex and intersecting power relations but as a binary phenomena carrying certain disadvantages for women and women alone.
>
> (Arora-Jonsson 2011, 750)

How can we ignore a debate on gender in the context of the Anthropocene? This chapter establishes the need to discuss the importance of structures and power in the context of gender. The concept of the Anthropocene (Steffen et al. 2011), which more or less explores human-environmental processes over time, must not ignore key issues of power, structures and agency. In this regard, Malm and Hornborg (2014, 66) argue that we need to address "realities of differentiated vulnerability on all scales of human society". While there is a large portion of the literature that generalises women as poor and most vulnerable, there needs to be a more nuanced understanding of the complex interactions and intersections of gender with other dimensions such as economy, class, caste and other such factors (Carr and Thompson 2014; Demetriades and Esplen 2008).

Much of the gender literature continues to focus on women (Schipper and Langston 2014), while what is needed for studying gender is an intersectionality approach (for example: gender and development; gender and poverty; gender and disasters in a much more disaggregated fashion). In our context, we need to look at the complex interactions of gender with disaster risk reduction, climate change adaptation and vulnerability. We see a slow change in gender research with new themes such as masculinities; sexual minorities; and the emergence of gender within climate change (Enarson, Fothergill, and Peek 2018). Within the larger literature on global environmental change, Ravera and colleagues (2016) argue that many have seen women as vulnerable compared to men, while not much literature within this field has addressed the intersections of gender and power. It could however also be that men and boys are vulnerable in different contexts. Men are seen as bread winners at times of hardship and may even constrain themselves from access to various services such as healthcare, as this may be seen as a weakness (Demetriades and Esplen 2008). Gender is important "because women and men play different roles in the household and because they must follow different gendered roles, they are differently affected by climate" (Schipper and Langston 2014, 4). However, there is a relatively sparse literature to guide a gender-based analysis, not least gender-based disaggregated data (Eklund and Tellier 2012).

With climate change, women will continue to bear disproportionate impacts (Panda, Shrivastava and Kapoor 2014). Arora-Jonsson argues that the majority of the literature on climate change and gender presents three arguments. "Firstly, that women need special attention because they are the poorest of the poor; secondly, because they have a higher mortality rate during natural calamities caused by climate change and thirdly because women are more environmentally conscious" (Arora-Jonsson 2011, 745). Here Arora-Jonsson argues for the need to differentiate between poverty and women (in this case gender), thus emphasising how vulnerability is a process affected by many different factors (Wisner et al. 2014).

This chapter draws on the vulnerability paradigm in the context of disasters. For operational purposes, "vulnerability refers to the propensity to be harmed, in this case by a hazard, and to be unable to deal with that harm alongside the social processes creating and maintaining that propensity. Vulnerability encompasses human decisions, values, governance, attitudes, and behaviour forming situations in which hazards could potentially cause harm" (Kelman et al. 2016, S130). It is argued that the political stance in the concept of vulnerability has been ignored or not emphasised enough (Gaillard 2018). The 1970s saw a turn in disaster studies with the marriage of the concept of vulnerability (from disaster studies) with that of international development (Kelman et al. 2016). In conjunction with the vulnerability paradigm, this chapter places root causes at the heart of the gender-disaster debate. Disasters are a result of root causes of vulnerability such as, for instance, development failures, larger structural agendas (such as the neo-liberal paradigms), and patriarchal structural contexts (see Kelman et al. 2016; Watts 1983; Wisner et al. 2014). Here, the Pressure and Release Model (Wisner et al. 2014, 51), highlights limited access to power, structures and resources, along with ideologies (political and economic), as the root causes in the progression of vulnerability. Gender is crucially situated within these structural elements. As mentioned above, the combination of different political and social structural factors affecting daily life is at the heart of addressing gender within an intersectionality approach.

This chapter is based on observations and field notes from various investigations in South Asia from 2009 to 2018. In 2008, I started investigating housing recovery in post-tsunami India and revisited some of these areas conducting vulnerability and risk assessments in the context of disasters in 2018. Between 2009 and 2012, I did a number of visits to Tamil Nadu to examine the nature of the recovery process in the post-2004 Indian Ocean tsunami. During this time, I explored how recovery posited a complex set of factors to be considered for long-term disaster risk reduction. I also draw from my work with students in Bangladesh in 2011 and 2012; and in Sri Lanka in 2017. The aim of these field investigations was to understand the particular factors of disaster risk in rural and urban settings that had contributed to vulnerability and risk assessments of the communities we were working with. In doing these assessments, gender was always highlighted as a key issue for failure or success of disaster risk reduction programming in South Asia. Against this backdrop, this chapter is situated around three major themes: (1) why gender in disasters?; (2) water, livelihoods and disaster risk reduction, and; (3) gender and participation, highlighted in the literature and from fieldwork experiences in South Asia over the past decade.

Why gender in disasters?

> Women disproportionately bear a heavier burden after disasters (e.g., care taking activities, organizing food, water and other basic necessities), but have fewer assets (e.g., cash, savings, house, land, livestock, tools and equipment).
>
> (Mukherji, Ganapati, and Rahill 2014)

To begin with, Bradshaw and Fordham (2015) argue that gender has not taken prominence in policy discussions with regard to disaster risk reduction. While literature points to women as being vulnerable and most affected during disasters, the political nature of gender has not been given much attention in disaster studies (similar to vulnerability as briefly referred to in the introduction). This is complemented by a lack of clarity about the vision of gender mainstreaming (Alston 2014, 290). A study in India found, for instance, that most schemes targeting climate change adaptation did not cover gender programming. Even more, disaster risk management took a low priority in four of the studied states with a very low budget for looking at disasters in the context of climate change (ibid.). As highlighted in the literature, women-headed households are a result of social processes and are rarely due to empowerment (Djoudi et al. 2016). These theoretical gaps give substantial reason to focus on gender in disasters.

In most of my post-tsunami fieldwork in Tamil Nadu in India (in 2008; 2010–11; and again in 2018), I found that women-headed households were either due to men dying during the tsunami or because of migration to urban areas in search of work rather than as a result of empowerment. This has also been highlighted as 'distress migration' where men migrate for work due to seasonal impacts of climatic events, while women take on the role as heads of the household (Bhatta et al. 2015). Even when women have migrated for work due to sprawling urban centres, they may end up in so called 'slums' or 'settlements' in which the built environment is not being safe, putting women at even higher risk (for example, lack of safety in public toilets) (Bhagat 2017). In Haiti for instance, as Lynn Horton (2012) notes, women-headed households had a double burden of sexual and societal pressures. This process thus puts additional pressure on many women, not only taking care of their family needs but also facing gendered constraints within their communities. These facets of gendered vulnerabilities are reproduced or exacerbated post-disasters, ranging from sexual harassment to taking care of family needs or even to an increase in domestic violence (Demetriades and Esplen 2008). During my fieldwork in 2018 in India, in Nagapattinam, I observed that some of the cyclone shelters had similar issues with regard to privacy and safety. In Haiti for example, after the 2010 earthquake, it was noted that the conditions in the relief camps undermined women's recovery. This was due to a lack of privacy and proper sanitation facilities and was worsened by illegal sexual behaviour towards women in the camps (Horton 2012). In Bangladesh, cyclone Sidr of 2007 showed higher incidences of early-marriage of girls. This, further resulted in many young women not pursuing higher education or dropping out of school (Horton 2012).

In one study, it was found that climate mitigation papers had less emphasis on gender compared to the adaptation literature due to mitigation being presumptuously considered male dominated (Djoudi et al. 2016). Further, while there is more emphasis for donors and different organisations on working with vulnerable groups, there seems to be less focus on moving towards transformational change (Djoudi et al. 2016). A specific example from the Indian Ocean disaster recovery in the post-tsunami context of 2004 had women included in the housing documents during reconstruction which was not common practice. This move is a major, although incremental, shift in landownership and power. However, this power of the name on the document does not necessarily change the power dynamics within the household, as highlighted by Juran (2012) in relation to the patriarchal post-tsunami structures. Similarly, in Sri Lanka women received less access to resources due to not being considered heads of households (Ariyabandhu 2006). Therefore, what we need is a much more radical shift of moving from resilience to transformation (Pelling 2010).

After the tsunami, more women than men lost their lives in Sri Lanka, which can be attributed to the socio-economic place of women in the affected communities (Enarson, Fothergill, and Peek 2018). Conducting fieldwork in Nagapattinam in Tamil Nadu in India 14 years after the Indian Ocean tsunami, revealed the absence of life saving skills, such as swimming, among women from coastal communities. This is a clear indicator of the need to address preparedness and Disaster Risk Reduction (DRR) through a gendered lens. Flooding has been one of the major disasters over the years in Sri Lanka. Recent research from Sri Lanka shows that women's needs were not included in flood recovery or risk reduction, highlighting how gender insensitive needs assessments have neglected many factors such as economic and livelihood assets of women (De Silva and Jayathilaka 2014). Further, this research also emphasised that disaster management projects tended to focus mostly on the technical aspects of the processes, while ignoring gender issues. While this is a common problem, it is not easily detectable. Addressing gender as a cross-cutting issue may, for several donors, be a mandatory requirement, but this does not automatically translate into gender taking a priority spot (although this is changing slowly).

Lahiri-Dutt (2012, 112) writes: "Human relationship with the environment is not gender-neutral; rural women in developing countries interact more directly with their local resource bases and are disproportionately affected adversely by degradation of these resources". In the analysis of Carr and Thompson (2014), four major themes emerge with regard to gender and agriculture: (1) gendered decision-making or lack of power to decide among women; (2) crops that women and men grow are different; (3) disparities in access to land; (4) failure to highlight women's activities in the context of gender, development and livelihoods. However, it is emphasised that more recent literature focuses on adaptive capacities and initiatives of women to battle climate change. Unfortunately, there appears to be a continual differential impact of disasters on not only women but also on other gender minorities across the world (see Gaillard, Gorman-Murray, and Fordham 2017).

The 1993 Latur earthquake displayed a range of patriarchal conventions. For example, it is estimated that more women died because of being trapped indoors, while their male counterparts could sleep outside (Seager 2014, 267). Similarly, in the North West Frontier Province in Pakistan, women did not leave their homes immediately after quakes (or as soon as they should have) as they would have to dress/cover in accordance with what is considered 'appropriate':

> Moreover, women are often discouraged from learning coping strategies and lifesaving skills, such as how to climb trees or swim. Both factors put them at a disadvantage when floods hit. Often women are not permitted to evacuate their homes without consent from their husbands or elder men in their families or communities. Gendered cultural codes of dress may inhibit their mobility during crises, resulting in higher disproportionate mortality during many disasters.
>
> (Nellemann, Verma, and Hislop 2012, 6)

In my fieldwork in 2018, I heard the story of a young girl who did not want to be rescued post-tsunami as she lost her clothes to the waves and did not want to lose her 'honour' if she came out without clothes. A community worker in Nagapattinam said that this story continues to live with her 14 years after the tsunami and expresses the need to address gender disparities in society. Even though Tamil Nadu is usually seen as a state with better development indices compared to other states in India, women are still controlled in different ways by family and relatives in the name of 'honour' (Carswell, Chambers, and De Neve 2018). Culture is here seen as both positive and negative with regard to vulnerability in disaster settings (Gaillard, Fordham, and Sanz 2015). In the context of South Asia, this plays out both ways as well. As emphasised above, while aspects of social capital come across strongly as an asset (Ganapati and Iuchi 2012), issues such as clothing are highlighted as a challenge.

A study in Dhaka in Bangladesh underlined that "when men plan space they seldom consider the advantages and disadvantages to women" (reported in Jabeen 2014, 114). Women were generally not consulted in planning or housing construction in urban settlements in Dhaka. This argument may be expanded to other contexts where women's participation in decision-making is not very high. Seager (2014) argues that early-warning systems have taken huge strides due to advanced technology. However, considering that every household has access to a mobile phone does not necessarily mean that every woman in the household has access to technology and more specifically to information. This puts them at the risk of information exclusion during disasters. In Bangladesh, Cannon writes that disasters are clearly linked to poverty and gender. Women in Bangladesh were seen to have lesser coping mechanisms due to larger societal/structural issues (Cannon 2002). More recently, Kabeer suggests that after the financial recession of 2008, women have had to accept low paid work for long hours while also being subjected to increased domestic violence (Kabeer 2015). With the recent Kerala floods in India in August 2018, a friend working in the humanitarian sector wrote me an email with his personal experience. With his permission, I quote him here:

> We went to the pockets in the villages where landless (without agricultural land) workers lived. On the visit, we came across some really tough stories of single women with children whose homes were destroyed. Homes which were badly constructed had collapsed, some women broke down telling us about it. They were eager to show their homes, it was basically a venting process for them, also we were the first outsiders to go that particular hamlet.

Intrigued by the real-time experience, I followed up with a phone call to understand the ground reality of some of the hardest hit regions and populations in Kerala. He explained that in some parts of Alappuzha in Kerala (also known as Alleppey), some of the old colonies built by the government many decades ago have been some of the worst affected by the flood. The toilets, usually constructed outside the house in this region, were heavily damaged. Speaking to the women in these colonies, he said, highlighted the immediate need to address water and sanitation issues. When asked questions about livelihood the women said, "we need to fix our toilets first". Privacy has currently become a major issue due to the sanitation problem. This highlights another recent story of South India where gender plays out differently and impacts women and other marginalised groups differently depending on the location of their housing and their socio-economic status in society.

Can disasters also provide an opportunity then? "Post-disaster sites may provide the focus for a renewed commitment to gender mainstreaming, not least because of the interactions between transnationals, state and local actors in sites where gender inequalities are transparent, where actions and resource allocations can inhibit or enhance gender equality and where the world's women wait" (Alston 2014, 293).

Water, livelihoods and disaster risk reduction

As water is central to livelihood debates and discussions, this section will outline some of the major constraints to gender and water issues in the context of disaster risk reduction. Results from a study in Accra and Cape Town "reinforce the importance of context to understand gender dimensions of water access and experience" (Harris et al. 2017, 573). Further, Carr and Thompson (2014) argue that groups within the same population can face an event differently or that some groups can even be worse affected by certain climate trends or during disasters (Gaillard 2010; Wisner et al. 2014). This makes the case for more gendered analyses and programming, particularly by looking at the large numbers of people involved in agrarian work in South Asia. This is supported by a study on climate-induced migration in South Asia that shows how "socio-economic, cultural and religious values in the region have resulted in its women and children being more vulnerable to adverse situations in comparison to men" (Bhatta et al. 2015, 15). Taking the example of the Assam and Orissa floods in India in 2012 and 2013, Krishnan and Twigg (2016) observed that women's menstrual hygiene needs were not considered in preparedness kits, nor were their needs met after the flood. Their research also shows how women had to commute long distances to not only fetch water but also for open defecation and to cater to menstrual hygiene needs. Further, the study highlights that women's access within and outside the household is restricted during menstruation. This increases women's vulnerabilities in conjunction with other factors such as access to water and access to markets. In Africa, it is noted that although women are generally responsible for bringing home water, access to water is restricted by male-dominated water user associations (Parker et al. 2016). This clearly highlights the intricacies of gender being a larger structural issue, where issues of root causes of vulnerability can also be seen in the light of justice (Parthasarathy 2018)

Studies highlight that "fetching drinking water befalling women in most South Asian societies would result in worsening the burden of procuring water when climatic changes result in changes in water quantity, quality, availability and seasonality in altering waterscapes" (Sultana 2014, 375). Women, however, are also often considered custodians of traditional knowledge and sustainable practices in many societies which contribute to building climate-smart households (Yadav and Lal 2018). Post-disaster fieldwork "highlights how the researcher's multiple identities could bring unique opportunities in terms of access to vulnerable groups after a disaster, such as women. Regardless of one's ethnic and gender positioning, it is important for the researcher to enter the field with an open mind and have the flexibility to make methodological or other changes in order to adapt to unknown challenges or to make the best of opportunities that might come up during fieldwork" (Mukherji, Ganapati, and Rahill 2014).

A study conducted in Bihar in India, Mehar, Mittal, and Prasad (2016), using large-scale surveys, concludes that men and women have differential coping strategies to climate change related events. "Capacities are often rooted in resources which are endogenous to the community and which rely on traditional knowledge, indigenous skills and technologies and solidarity networks" (Gaillard 2010, 220). Coping is closely linked with existing capacities of communities. For example, the UNISDR definition highlights that coping capacities are crucial not only during disasters but also in everyday life. "Coping capacity is the ability of people, organizations and systems, using available skills and resources, to manage adverse conditions, risk or disasters. The capacity to cope requires continuing awareness, resources and good management, both in normal times as well as during disasters or adverse conditions" (UNISDR 2009). Further, Mehar, Mittal, and Prasad (2016) also highlight how men play a much more crucial role in decision-making with regard to adaptation mechanisms. This can be attributed to the social structure within a household in relation to power and decision-making. Sugden et al. (2014) note that in Bangladesh, irrespective of the nature of disasters and water contamination, fetching water becomes the sole responsibility of the women. While conducting vulnerability and risk assessments (with a group of students) in Sathkira region in 2011 during my fieldwork, I found that women walk as far as 10 kilometres a day to fetch drinking water. This is also confirmed in the study by Sugden et al. (2014). Moreno-Walton and Koenig (2016) argue that women's additional responsibility of needing to care for the family and children puts an additional burden on them, thereby reducing their coping mechanisms for an early recovery.

Women working in more informal employments (Nelson et al. 2002) may witness increased/exacerbated disaster impacts. For example, after the Indian Ocean tsunami of 2004, my fieldwork identified how many female street vendors did not have any documentation to prove the physical assets they possessed before the disaster. Similarly, women working in small food stalls or working as wage labourers on agricultural farms tended to lose their jobs due to a possible lack of commercial activities in the immediate aftermath of disasters. The reasons for these increased hardships are not only economic, but also rooted in socio-cultural structures of society. Research shows that there are major barriers to risk reduction and adaptation connected to land in agricultural communities. These include a lack of inheritance of family property which in turn results in reduced incomes for women. Furthermore non-ownership of land relates to reduced or no access to information and technology in agriculture as information is provided to those who are formal landowners. Also, lack of information on new technology or adaptation strategies leads to a higher dependence on natural resources for women. In times of disasters, this dependence on natural resources puts certain groups at a higher disadvantage (Sugden et al. 2014).

Participation and gender

"Vulnerability is not a natural attribute of women, but rooted in gender inequality" (Smyth and Sweetman 2015, 415). Women's access to public institutions is problematic in many places due to male-dominated office spaces. Even if the male staff are able to communicate with women on the field, male-dominated public spaces affect participation and influence (Sugden et al. 2014). This was exactly my experience in Sathkira while conducting fieldwork in 2011 and I had similar experiences during fieldwork in the settlements in Dhaka in 2012. In Sathkira, over a span of three days, different modes of conducting focus groups were tried in order to ensure participation of all groups and hear everyone's concerns regarding disaster risk reduction. In the first instance, a mixed focus group would have failed completely as it would have been unable to discuss all the planned themes due to cultural sensitivity. It is important to note that "focus groups often produce data that are seldom produced through individual interviewing and observation and thus yield particularly powerful knowledges and insights" (Kamberelis and Dinitriadis 2011, 558). As the aim of the focus group was to capture disaster risk issues facing the community, we did not choose a mixed group as that could easily have become dominated by powerful members from the village council (predominantly men). After identifying different disaster risk issues facing the community, this was to be followed with community disaster mapping exercises to get the views of different members of the community. Also, there was a risk of women becoming mute spectators in the focus group. Therefore, an all women focus group with a female interpreter was tried. However, this format did not yield significant participation of all women due to the presence of women from dominant families (who were seen as associated with families of local leaders). This diluted the conversation regarding access to basic resources, such as water and other governmental schemes and programmes. The field hence presents a number of challenges and opportunities for addressing gender. One definitely needs to be mindful of not creating internal conflicts within or between different groups in the community. One's own identity as a researcher, as well as questions of ethnicity and individual gender perspectives, play a crucial role in asking questions and seeking for broader knowledge (for example, see Kunze and Padmanabhan 2014). What Gaillard (2018) argues for is the need for local scholars to take over disaster research in their respective countries. Who else knows the context better? Often, we notice a powerful 'scholarly play' (including disaster studies), that limits the levels of participation where gender certainly stands as an important factor.

In Nagapattinam, in Tamil Nadu, India in 2008, while conducting Vulnerability and Capacity assessments four years after the tsunami, one of the community leaders (male) suggested that he could draw community maps and discuss vulnerability issues on behalf of everyone. Being my first time in the field conducting such an exercise, I did not know how to respond to this situation. However, it taught me to be patient but firm, asking the local leader to engage other people in the village to do this mapping. Ten years later, in 2018, in a different community in the same region, I encountered similar experiences but of a slightly different character while conducting participatory vulnerability assessments. We divided the men and women to discuss separately and map out risks to disasters in the community. In the male group, one person who was not only holding a position of power in the community but had experience conducting such exercises in the past, dominated the entire exercise. This raises questions of whether exercises such as these can ever be truly 'participatory'. At the same time, within the women's groups, all the women participated after some nudging. Discussions of what risks were ranked higher than others were of course very different in the two groups. While the men were concerned with droughts and their impacts on livelihoods; the women were more concerned with access to water and the long distances they would have to commute to fetch water. There are both structural and political interesectionalities at play here (Crenshaw 1991). The structural aspect of women being a different social group and the political aspect of power, institution and agency make the situation worse. This issue of structural versus political intersections needs to be investigated further as it is theoretically unclear how this can be resolved (Walby, Armstrong, and Strid 2012).

Austin and McKinney (2016) argue that increased access to economic facilities such as credit sources can help women reduce vulnerabilities during disasters. Economic empowerment continues to be a major challenge in many parts of South Asia, especially for women. In 2011–2012, I was conducting interviews for a disaster recovery coordination study in the light of the Indian Ocean tsunami of 2004. The aim of the study was to explore different recovery approaches after the tsunami. During this field study, I interviewed professionals from different organisations (government and non-governmental organisations) who were actively engaged in post-tsunami disaster recovery. Even today, the case of women vendors is very fresh in my memory. In Nagapattinam, after the Indian Ocean tsunami, women vendors (who sold food along the coast) were largely neglected in the initial phases of the recovery process due to lack of documentation to prove their livelihood assets (such as stoves, carts etc.). The civil society organisations working with the local leadership managed to raise attention to this matter (Fieldwork 2011–2012). This highlights the importance of inclusion of different groups of affected people in all aspects of disaster risk management.

Addressing gender issues in the context of disaster risk reduction and climate change adaptation does not merely refer to the inclusion of representatives from all groups or ensuring a head count from different groups. Addressing gender is to address deep-rooted socio-structural issues, such as patriarchal power dynamics and access to resources (such as economic credit). While synergies may be drawn between climate change adaptation and disaster risk reduction strategies, both these processes are challenged with questions of gender inequalities on the ground. Policy dialogues on gender mainstreaming have been in vogue for a long time now. However, change with regard to gender aspects must be seen in the light of power shifts and institutionalisation of gender as a key issue in larger development processes, which in turn feed into Disaster Risk Reduction (DRR) and Climate Change Adaptation (CCA).

Cutter (2017) argues that there has been an increase in understanding gender issues and disparities. However, there is more to be done with regard to loss and damage data in the context of CCA and disaster risk management (Cutter 2017; Cutter and Gall 2015). Here, even anecdotal evidence is important, as it highlights individual and context specific analyses that help us understand the complexities of context and culture. At the policy level, Nagel (2016) argues that women have been underrepresented in climate negotiations, such as the global climate change conferences (COPs). These are clear signs of male-dominance in both disaster and climate sectors. Looking at the United Nations Framework Convention on Climate Change (UNFCC), Hemmati and Röhr (2009, 26) write, "At the highest level – that of heads of delegations – women are substantially less well-represented". Further, in the context of national disaster management agencies traditionally belonging to the civil protection units, these agencies have less women than men involved. While there is increased recognition of this problem, my observation is that we have not achieved a gender balance at high-level policy and ministerial meetings on disasters and climate change because "there is no true justice without gender justice" (Hemmati and Röhr 2009, 30). How do we overturn this to steer active transformation with regard to gender?

It is argued that for successful gender mainstreaming, along with upright use of institutions, there needs to be efficient use of available gender disaggregated data (Dankelman 2010). One should be constantly aware of gender as being all-inclusive and not only a woman issue. As Eklund and Tellier argue in terms of "'invisible men', that is, that men's vulnerabilities are not well captured in existing data sets" (2012). Further, it is argued that although women and children may continue to undergo shocks and stresses of disasters and environmental change, "we don't know the magnitude or geographic extent of the burden because there is no legacy of data reliability to consistently document these injustices" (Cutter 2017, 117). Active participation of different marginalised groups cannot be a reality until we know the depth of the problem. Hence, to argue for it we need to have sufficient data at all levels of existing policy domains. Until then, participation will only be that of the powerful representing the voiceless. The larger question remains in terms of how gender mainstreaming siloes can be challenged and how not to see gender as an entirely separate policy domain (Le Masson, Norton, and Wilkinson 2015). Rather, gender should be effectively brought into all programming for disaster risk reduction and climate change adaptation.

Conclusions

This chapter has primarily been a reflection on working with disaster risk reduction and climate change adaptation issues in South Asia over the past many years.

As Helen Clark notes "any serious shift towards more sustainable societies has to include gender equality".[1] This is no longer a debate about women's issues but about addressing higher orders of power structures; thus moving towards fundamental transformational change. These changes would go a long way in bridging the gaps to achieve equity. Thinking of my mother's words in lay terms, 'who cared', is almost resembling the theoretical literature and its lack of priority to gender in policy issues. I am not sure if my mother even considered the drought that affected her village as a disaster, but she certainly continues to have strong impressions on gender and the drought. This is similar to the anecdotes cited in this chapter in the context of climate change and disasters. As Sarah Bradshaw (2014) argues, unless gender issues are specifically addressed and not merely including women, root causes of vulnerability (mentioned in the introduction) will remain untargeted. A key learning from current research is the importance of tackling gender issues as a fundamental rule to participation.

Furthermore, as Kelman et al. (2012) indicate, the lack of tools for dialogue and for mainstreaming disaster risk, means that gender as a subject has been given only 'token' attention. Here, the question of why gender issues tend to remain a responsibility of 'gender experts' and not everyone's business, ought to be asked. "The conceptualization of disasters and risks as objective things rather than as inherent characteristics of evolving processes and relationships fosters technocracy and technocratic and bureaucratic approaches to DRR" (Lavell and Maskrey 2014, 273). These issues of gender and disasters must also be seen in the light of justice (Parthasarathy 2018) and beyond traditional means and notions of disaster governance. Therefore, addressing gender in the context of disaster risk reduction and climate change adaptation ought to include everyone working in this sector/field, including relevant institutions, in order to move towards transformational change.

Note

1 www.undp.org/content/undp/en/home/presscenter/events/2012/february/csw56.html

References

Alston, Margaret. 2014. "Gender Mainstreaming and Climate Change." *Women's Studies International Forum* 47: 287–94.

Ariyabandhu, M. 2006. "Gender Issues in Recovery from the December 2004 Indian Ocean Tsunami: The Case of Sri Lanka." *Earthquake Spectra* 22, no. 3: 759–75.

Arora-Jonsson, Seema. 2011. "Virtue and Vulnerability: Discourses on Women, Gender and Climate Change." *Global Environmental Change* 21, no. 2: 744–51.

Austin, Kelly F., and Laura A. McKinney. 2016. "Disaster Devastation in Poor Nations: The Direct and Indirect Effects of Gender Equality, Ecological Losses, and Development." *Social Forces* 95, no. 1: 355–80.

Bhagat, R. 2017. "Migration, Gender and Right to the City – The Indian Context." *Economic and Political Weekly* II, no. 32: 35–40.

Bhatta, Gopal Datt, Pramod K. Aggarwal, Santosh Poudel, and Anne Debbie Belgrave. 2015. "Climate-Induced Migration in South Asia: Migration Decisions and the Gender

Dimensions of Adverse Climatic Events." *Journal of Rural and Community Development* 10, no. 4: 1–23.

Bradshaw, Sarah. 2014. "Engendering Development and Disasters." *Disasters* 39, no. S1: 54–75.

Bradshaw, Sarah, and Maureen Fordham. 2015. "Double Disaster: Disaster through a Gender Lens." In *Hazards, Risks and Disasters in Society*, edited by Andrew E. Collins, Samantha Jones, Bernard Mayena, and Janaka Jayawickrama, 233–51. London: Elsevier.

Cannon, Terry. 2002. "Gender and Climate Hazards in Bangladesh." *Gender and Development* 10, no. 2: 45–50.

Carr, Edward R., and Mary C. Thompson. 2014. "Gender and Climate Change Adaptation in Agrarian Settings: Current Thinking, New Directions, and Research Frontiers." *Geography Compass* 8: 182–97.

Carswell, Grace, Thomas Chambers, and Geert De Neve. 2018. "Waiting for the State: Gender, Citizenship and Everyday Encounters with Bureaucracy in India." *Environment and Planning C: Politics and Space* 0, no. 0: 1–20. DOI: 10.1177/0263774X18802930.

Crenshaw, KW. 1991. "Mapping the Margins: Intersectionality, Identity Politics, and Violence against Women of Color." *Stanford Law Review* 43, no. 6: 1241–99.

Cutter, Susan L. 2017. "The Forgotten Casualties Redux: Women, Children, and Disaster Risk." *Global Environmental Change* 42: 117–21.

Cutter, Susan L, and Melanie Gall. 2015. "Sendai Targets at Risk." *Nature Climate Change* 5, no. 8: 707–09.

Dankelman, Irene. 2010. "Introduction: Exploring Gender, Environment and Climate Change." In *Gender and Climate Change: An Introduction*, edited by Irene Dankelman, 1–20. London: EarthScan.

Demetriades, Justina, and Emily Esplen. 2008. "The Gender Dimensions of Poverty and Climate Change Adaptation." *IDS Bulletin* 39, no. 4: 24–31.

De Silva, Kushani, and Ramanie Jayathilaka. 2014. "Gender in the Context of Disaster Risk Reduction: A Case Study of a Flood Risk Reduction Project in the Gampaha District in Sri Lanka." *Procedia Economics and Finance* 18: 873–81.

Djoudi, Houria, Bruno Locatelli, Chloe Vaast, Kiran Asher, Maria Brockhaus, and Bimbika Basnett Sijapati. 2016. "Beyond Dichotomies: Gender and Intersecting Inequalities in Climate Change Studies." *Ambio* 45, no. S3: 248–62.

Eklund, Lisa, and Siri Tellier. 2012. "Gender and International Crisis Response: Do We Have the Data, and Does It Matter?" *Disaster* 36, no. 4: 589–608.

Enarson, Elaine, Alice Fothergill, and Lori Peek. 2018. "Gender and Disaster: Foundations and New Directions for Research and Practice." In *Handbook of Disaster Research*, edited by Rodríguez Havidan, William Donner, and J. Joseph Trainor, 205–23. Cham: Springer.

Gaillard, J. C. 2010. "Vulnerability, Capacity and Resilience: Perspectives for Climate and Development Policy." *Journal of International Development* 232: 218–32.

Gaillard, J. C. 2018. "Disaster Studies Inside Out." *Disasters* 43, no. S1: S7–S17. DOI: 10.1111/disa.12323.

Gaillard, J. C., Maureen Fordham, and Kristinne Sanz. 2015. "Culture, Gender and Sexuality: Perspectives for Disaster Risk Reduction." In *Cultures and Disasters: Understanding Cultural Framings in Disaster Risk Reduction*, edited by Greg Kruger, Greg Bankoff, Terry Cannon, Benedikt Orlowski, and Lisa Schipper, 222–34. Abingdon: Routledge.

Gaillard, J. C., Andrew Gorman-Murray, and Maureen Fordham. 2017. "Sexual and Gender Minorities in Disaster." *Gender, Place & Culture* 24, no. 1: 18–26.

Gaillard, J. C., and Jessica Mercer. 2012. "From Knowledge to Action: Bridging Gaps in Disaster Risk Reduction." *Progress in Human Geography* 37, no. 1: 93–114.

Ganapati, N. Emel, and Iuchi, K. 2012. "In Good Company: Why Social Capital Matters for Women During Disaster Recovery." *Public Administration Review* 72, no. 3: 419–29.

Harris, Leila M., Danika L. Kleiber, Jacqueline Goldin, Akosua Darkwah, and Cynthia Morinville. 2017. "Intersections of Gender and Water: Comparative Approaches to Everyday Gendered Negotiations of Water Access in Underserved Areas of Accra, Ghana and Cape Town, South Africa." *Journal of Gender Studies* 26, no. 5: 561–82.

Hemmati, Minu, and Ulrike Röhr. 2009. "Engendering the Climate-Change Negotiations: Experiences, Challenges, and Steps Forward." *Gender and Development* 17, no. 1: 19–32.

Horton, Lynn. 2012. "After the Earthquake: Gender Inequality and Transformation in Post-Disaster Haiti." *Gender and Development* 20, no. 2: 295–308.

Jabeen, Huraera. 2014. "Adapting the Built Environment: The Role of Gender in Shaping Vulnerability and Resilience to Climate Extremes in Dhaka." *Environment and Urbanization* 26, no. 1: 147–65.

Juran, Luke. 2012. "The Gendered Nature of Disasters: Women Survivors in Post-Tsunami Tamil Nadu." *Indian Journal of Gender Studies* 19, no. 1: 1–29.

Kabeer, Naila. 2015. "Gender, Poverty, and Inequality: A Brief History of Feminist Contributions in the Field of International Development." *Gender and Development* 23, no. 2: 189–205.

Kamberelis, George, and Greg Dinitriadis. 2011. "Focus Groups: Contingent Articulations of Pedagogy, Politics, and Inquiry." In *The Sage Handbook of Qualitative Research* (4th ed.), edited by Norman K. Denzin and Lincoln Yvonna, 545–61. Thousand Oaks: SAGE Publications.

Kelman, Ilan, J. C. Gaillard, James Lewis, and Jessica Mercer. 2016. "Learning from the History of Disaster Vulnerability and Resilience Research and Practice for Climate Change." *Natural Hazards* 82, no. 1: 129–42.

Krishnan, Sneha, and John Twigg. 2016. "Menstrual Hygiene: A 'Silent' Need During Disaster Recovery." *Waterlines* 35, no. 3: 265–76.

Kunze, Isabelle, and Martina Padmanabhan. 2014. "Discovering Positionalities in the Countryside: Methodological Reflections on Doing Fieldwork in South India." *Erdkunde* 68, no. 4: 277–88.

Lahiri-Dutt, Kuntala. 2012. "Climate Justice Must Integrate Gender Equality: What Are the Key Principles?" In *Impact of Climate Change on Water and Health*, edited by Velma I. Grover, 106–19. Boca Raton: CRC Press.

Lavell, Allan, and Andrew Maskrey. 2014. "The Future of Disaster Risk Management." *Environmental Hazards-Human and Policy Dimensions* 13, no. 4: 267–80.

Le Masson, Virginie, Andrew Norton, and Emily Wilkinson. 2015. "Gender and Resilience." Working Paper. BRACED Knowledge Manager. https://europa.eu/capacity4dev/file/27825/download?token=hSLdYcBs.

Malm, Andreas, and Alf Hornborg. 2014. "The Geology of Mankind? A Critique of the Anthropocene Narrative." *The Anthropocene Review* 1, no. 1: 62–69.

Mehar, Mamta, Surabhi Mittal, and Narayan Prasad. 2016. "Farmers Coping Strategies for Climate Shock: Is It Differentiated by Gender?" *Journal of Rural Studies* 44: 123–31.

Moreno-walton, Lisa, and Kristi Koenig. 2016. "Disaster Resilience: Addressing Gender Disparities." *World, Medical & Health Policy* 8, no. 1: 46–57.

Mukherji, Anuradha, N. Emel Ganapati, and Guitele Rahill. 2014. "Expecting the Unexpected: Field Research in Post-Disaster Settings." *Natural Hazards* 73, no. 2: 805–28.

Nagel, Joane. 2016. *Gender and Climate Change Impacts, Science, Policy*. New York and London: Routledge.

Nellemann, Christian, Ritu Verma, and Lawrence Hislop. 2012. "Women at the Frontline of Climate Change: Gender Risks and Hopes." Rapid Response Assessment. United Nations Environment Programme (UNEP), GRID-Arendal. Norway: Birkeland Trykkeri AS. www.grida.no/publications/rr/women-and-climate-change/.

Nelson, Valerie, Kate Meadows, Terry Cannon, John Morton, and Adrienne Martin. 2002. "Uncertain Predictions, Invisible Impacts, and the Need to Mainstream Gender in Climate Change Adaptations." *Gender and Development* 10, no. 2: 51–59.

Panda, Gyana Ranjan, Saumya Shrivastava, and Aditi Kapoor2014. "Climate Change and Gender: Study of Adaptation Expenditure in Select States of India." In *Handbook of Climate Change Adaptation*, edited by Walter Leal Filho, 765–84. Berlin and Heidelberg: Springer.

Parker, Helen, Naomi Oates, Nathaniel Mason, and Roger Calow. 2016. "Gender, Agriculture and Water Insecurity." Policy Brief. www.odi.org/sites/odi.org.uk/files/resource-documents/10533.pdf.

Parthasarathy, Devanathan. 2018. "Inequality, Uncertainty, and Vulnerability: Rethinking Governance from a Disaster Justice Perspective." *Environment and Planning E: Nature and Space* 1, no. 3: 422–42.

Pelling, Mark. 2010. *Adaptation to Climate Change: From Resilience to Transformation*. London: Routledge.

Ravera, Federica, Irene Iniesta-Arandia, Unai Pascual, Purabi Bose, and Berta Martı. 2016. "Gender Perspectives in Resilience, Vulnerability and Adaptation to Global Environmental Change." *Ambio* 45, no. 3: 235–47.

Schipper, Lisa, and Lara Langston. 2014. "Gender Equality and Climate Compatible Development: Drivers and Challenges to People's Empowerment." Literature Review. London Climate Development Knowledge Network/Overseas Development Institute (ODI).

Seager, Joni. 2014. "Disasters Are Gendered: What's New?" In *Reducing Disaster: Early Warning Systems for Climate Change*, edited by Zinta Zommers and Ashbindu Singh, 265–82. Dordrecht, Heidelberg, New York, and London: Springer.

Smyth, Ines, and Caroline Sweetman. 2015. "Introduction: Gender and Resilience." *Gender and Development* 23, no. 3: 405–14.

Steffen, Will, Jacques Gribevald, Paul Crutzen, and John McNeill. 2011. "The Anthropocene: Conceptual and Historical Perspectives." *Philosophical Transactions of the Royal Society A* 369: 842–67.

Sugden, Fraser, Sanjiv De Silva, Floriane Clement, Niki Maskey-Amatya, Vidya Ramesh, Anil Philip, and Luna Bharati. 2014. "A Framework to Understand Gender and Structural Vulnerability to Climate Change in the Ganges River Basin: Lessons from Bangladesh, India and Nepal." Working Paper 159. International Water Management Institute (IWMI). https://ccafs.cgiar.org/publications/framework-understand-gender-and-structural-vulnerability-climate-change-ganges-river.

Sultana, Farhana. 2014. "Gendering Climate Change: Geographical Insights." *Professional Geographer* 66, no. 3: 372–81.

United Nations International Strategy for Disaster Reduction (UNISDR). 2009. "Terminology." www.unisdr.org/we/inform/terminology.

Walby, Sylvia, Jo Armstrong, and Sofia Strid. 2012. "Intersectionality: Multiple Inequalities in Social Theory." *Sociology* 46, no. 2: 224–40.

Watts, Michael. 1983. "On the Poverty of Theory: Natural Hazards Research in Context." In *Interpretations of Calamity: From the Viewpoint of Human Ecology*, edited by Kenneth Hewitt, 230–61. Boston, London, and Sydney: Allen & Unwin Inc.

Wisner, Ben, Piers Blaikie, Terry Cannon, and Ian Davis. 2014. *At Risk: Natural Hazards, People's Vulnerability and Disasters*. London: Routledge.

Yadav, S. S., and Rattan Lal. 2018. "Vulnerability of Women to Climate Change in Arid and Semi-Arid Regions: The Case of India and South Asia." *Journal of Arid Environments* 149: 4–17.

8 #leavenoonebehind

Women, gender planning and disaster risk reduction in Nepal

*Katie Oven, Jonathan Rigg, Shubheksha Rana,
Arya Gautam and Toran Singh*

Introduction

The commitment to 'leave no one behind' is a foundational element of the 2030 Agenda for Sustainable Development and a response to the growing recognition that aggregate data sets, commonly used when reporting against international development targets such as the Millennium Development Goals, mask serious problems (Manuel et al. 2018). While development in the aggregate may be achieved, groups that are commonly left behind include, among others, the elderly, people with disabilities, ethnic and religious minorities, sexual minorities, and women and girls. In 2018, the Overseas Development Institute (ODI) undertook a review of country-level progress against their commitment to meet the 'leave no one behind' agenda, in the context of disaster and climate change resilience (Manuel et al. 2018).

Based on their analysis, 64 percent of countries who submitted voluntary national reviews were deemed 'ready' to meet their commitment, 28 percent were partially ready, and 8 percent were either not ready or there were insufficient data available to determine their progress (Manuel et al. 2018). Nepal, the focus of this chapter, was deemed to be partially ready. While the necessary data are being gathered to identify those at risk of being left behind, Nepal does not have key policies in place, for example, in relation to women's access to land and employment, or universal access to health. Further, the analysis concluded that the Government of Nepal was not investing enough in education, health and social protection, sectors recognised as critical for supporting those most at risk.

This chapter focuses on social resilience which has been defined by Adger (2000, 347) as "the ability of groups or communities to cope with external stresses or disturbances as a result of social, political or environmental change". More specifically, we explore attempts to build social resilience in the context of flooding through donor-funded development projects using Nepal as a case study. We therefore situate this work within wider discussions on the Anthropocene and respond critically to the science/engineering 'fix' which is often seen as the solution (see, e.g., Crutzen 2002 and for a critique Rigg and Mason 2018). As some others have done, we explore the intersectionality between gender and other social factors including caste and ethnicity. However, we take the discussion further by asking how these historical and inherited patterns of gendered relations and processes of marginalisation intersect with contemporary transformations in economy and society. The chapter argues that while there are tangible benefits to resilience building initiatives, for example, through the development of flood early-warning systems, projects rarely address the root causes of disasters which give rise to household vulnerability in the first place. This reflects, at least in part, the 'projectisation' of resilience building (cf. Li 2016).

The first section of the chapter explores the concepts of social vulnerability and resilience in the context of disaster risk reduction (DRR) and makes the case for a gendered perspective, with a specific focus on women. The next section contextualises the study which was conducted at a time when Nepal was undergoing a period of significant political change, with implications for how DRR is governed, as well as women's rights and citizenship. We then go on to introduce the study on which this paper draws and the methodological approach adopted. The findings and discussion section explores how women are being incorporated into resilience planning and how this is playing out on the ground. In the final section, we reflect on some of the wider implications of the findings for disaster resilience policy and practice and the 'leave no one behind' agenda.

Understanding disaster vulnerability: a gender and development perspective

Unsurprisingly, perhaps, there is no single accepted definition of vulnerability, reflecting its use in different fields of study across the natural and social sciences. O'Brien et al. (2007) helpfully identify two different interpretations of vulnerability in the context of climate change adaptation, which they summarise as 'outcome vulnerability' and 'contextual vulnerability'. "Outcome vulnerability is considered a linear result of the projected impact of climate change on a particular exposure unit (which can be either biophysical or social), offset by adaptation measures" (O'Brien et al. 2007, 75). In contrast, "contextual vulnerability [. . .] is based on a processual and multidimensional view of climate-society interactions. Both climate variability and change are considered to occur in the context of political, institutional, economic and social structures and changes, which interact dynamically with contextual conditions associated with a particular 'exposure unit'" (O'Brien et al. 2007, 76). How vulnerability is interpreted, therefore, has important implications for DRR policy and practice (Nagoda 2015). In this study, we seek to understand why some social groups are more vulnerable than others to disasters, in particular flood disasters, and the intersecting social, cultural, political and economic factors that affect vulnerability and resilience including caste, ethnicity, gender, income levels and political party networks (Nagoda and Nightingale 2017). We, therefore, in O'Brien et al.'s (2007) terms, take a contextual approach to understanding disaster vulnerability.

There is a growing body of work exploring how gender shapes vulnerabilities and resilience in the context of disasters and climate change (see, e.g., Bradshaw 2014; Cupples 2007; Fordham 1998; Sultana 2013). Single women, woman-headed households, women living with disabilities, pregnant and lactating women, adolescent girls, senior citizens, children, and caste and ethnically based minorities, are often cited as some of the most vulnerable groups in the event of a disaster (Gender Equality 2017). For example, the 2005 Indian Ocean Tsunami killed four times more women than men in Tamil Nadu (India), Sri Lanka and Thailand, mainly for the reason that gender divisions of labour and timing of the tsunami meant that women were at greater risk of exposure (Oxfam 2005; Falk 2012). Some sources indicate that the 2015 Gorkha earthquake in Nepal also killed significantly more women than men, with 55 percent of those who died being women (with the figure ranging from 49 percent to 64 percent depending upon the district) (GoN 2015 cited in Petal et al. 2017, 33).[1] This may reflect the high rates of outmigration of working-aged men from the earthquake-affected districts for employment, and the gendered division of labour, with more women indoors cooking and caring for young children at the time of the earthquake (Petal et al. 2017). However, as noted by Bradshaw (2014), in general data to support the overall assertion that women suffer the most from the impacts of disasters, is sorely lacking. Further, Nightingale (2015a, 221) makes the case that we should pay attention to social justice issues in relation to climate change adaptation "not because of [the] disproportionate impacts on women, but because it is impossible to produce durable adaptation programmes without accounting for socionatures and power".

As noted by Wisner, Berger, and Gaillard (2017, 27), so-called 'vulnerable groups' such as women are often, over time, seen "as groups of people with agency, with 'special contributions' and with capacities in relation to disaster risk reduction". Realising this potential can, however, be challenging in patriarchal societies with complex power relations between different social groups. It can also lead to a 'feminisation of responsibility' that reinforces rather than challenges gender relations (Bradshaw 2014). Cupples (2007, 155) highlights a "danger of oversimplifying how gender shapes responses to disaster or is responsible for generating certain kinds of vulnerabilities or strengths". She notes, as others have done, a tendency "to homogenise and essentialise women, and ethnocentrically treat them as victims of their culture" (Cupples 2007, 156). Indeed, Cupples' (2007) research with women in Nicaragua impacted by Hurricane Mitch in 1998 showed how hurricane survivors adopted a variety of subject positions that were sometimes contradictory. She therefore argues that "rather than concentrat[ing] exclusively on what a disaster *does* to women, it is more productive to focus on how women feel about and reflect on the spatial realignments resulting from these events or processes and on the ways in which they constrain and facilitate" (Cupples 2007, 170, emphasis original). These recommendations are also relevant to community-based disaster risk reduction (CBDRR) and resilience building initiatives, which we discuss further below.

CBDRR: from radical roots to projectisation

CBDRR has become a cornerstone of DRR programming, promoted by donor organisations, NGOs and, increasingly, governments internationally (Maskrey

2011; see also Twigg 2015). Rooted in community and participatory development, the practice of CBDRR emerged in the 1970s, and was formalised and promoted in the 1980s by NGOs and civil society organisations (Delica-Willison and Gaillard 2012). CBDRR approaches were designed to reduce vulnerability and increase disaster resilience by responding to local problems and needs, capitalising on local knowledge and expertise and strengthening communities' technical and organisational capacities to prepare and respond (Wisner 2006). These early approaches, we argue, were based on a 'contextual' understanding of disaster vulnerability that recognised the multidimensional causes of vulnerability (O'Brien et al. 2007).

However, while CBDRR may be rooted in neo-populism, the idea being that DRR is best built through local knowledge, intermediate technologies, indigenous wisdom and local resources and, to some extent neo-Marxist thinking, which recognises the role of the state in protecting its population (Rigg and Oven 2015), such approaches have undoubtedly been neoliberalised or 'projectised'. This builds upon arguments put forward by Tania Murray Li (2007, 2016) whereby development challenges – in this case DRR – are 'rendered technical'. In such instances, the root causes of poverty and social inequality (for example, the caste system and associated marginalisation, gender divisions in society, feudalism, dispossession of land and market dependencies) are largely sidelined and the problem re-posed in technical terms which can be addressed through a development intervention (see also Ojha et al. 2016; Ribot 2014). Inevitably, the identification of a problem is intimately linked to the availability of a solution, in this case disaster risk management plans, flood early-warning systems, raised water pumps and grain stores. The projects are therefore largely apolitical which is welcome by recipient governments as they do little to challenge the status quo. For O'Brien et al. (2007), this would be viewed as an 'outcome vulnerability approach', whereby vulnerability is attributed to the disaster event, rather than the underlying root causes.

Worryingly, perhaps, for Li (2016, 83), "the main effect of these projects is to diminish the role of policy and to limit (but not quite eliminate) political debate". NGOs find themselves straddling an awkward project/politics divide (Edwards and Hulme 1996, cited in Li 2016). They must foreground expertise to justify their intervention, and are limited in terms of what they can do by way of policy dialogue (Li 2016). Donors also place emphasis on what can be measured (Kleinman 2017), which significantly impacts on an NGO's 'logic of operation' (Mueller-Hirth 2012) as they become focused on how many disaster management committees they can establish, and how many women and girls they can reach, rather than longer-term DRR outcomes.

Community-based participatory approaches to development have been subject to much and sustained criticism (see e.g. Hickey and Mohan 2006; Cooke and Kothari's 2001 'tyranny of participation' debate; Mansuri and Rao's 2013 critical take on participatory development for the World Bank), with many of these debates relevant to 'projectised' CBDRR. CBDRR projects are often implemented at the community level rather than being community owned and led (Maskrey 2011). Their short-term nature can impact negatively on their sustainability, particularly when communities are not empowered through skills training and funding to take the projects forward (Delica-Willison and Gaillard 2012). CBDRR can be an added burden for householders, especially when unaccompanied by either additional resources or greater decision-making powers (Allen 2006). Communities may also be reluctant to invest their own time and resources when they have other priority concerns (Jones, Aryal, and Collins 2013). CBDRR projects "can also have the effect of unintentionally naturalizing vulnerability and individualizing responsibility for self-securitization in the name of empowerment . . . [and] provide an excuse for government's neglect of marginalised citizens" (Gladfelter 2018).

While programmes may be designed to benefit communities at large, patronage and power can lead to elite capture (Delica-Willison and Gaillard 2012; Pelling 2007), while embedded and hard-to-shift participatory exclusions can keep the most vulnerable at the margins (Nagoda and Erikson 2015). As a result, the impacts of development projects can vary greatly between households within the same community (Nagoda 2015). Further, as Nightingale and Rankin (2014) remind us from the perspective of feminist scholarship, the household is often the origin of these inequalities. These critiques have important implications for DRR and the 'leave no one behind agenda'.

The hazard and vulnerability context in Nepal

We focus this particular chapter on Nepal, a country highly susceptible to a range of geophysical and hydrometeorological hazards including earthquakes, landslides and floods. Recent events include the 2015 Gorkha earthquake and subsequent aftershocks, which killed around 9,000 people in Central and Western Nepal; and the 2017 floods which affected 35 districts, killed 134 people, and displaced tens of thousands more. In addition to these rapid onset events, Nepal is susceptible to slow-onset disasters such as drought resulting in significant food insecurity (Nagoda and Nightingale 2017). Nepal is a low-income country ranked 144 out of 188 countries in the composite Human Development Index (UNDP 2016), with a predominantly rural population (80.66 percent) (World Bank data).[2] While money metric poverty has fallen in recent years from 46.1 percent of the population at $1.90 per day to 15 percent in 2010 (World Bank data),[3] social inequality based on caste, ethnicity and gender remains severe, with wide variations in human development across population groups (UNDP 2014). International migration has become the key to sustaining livelihoods due to the paucity of opportunities for income generation locally (Sugden et al. 2014; Sunam 2014). In 2017, personal remittances accounted for almost 30 percent of the country's GDP (World Bank data).[4]

Politically, Nepal is undergoing a complex transition following the establishment of multi-party democracy in the 1990s and a decade-long civil conflict fuelled by severe poverty, pronounced socio-economic inequality and the marginalisation of indigenous groups (Rankin et al. 2018). Following a protracted constitution process which began at the end of the civil conflict in 2006, a new constitution was finally adopted five months after the 2015 Gorkha earthquake. For some, the new constitution is less progressive than had been hoped and significant concerns remain regarding the rights of ethnic minorities and women, particularly concerning political representation and citizenship (Muni 2015; Rankin et al. 2018). Indeed, overall, "despite the social and legislative changes that have taken place in the transition to democracy, exclusionary policies mean that many of the population remain unequal citizens including women, dalits [scheduled castes] and indigenous ethnic groups" (Richardson et al. 2016).

While the social norms relating to women differ to some extent across Nepal, women are rarely considered village leaders or have privileged access to community decision-making processes (Nightingale 2002, 19). Pant and Standing (2011) identified three key factors working together to exclude women from participation in Nepal: men, society and the government's discriminatory practices. Their research, which explored women's participation in an NGO-sponsored irrigation and drinking water project, found that women were often excluded from decision-making processes. This was attributed, at least in part, to the NGO failing to recognise women's potential valuable contribution. While many development projects are now focused on women's empowerment and inclusion in line with the 'Gender and Development' approach within development, pre-existing power relations continue to exist (Nagoda and Nightingale 2017). As summarised by Pant and Standing (2011, 413): "Mandatory female representation [in local level decision-making bodies] is a good start, but women are rarely in executive positions of power, and further sensitisation of women's issues remains necessary". Indeed, "while gender is a key axis of power in rural communities, it is not the only important aspect of social difference" (Nightingale 2002, 17). It is therefore necessary to examine how gender intersects with other axes of marginalisation such as ethnicity and caste which "continue to be very significant social relations that shape the division of labour as well as a variety of spatial and bodily practices associated with ritual pollution" (Nightingale 2005, 586).

CBDRR policy in Nepal

The National Strategy for Disaster Risk Management in Nepal was developed between 2007 and 2008 with funding from the European Commission for Humanitarian Aid. Heavily influenced by the 2005 Hyogo Framework for Action (HFA), the Strategy was intended to "facilitate [the] fulfillment of the commitments made by Nepal through various international conventions and forums towards DRR" (NSDRM 2008, iv). The Strategy, which was developed through a highly consultative and participatory process, gained significant traction amongst national and international stakeholders engaged in DRR in Nepal (Jones, Aryal, and Collins 2014). The document itself showed awareness and knowledge of contextual vulnerability, for example: "To take a basic example of floods, which is the recurrent hazard in Nepal, the problem tree would show poverty and disempowerment among the fundamental causes and lack of land tenure in safe areas as one of the underlying causes" (NSDRM 2008, 6). The Strategy also included a specific section on 'Gender and Social Inclusion' which recognised the need to "mainstream gender and social inclusion into all stages of disaster risk management" (NSDRM 2008, 25). However, the action plan itself was focused more towards outcome vulnerability, with the suggested strategies mapping onto the HFA priorities for action.[5]

In line with the National Strategy, the LDRMP was developed by the then Ministry of Local Development in 2011, with support from the INGO Oxfam, with the aim of institutionalising DRR at the local level through Village Development Committees (VDCs)[6] and municipalities (MoLD 2011). Alongside the LDRMP, and in response to the large number of CBDRR projects being implemented by the international development community, the Ministry of Local Development also developed the "9 Minimum Characteristics of a Disaster Resilient Community" in collaboration with (I)NGO partners, with the aim of ensuring a "more consistent, holistic and harmonised approach to CBDRR at the local level" (Oven et al. 2017, 18). The LDRMP process began at the district level with the identification of vulnerable VDCs/municipalities within them. This was followed by the establishment of disaster management committees, the completion of hazard and vulnerability assessments using a range of participatory tools, the identification and prioritisation of disaster risk management activities, and the development of a plan which was (in theory) implemented following approval by the VDC or municipality (MoLD 2011). The 9 Minimum Characteristics (Figure 8.1), map onto the LDRMP and include, for example, the establishment of an organisational base, the completion of a risk and capacity assessment, and the development of a disaster risk management plan. The characteristics also include more general points such as access to DRR resources and the identification of risk and vulnerability reduction measures. Along with the characteristics, a series of output indicators were developed to enable (I)NGOs to monitor project impact (Flagship 4 2013). For example, one-third of local disaster risk management committee members were to be drawn from so called 'vulnerable groups' (ibid).

Overall, progress in terms of technical outputs was encouraging with more than 800 LDRMPs developed by December 2015 (Oven et al. 2017). However, the Government of Nepal identified a number of barriers to effective implementation including the absence of a specific budget at the district, VDC and municipal levels, and legitimate local governance structures accountable for implementation (MoHA 2015).[7]

The next section of the chapter explores how CBDRR policy is playing out on the ground in this post-conflict, transitional state, from the perspective of householders themselves. For this we draw on a case study which is part of a wider project exploring the relevance and effectiveness of Nepal's 9 Minimum

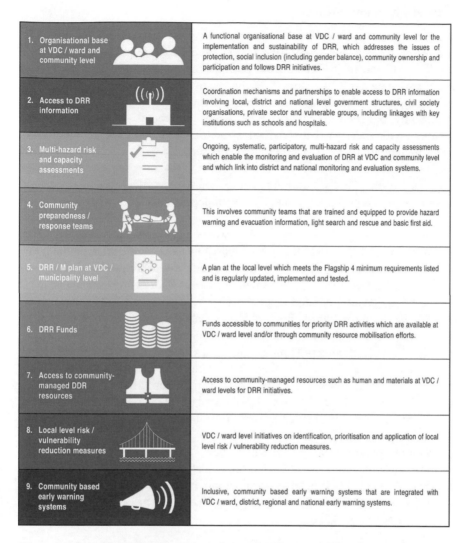

Figure 8.1 Nepal's 9 Minimum Characteristics of a Disaster Resilient Community
(Source: Adapted from Flagship 4 2013)

Characteristics of a Disaster Resilient Community, funded by the UK DFID's South Asia Research Hub (see Oven et al. 2017 for an overview of the wider study). While the focus was not on gender or indeed women specifically, the high level of male outmigration for employment meant that many of the participants were women enabling us to explore CBDRR and the inclusivity agenda through the lens of the women involved in the study themselves.

Case study communities

We focus this chapter on Bardiya District, in Mid-Western Nepal.[8] Bardiya is located on Nepal's low-lying *Terai* plains, bordering the Indian state of Uttar Pradesh. Several major rivers flow through the district including the Babai River and branches of the Karnali River, one of Nepal's largest river systems, rendering the district highly susceptible to flooding. Bardiya is home to a predominantly indigenous (*adivasi*) population comprising *Tharu* and *Sonaha* peoples, with their particular social structures and histories of subordination (Sugden et al. 2014). The research focused on four flood-prone wards in Rajapur Municipality, which was established in 2014 following the amalgamation of six former VDCs. Each ward comprised several villages, which were all rural in nature, with a predominantly *Tharu* population:

- Ward A, was the largest of the four wards and comprised 12 villages and more than 1,000 households. The research focused on one village within the ward where the majority of the 635 households were former bonded labourers (*Mukta Kamaiyas*),[9] with a small number of *Dalit* householders who used to be agricultural bonded labourers (*Haliyas*). They moved to the area in 2005/6, following the abolition of bonded labour in 2000, having been given 5 kathas of land (1 katha = 0.03 ha) by the government.
- Ward B comprised approximately 500 households across five villages, three of which were prone to flooding. In addition to the dominant *Tharu* population, the ward was also home to *Sonahas* (a minority fishing community)[10] and *Dalit* households. The research focused on all three flood prone villages.
- Ward C was smaller, comprising around 350 households. Two villages made up the ward, although some *Tharu* householders also had links with a village in a neighbouring ward from where they were displaced following a flood event in 1983. We focused the research on all three villages.
- Finally, ward D was a small ward comprising two villages and around 350 households, the majority of which were *Tharu*. The research focused on the most flood prone village.

The main livelihood activity for the householders involved in the study was agriculture, with some landless householders engaged in sharecropping arrangements with other landowners. Employment opportunities locally were limited. Some women engaged in day wage labour locally, while most men migrated to the cities of Nepalgunj or Birgunj on the Nepal–India border, to other cities in the Terai, or to India for seasonal employment. As a result, there were very few active men in the villages. In one focus group discussion (FGD) with *Tharu* and *Dalit* women in Ward C, 9 of the 12 women present had husbands working in India. A small number of householders had set up small businesses. For example, in Wards C and D there had been a move towards commercial vegetable farming, linked to the establishment of the Women's Agricultural Group with NGO support.

The four case study wards experienced seasonal floods every year, which damaged and destroyed farmland and crops. Major flood disasters occurred in 2005–6 (2062–63 BS), resulting in a large number of human as well as livestock casualties; and in 2014 (2071 BS), when the floods destroyed houses and farmland but no human casualties were reported. Since the research was undertaken, a major flood occurred in August 2017, which resulted in four fatalities and affected nearly 135,000 people in Bardiya District alone (NPC 2017, 2).

Householders in all four of the case study wards have been the target beneficiaries of various NGO-led projects focusing on disaster preparedness and response, livelihood strengthening, community forestry and microfinance. The disaster preparedness and response projects in particular began in 2011 and 2013 respectively, heavily shaped by the vision of the international development community, and later the LDRMP Guideline (or 'The Pink Book' as it was commonly referred to in FGDs) and the 9 Minimum Characteristics of a Disaster Resilient Community.

Methods

The research involved a series of key informant interviews with NGO representatives responsible for implementing CBDRR projects, with a view to understanding how the 9 Minimum Characteristics have been used by implementing partners; and with other key community-level stakeholders including the head of the village and the Chairperson of the Ward Citizen Forum[11] to gather background information on the case study villages. The interviews were followed by a series of FGDs with community groups established as part of the CBDRR intervention, for examples the Community Disaster Management Committees (CDMCs) and associated task forces (e.g., first aid, search and rescue); and with pre-existing community groups, including the Ward Citizen Fora and women's groups. The aim of the FGD was to explore people's perspectives on their own vulnerability and resilience, and on the usefulness of the 9 Minimum Characteristics in supporting the community to prepare for, respond to, and recover from, a disaster. We used the ward, the lowest administrative unit in Nepal, as the means of entry into the community. In doing so we acknowledged the complexity of the term 'community' (Titz, Cannon, and Kruger 2018) and remained aware that "people's sense of themselves and their place is often in friction with how resilience planners imagine they might foster social capital and harness networks into their disaster risk management schemes" (Nightingale 2015b, 183). See also Shneiderman (2015) for a critique of "the village" as a set of social relations in Nepal. A 'snowballing approach' was adopted to ensure we gained a range of perspectives.

A total of eight interviews and 16 FGDs (Tables 8.1 and 8.2) were undertaken at the community level across the six case study wards in March 2016. Interviews were also undertaken with the project officer and social mobiliser from the local NGOs implementing the CBDRR projects, and with the Executive Office in the Municipality.[12] The interviews were undertaken in Nepali and Tharu language as required by Gautam and Singh, and were digitally recorded with detailed

Table 8.1 Interviews with Community Members across the Four Case Study Wards

	Description of interviewee	*Gender and ethnicity*
Ward A	Primary school principle	Male, Tharu
	Member of the Search and Rescue Committee	Male, Tharu
	Bhadhgar/Chairperson of the Mukta Kamaiya Society	Male, Tharu
Ward B	Coordinator of the Ward Citizen Forum	Female, Tharu
	Task force members	Female (x 2), Tharu
Ward C	Principle, Lower Secondary School	Chhetri
Ward D	Bhadhgar and Secretary of the CDMC	Male, Tharu
	Coordinator of the Ward Citizen Forum	Male, Tharu

Table 8.2 FGDs Undertaken with Formal and Informal Community Groups in the Four Case Study Wards

	Description of group	*Total no. of participants (no. female participants in parentheses)*	*Caste/ethnicity of participants*
Ward A	CDMC	10 (6)	Tharu
	Dalit householders	3 (1)	Dalit
	Task force	3 (3)	Tharu
Ward B	Tharu Women's Community	14 (14)	Tharu
	Forestry Group	11 (8)	Tharu, Sonaha
	CDMC	6 (5)	Tharu, Chhetri, Dalit
	CDMC	7 (7)	
	Women's Agriculture Group		
Ward C	CDMC	10 (6)	Tharu
	First Aid, Search and Rescue	4	Tharu
	Committee	11 (10)	Tharu, Dalit
	Sub-community	10 (10)	Tharu
	Women's Agriculture Group		
Ward D	CDMC	14 (11)	
	Saving and Cooperative Group	11 (0)	Tharu
	Mothers' Group	15 (15)	Tharu, Brahmin
Total	16	129 (96)	

summaries of the discussions prepared based on the recording.[13] A grounded theory approach was adopted, with the notes reviewed and coded thematically by Oven and Rana. The codes were then triangulated with the findings from a two-day debrief workshop.

Findings and discussion

We structure the findings from the empirical research around three key themes. The first explores how communities are defined from the perspective of local people themselves, and the tensions and incongruities that exist between these local framings and the framings commonly used in donor-funded CBDRR projects. The second theme explores gender relations and the specific role of women in decision-making within the household and community. The third theme explores the impacts and implications of a technocratic, projectised approach to CBDRR with a specific focus on women. Across all three themes, we pay attention to the intersectionality between gender and other social factors including caste and ethnicity, and how these historical and inherited patterns of gendered relations and processes of marginalisation intersect with contemporary transformations in economy and society.

Defining a community: associations and tensions within and between villages

CBDRR initiatives in Nepal commonly involve the establishment of institutions at the local level in the form of CDMCs. These CDMCs are territorially based, commonly established at the ward or sub-ward level, with the aim of serving particular villages. This can be problematic as such models fail to take into account networks and associations, as well as tensions, within and between villages. For example, three villages made up Ward C, one of which included both *Dalit* and *Tharu* households. The *Dalit* households in this particular village had close ties with households in a neighbouring village in the same ward, while the *Tharu* households were closer to households in a village in a neighbouring ward where their deity resided and where they used to live before they were displaced by a flood in 1983. However, government and NGO programmes, including CBDRR-related activities, were largely organised on a ward-by-ward basis. Officially, residents could only attend development programmes in the ward where they lived but this ignored the strong networks and social ties that residents relied upon in the event of a disaster. Indeed, these networks and ties commonly reflected the complex intersection of exclusions at the community level, which we interrogate further below. What is clear from the findings is that imposing a new governance structure that is counter to the way communities organise themselves is unlikely to be effective.

Participants spoke of people helping each other to prepare for, and respond to, flood events. For example, in Ward B, "People from [one village] usually bring all their grains and belongings to [our village] before the flood. We help them to transfer all of the stuff in tractors and carts. Task force members went to [their village] to rescue people and they also helped in moving belongings in their own village" (CDMC member, Ward B). Similarly, in Ward C, teams of four to five people formed to rebuild damaged houses once the floodwaters receded; while in Ward A, participants described the tole-level[14] patrols that had been established with the aim of preventing elephant attacks. The patrols were often made up of women, with households fined if they did not turn up, as per the community rules. However, while there was evidence of community self-help and reciprocity, some people questioned the intentions and motivations of the people coming to help possibly indicating underlying tensions, and power dynamics, which CBDRR practitioners may be unaware of.

In Ward A, some participants spoke of the discrimination that they had experienced by neighbouring *Tharu* people who resented the *Mukta Kamaiyas* for clearing and occupying land that was formerly national forest. Some *Dalit* householders also spoke of the discrimination that they had experienced by the *Mukta Kamaiyas* within the same ward, and the *Badhgahr* (village headman) who was *Tharu*. As agricultural bonded labourers (*Haliyas*), they had a different social status to the *Mukta Kamaiyas* and were referred to as 'squatters'. They felt looked down upon by the *Mukta Kamaiyas* despite facing similar political struggles, and were given land of poorer quality, which was unsuitable for farming. As one FGD participant explained, "We came here because the political leaders convinced us to move here. We settled here with their permission. But now no one cares about us, we ended up being their vote bank" (Direct quotation, from a *Dalit* householder, FGD, Ward A).

In Ward B, tensions were reported between the toles that made up the ward, with many of the development programmes centred on one tole in particular. The participants in one FGD, who were mostly women, attributed this to the strong leadership in that particular tole which enabled them to engage with NGOs and government, and to make the case for the support they needed. They noted that in other toles leadership was weak, which some attributed to the absence of men in the community, perhaps reflecting that men are often more politically connected.

These 'community' conditions highlight three issues that are also relevant when it comes to considering gender relations and DRR. Firstly, the bureaucratisation of space reflected in formal administrative divisions do not, often, map onto more organic community forms. In time, such formal units may create their own community logics but until that point there will be a mismatch between the space of administration and spaces of community (see also Shneiderman 2015). The second issue relates to these organic, often styled 'community' spaces. Just as gender theorists in the 1980s (see Folbre 1986; Moser 1989) enjoined scholars to look into the black box of the household, so too with community, with the community itself being both constructed and imagined (Titz, Cannon, and Kruger 2018). Thirdly, where CBDRR projects land is commonly neither a neutral nor a harmonious space in terms of local politics and decision-making. Understanding these tensions, how they are produced and reproduced, and their gendered manifestations, is therefore essential if we are to understand why apparently locally engaged and sensitive interventions do not necessarily engage and empower all community members. We expand on this further below, with a specific focus on women and gender relations at the village level.

Gender relations at the village level and the role of women in decision-making

As noted elsewhere (Sugden et al. 2014; see also Allendorf 2007; Gurung and Bisht 2014), women form the backbone of the agrarian workforce in Nepal as well as being responsible for the reproductive tasks of the household including child rearing, food preparation and the collection of firewood and fodder. Men also engage in agriculture, but as the predominant wage earners, commonly live away from the village. As a result, there were a number of *de facto* women-headed households in the case study wards. With few opportunities for day wage labour locally, and their reproductive tasks within the home, most women in the case study wards relied on agriculture and remittance income from male members of the household.

There were examples of women's empowerment captured in the interviews and FGDs. For example, discussions with some FG participants in Ward A suggested that the situation within the household was changing: "we are free to move around, things have changed. Men understand more now, it's not like before" (FGD with the Women's Agriculture Group, Ward A). This awareness may be attributed to education; experiences of living in other places (and countries); the promotion of women's empowerment by donor-funded projects; as well as the impacts of the Maoist insurgency which, it could be argued, "create[d] openings for destabilizing existing caste, gender or class-based identities" (Nightingale and Rankin 2014, 110; see also Yadav 2016).[15]

In Ward B, however, some of the women who participated in the FGDs explained that they were not allowed to move freely and were in fact restricted in their movements by their husbands and other men in the community. In Ward C, the issue of gender-based violence was raised: "Violence against women is still a grave issue in this community. Women are not treated well in the families and there have been cases of sexual violence. Single women are more vulnerable in the community. Further, the dowry system is deeply rooted in the community. The community has changed over the years but there are still problems" (FGD with task force members, Ward C). This comment reflects the patriarchal, patrilocal and patrilineal households in Nepal (see Nightingale and Rankin 2014). Indeed, while the FGDs were underway in Ward D, an awareness-raising programme was being run by the Area Police Office on the subject of increasing suicide rates among women in Rajapur. While the issue of suicide was not raised directly by the research participants, research elsewhere has highlighted suicide as a leading cause of death for women of reproductive age in Nepal (Pradhan et al. 2011; see also Thapaliya, Sharma, and Upadhyaya 2018). This has been linked to pressures to marry and bear children, often at a young age, as well as domestic violence (Pradhan et al. 2011). With the household commonly the origin of inequalities (Nightingale and Rankin 2014), it is perhaps unsurprising that these gender-based inequalities were very much present within the community and impacting community-level decision-making processes. We explore this further below.

In the absence of locally elected representatives when the study was undertaken, the Ward Citizen Forum (WCF) was the main link between the ward and the municipality. The bottom-up planning process began at the village level, and was usually led by the *Badghar*, the elected chief of the village or tole,[16] who was responsible for taking village issues and concerns to the WCF. The 22 *Badghars* from the 22 toles in Ward A met every month to discuss the issues raised. At the ward level there was much debate as every village had its own demands and needs. As one village headman explained, it was easier to pass plans that benefitted more than one community, with physical infrastructure (electricity and gravel roads) identified as priorities for all four wards. In Ward B, the annual budget in 2015/6 was spent on the construction of culverts to manage the floodwater and gravel roads, with clear links to DRR, although these links were not explicitly made in the planning documents. The case study village in Ward A seemed happy with the budget they were receiving, which the *Badghar* reported to be three times that of some other villages in the ward. However, in light of the governance gap, some villagers questioned if the issues raised at the WCF meetings actually reached the municipality.

There were examples of women taking on leadership roles. As one interviewee explained, "It used to be men who coordinated the WCF but last year the women raised the issue of equal representation and participation. I was chosen as coordinator for one year. However, the voices of women are heard less and pressure from men in decision-making is extremely high" (Interview with the Coordinator of the Ward Citizen Forum in Ward B, female, former Maoist combatant). Indeed, while women can be *Badghars*, men continue to have a stronger voice in decision-making. As the female members of a CDMC in Ward B explained:

> [W]hen we took this plan [village development plan made with NGO support] to the [WCF] meeting no one listened to us. Our field facilitator urged us to submit it and we spoke with the WCF coordinator and he finally agreed to take it. They wanted a plan that was made by the *Badghars* and men but we told them none of them took charge of it and we did. But the WCF coordinator was smart and kept his village's plan on the top and ours at the end. No one looked at our plan. It is very difficult for women to be heard in these meetings. We complained but they told us that the *Badghar* had to propose it, so later we asked one of the men in the village to come. Finally they took the plan.

This highlights some of the entrenched cultural barriers which quotas for female participation, alone, are unlikely to address.[17] Indeed, as noted elsewhere in the context of community forestry, attendance is often equated with participation despite a good deal of research suggesting that often this is not the case (Nightingale 2002). Even if the most vulnerable are included, they commonly have negligible influence on decision-making processes (Nagoda 2015). We argue that there is a three-fold series of challenges to navigate if women's involvement is to be meaningful. First, it is necessary to achieve attendance – mere physical presence at key meetings, for instance. Second, to ensure that attendance is translated into participation, in terms of women being active and engaged in meetings and other activities. And third, to make this participation meaningful and transformative so that women's voices are truly heard. In a similar vein to analyses of 'participation' which draw a distinction between nominal, instrumental and transformative participation (see Cornwall 2003), there is a world of difference between nominal 'listening' and truly hearing, in a transformative sense. Being truly heard requires that we really listen to women's diverse experiences by engaging women on their own terms and not attempting to reduce their experiences to the universal and the understandable (Bohler-Muller 2002; see also Cupples 2007). Meeting each of these challenges becomes progressively more difficult because they challenge embedded behavioural norms and, more importantly, entrenched interests. We illustrate this point further below, drawing on specific examples from the CDMCs.

Like other user groups, CDMCs rely upon the 'equitable engagement' of committee members with the aim of overcoming gender, caste, ethnicity and other socially defined inequalities (Nightingale 2015b, 199). The LDRMP guidelines recognise the importance of social inclusion and participation. For example, in the context of the formation of the Local Disaster Management Committee: "this should ensure inclusion of women and Dalit[s] in the committee" (15). In the identification and prioritisation of risk management activities: "the participation of the vulnerable community, the poor, People with Disability (PWD), women, dalits and ethnic groups shall be ensured" (9). Similarly, "When choosing task-force members, prioritisation should be given to women, Dalits, ethnic community, PWDs and representatives of highly vulnerable communities with relevant expertise" (40). This shows that there is some awareness of the social and political context in which vulnerability is created in policy documents, although little if any action to address root causes. These root causes reflect both inherited exposure (for example, the caste system and associated marginalisation, gender divisions in society and feudalism) and precarity or produced exposure (resulting from, for example, growing market dependencies) (Rigg et al. 2016, 66). We return to this in the next section of the paper.

CDMCs were established across the four case study wards starting in 2011 with donor funding and NGO support. Given the traditional social structure of *Tharu* communities involving an elected chief (*Badghar*) and an assistant or village messenger (*Chaukidar*), it was unclear how this more 'Western' mode of community organisation fitted with inherited and established norms. What was clear from the interviews and FGDs was that decisions regarding DRR were taken by the *Badghar* and were communicated via the *Chaukidar*. In Ward D, when the CDMC was established in 2011, "[a] big meeting was called . . . but not many people attended this as they didn't care about it. The problem was that it was a meeting called by normal people and not the *Badghar*. Until the *Badghar* calls for such meetings, people won't give much importance" (FGD with CDMC members, mainly female, Ward D). The *Badghar* was, at the time, the Secretary of the CDMC although the interview revealed that he knew very little about DRR. Working through the governance structures of a particular ethnic group, in this case *Tharu* people, may initially make sense. However, such structures potentially marginalise others, for example, *Sonaha* and *Dalit* householders.

In Ward B, all three flood-prone communities had their own CDMC. One of the CDMCs included representatives from the *Tharu* and *Sonaha* communities but there were clear tensions between the two groups. Importantly, despite the three toles helping each other in the event of a flood, no formal coordination between the CDMCs was reported. Indeed, according to the Coordinator of the WCF, none of the three CDMCs was even functional.

In Ward A, despite the village *Badghar* claiming that the whole of the tole was involved in preparing the disaster management plan, members of the *Dalit* community claimed that they were not represented on the CDMC or within any task groups. The *Dalit* householders interviewed were aware to some extent of the DRR activities underway within their community, but were not actively involved. This, at least in part, reflects their absence from the community for day-wage labour rather than caste-based discrimination, although this should not be disregarded completely. The lack of participation of *Dalit* householders in development projects was reported in other wards too. According to the coordinator of the WCF in Ward D who is *Tharu*, *Dalit* households do not attend many ward level programmes because they claim that they do not have time.

In all of the case study villages, women were actively involved in the CDMCs and task forces, such as first aid and search and rescue, and had participated in trainings. Further, all communities had established groups of female volunteers who were responsible for sharing information with householders in the community, for example, they distributed calendars which contained important DRR information and contact numbers. These activities were undertaken alongside existing reproductive and productive tasks. As participants from a women's agricultural group in Ward B explained: "none of the men in the community agreed to be in the committee, so women came forward. Most men don't attend any such meetings. Although they meet every month to discuss what to do, make plans and also discuss where to use the emergency funds, the men in the DMC never attend such meetings" (Women's agriculture Group, Ward B). A *Badghar* in Ward C explained why the participation of men was so low in his village: "men are mostly involved in major decision-making processes. Men don't have interest in DRR activities. It is usually women who attend most of the meetings" (Interview with *Badghar* and Secretary of the CDMC, Ward D). This further suggests that DRR was not a priority concern for male members of the community. Women showed a greater level of concern perhaps reflecting that it is the women who have to deal with the floods, look after the house, their children and elderly family members, as well as tending the farmland and protecting the crops. Indeed, as noted by Moser (1989, 1801) men are "involved in community activities in markedly different ways from women [. . .] women most frequently make up the rank and file voluntary membership while men are only involved in positions of direct authority and work in a paid capacity". It was therefore not uncommon for women to be members of multiple committees.

In addition to these traditional caste and gender-based exclusions, then, we also see the impacts associated with contemporary transformations in economy and society. For example, in Ward C, participants reported that while a CDMC had been established, it was largely inactive as most of the members were away from the village for work. The CDMC was subsequently reformed with NGO support and members who could give time and who were more likely to remain in the village were recruited. As the FGD participants explained, "it was difficult to organise any meeting as most people were away for work. People from [village name] which is at high risk [of flooding] are not in the CDMC. As they were never regular in the meeting, CDMC had to replace these members. They go for labour work" (Members of the CDMC, Ward C).

Similarly, in Ward C, people were unaware of who exactly was involved in the CDMC. As the women from a sub-community explained: "A lot of people were asked to be members of the CDMC from this community but no one agreed. There are currently two people from this community who are on the committee but one person is currently in India. It has been two to three months since she last attended the meeting. People are busy in labour work so that is why no one is interested in being part of the committee" (FGD with a sub-community in Ward C). Evident in these responses is the need to not only make such groups socially inclusive, but also to recognise the broader context within which they operate. Many villages' local economies and available opportunities are not sufficient to meet residents' needs – localities are *de facto* sub-livelihood. This creates a situation in which many women and men are simply absent from their communities, and no matter how inclusive a CDMC may be, the practicalities of filling positions are hard to surmount.

CBDRR rendered technical

Given the nature of CBDRR, which has been 'rendered technical', it is perhaps unsurprising that the community disaster risk management plans prepared by the CDMCs with NGO support detailed standard technical interventions. For example, Ward B's plan included the construction of a community building to provide a space for the delivery of training (for example, in agriculture, savings and credit, and disaster preparedness) and for the community and people in nearby villages to gather during a flood, as well as raised toilets and hand pumps, echoing the technical expertise of the NGOs implementing the projects. As argued by Li (2016) in the context of development projects in Indonesia, it could be argued that the CBDRR projects were 'pragmatically accepted' by the communities in Bardiya. They did not necessarily address the underlying issues of concern for the householders (for example, citizenship in the case if the *Mukta Kamaiyas*), but in many cases the projects led to concrete deliverables, many of which were useful in the context of seasonal and extreme flooding.

A vulnerability and capacity assessment was undertaken with NGO support in Ward A in 2014, and we were told that the CDMC still had a copy. While this particular document was not viewed by the research team, experience suggests that such documents are, at best, a light touch overview of what are deeply embedded social exclusions. As noted by Ojha et al. (2016, 424) in the context of climate change adaptation in Nepal, such efforts to promote participation are often "too technocratic to empower the local groups most at risk . . . with little consideration given to how the underlying power relations are addressed". It was also unclear if the report had ever reached the municipal council or indeed, if the assessment fed into the Local Disaster Risk Management Plan (LDRMP) at the municipal level. There was a sense, then, that bottom-up planning processes were not really bottom-up at all.

Mock drills for flood preparedness were organised every year and there were clear protocols during the monsoon whereby the *Badghar* or the CDMC

Chairperson telephoned the government's gauge reader who lived upstream. There were also multiple donor-funded projects supporting the construction of *machans* (raised platforms that were constructed to keep people, animals, grains and valuables safe from floodwaters), raised grain storage units, and raised hand pumps (Figures 8.2 and 8.3). In Ward D, the case study community bought land to construct the community shelter house with support from the municipality, international and local NGOs. The case study community in Ward A requested the establishment of an emergency fund through the WCF. Such a fund was also requested by other wards and was approved, with 5 lakh [NPR 500,000 or US$4,250] allocated by the municipality for DM in 2016 (FGD, CDMC, Ward A). The CDMC also had its own fund (NPR 18,000 or US$160), which was used to repair equipment such as sirens and boats, among other things. In other villages, for example in Ward B, sustaining a community-level emergency fund was more of a challenge as households could not afford to donate money.

The CBDRR projects were very good at addressing outcome vulnerability. As explained by members of a Mothers' Group, the group had identified households where there were pregnant women and newly born babies so they could inform the CDMC in the event of a flood and extra assistance could be provided. The female participants particularly valued the early-warning system, which gave them three hours advanced warning that the floodwaters were on their way; and

Figure 8.2 Community Building or *Machan* in a Case Study Ward in Rajapur Municipality, March 2016

Figure 8.3 Raised Water Pump in a Case Study Ward in Rajapur Municipality, March 2016

the *machan* which provided a safe place for disabled people and pregnant women. Boats were also important for transporting belongings: "at the end of the day the most important for us is a way to store our grains. The mud one [grain store] that we make gets wet during flood" (FGD with a women's agriculture group, Ward C). Other female participants highlighted that "there is no point in just saving our life. What will we do after that [without food and housing]?" (Direct quotation, FGD with a sub-community, Ward C). In one ward, however, the seasonal floodwaters were too high so *machans* and raised hand pumps were of no use. In another village, funds were the issue, "A *machan* is needed but there is no space to build it . . . and we don't have the resources to purchase such land" (FGD with a women's agricultural group, Ward B).

Participants in all FGDs welcomed the embankment, which was part of a larger-scale DRR project: "We cannot say that the flood won't come this year; we always live in that fear. But maybe this year it won't come as there is embankment along the river. We feel much [more] secure and this is because of embankment" (Direct quotation, FGD with a women's agricultural group, Ward B). For some the "Embankment is the only long-term solution for disaster preparedness. People are already doing what they can right now, so there has to be concrete solutions for this" (Coordinator of the WCF, male, Ward D). Some participants were more sceptical, however. This scepticism reflected, at least in part, the fact that people had to donate their already small land holdings without compensation. There was also some concern that because the embankment was only being built along certain sections of the river that this could make the flooding worse in other areas. As members of a FGD with a male cooperative group in Ward D explained: "This [the embankment] would be really helpful. However, the construction of the embankment has been in parts. If there is flood now, none of these houses would be safe. Because only certain parts of the river has embankment, the water flow is going to be high in areas where there is no embankment".

While this technical approach to CBDRR brings some concrete and tangible benefits to the villages involved, it also exercises two important sets of issues. To begin with, it obscures the deep-seated cultural and social inequities that create and sustain vulnerability in the first place, and may also undermine efforts to achieve DRR. However, these are, as noted, often the most invidious and intractable because they speak to the essential nature of communities. Second, this process of rendering technical depoliticises CBDRR. This is not to say that the NGOs implementing the projects are unaware of the wider contextual vulnerability and root causes of disasters. Indeed, for the NGOs involved in this study, advocacy and lobbying for the citizenship rights of the *Mukta Kamaiyas* was central to the work they do. However, as noted by Li (2016), there is no way to render these issues technical or indeed to resolve them as wider structural reform is needed. It is, however, important to note that while the projects themselves may depoliticise CBDRR, they land in highly politicised contexts resulting from ongoing demands for social and political inclusion, challenging the idea of the passive beneficiary.

It is perhaps unsurprising that according to the CDMC representatives in Ward B, the biggest need was to improve the living standards of the poor. If they had employment opportunities, so it was said and assumed, they would automatically be resilient. For the female participants involved in the women's agriculture group, a good income would enable them to build concrete houses (FGD, women's agricultural group, Ward B). FGD participants in Ward A concurred with this, "If you have employment, your source of income expands and increases. It's after that people can decide what kind of houses to build and what kind of land to build it on" (Direct quotation, FGD with the CDMC, Ward A).

Conclusions

This chapter has critically explored attempts to build social resilience to disasters through donor-funded CBDRR projects, with a specific focus on women in four case study wards in Mid-West Nepal. Our findings suggest that CBDRR projects informed by the 9 Minimum Characteristics of a Disaster Resilient Community addressed some of the practical needs of women (for example, through the development and implementation of early-warning systems and raised water pumps). However, this was limited to some extent by the lack of consideration given to the 'triple role' played by women in terms of reproduction, production and community management (Moser 1989). As a result, significant expectations were placed on women in the community, who often found themselves attending rather than contributing to CBDRR, reflecting historical and inherited gendered relations and processes of marginalisation. Some positive examples were, however, noted with women contributing to decision-making processes in a more meaningful way reflecting changes in gender relations within the household and the wider community.

Given the technocratic nature of CBDRR projects, it is perhaps unsurprising that limited action was taken in relation to women's strategic gender needs which emerge from wider issues of subordination. Here the findings remind us that gender is just one form of marginalisation, often working in tandem with other forms at the community level, in particular, caste and ethnicity, which CBDRR initiatives have a tendency to mask. Moving forward, Moser's (1989) recommendations for gender planning continue to hold true. CBDRR projects should try to meet practical gender needs not only through technical interventions but by focusing "on the domestic arena, on income earning activities, and also on community-level requirements of housing and basic services" (Moser 1989, 1803). However, to truly 'leave no one behind' in the context of disasters and climate change, requires that strategic needs and root causes of vulnerability are addressed which requires more significant political transformation. While there is much optimism here, as Tamang (2018) continues to remind us, there remain significant barriers to overcome.

Acknowledgements

This research was carried out as part of a *Review of Nepal's 9 Minimum Characteristics of a Disaster Resilient Community* funded by the UK Department for International Development's South Asia Research Hub. The authors would like to thank the wider research team including Dr Shailendra Sigdel at the Foundation for Development Management (FDM) in Kathmandu, Dr Ben Wisner (Oberlin College, Ohio), Dr Samantha Jones (Northumbria University), Prof. Alex Densmore (Durham University) and Ajoy Datta (Overseas Development Institute). Thanks go to our collaborators at the IFRC, Krishna KC and Nikhil Shrestha, and the former Ministry of Federal Affairs and Local Development, Government of Nepal, Mr Gopi Krishna Khanal, Mr Rishi Raj Acharya and Mr Purusottam Subedi. Special thanks go to the international and national NGOs involved in the study and to the individuals who kindly participated in the research; we are incredibly grateful for your time and willingness to engage. Finally, we are grateful to Helle Rydstrom and Catarina Kinnvall for the opportunity to contribute to this volume, as well as their valuable comments on an earlier draft.

Notes

1 It should be noted, however, that Petal et al.'s (2018, 6) forensic study on the causes of deaths and injuries in the 2015 Gorkha earthquake found no significant difference in the rate of death for men and women in their sample.
2 https://data.worldbank.org/indicator/SP.RUR.TOTL.ZS

3 http://databank.worldbank.org/data/download/poverty/33EF03BB-9722-4AE2-ABC7-AA2972D68AFE/Global_POVEQ_NPL.pdf
4 https://data.worldbank.org/indicator/BX.TRF.PWKR.DT.GD.ZS
5 The five priorities of the HFA were: (1) Ensure that DRR is a national and a local priority; (2) Identify, assess and monitor disaster risks and enhance early-warning systems; (3) Use knowledge, innovation and education to build a culture of safety and resilience at all levels; (4) Reduce the underlying risk factors; (5) Strengthening disaster preparedness for effective response at all levels (www.unisdr.org/we/coordinate/hfa).
6 VDCs were the lowest administrative level of the Ministry of Federal Affairs and Local Development. Each district comprised several VDCs. VDCs were dissolved in March 2017 and replaced by *gaunpalikas* or rural municipalities.
7 It is important to note that the governance context in Nepal has changed significantly since this research was undertaken in 2016. Provinces, and urban and rural municipalities, have been established under the new federal state structure, local elections have taken place for the first time in 20 years, and the new Disaster Management Act has been ratified. It remains unclear if and how CBDRR will be implemented moving forward (Oven 2019). We focus here on the situation as it was in 2016.
8 Under federal restructuring, Bardiya District now falls within Province 5.
9 Adhikari (2014) provides a brief historical overview of the dispossession of *Tharu* people in Bardiya District. Historically, the *Tharus* inhabited the malarial forests in the western Terai plains. The forests were cleared by the government in the 1950s and people from other regions were encouraged to settle there. The new settlers included upper-caste people from the hills, who gradually established ownership of the land. This resulted in the dispossession of a large number of *Tharus*, many of whom were forced into bonded labour.
10 The *Sonahas* are a traditional group living near the Karnali River, engaged in fishing and gold mining. They have been campaigning to be recognised as an indigenous or Janajati group like the Kamaiyas. Without this status they claim that they are deprived of government facilities (see Panthi 2018)
11 The WCF was a local level decision-making body created under the Local Governance Community Development Programme, in the absence of locally elected government.
12 We acknowledge here the significant contribution of Arya Gautam and Toran Singh who undertook the interviews and facilitated the focus groups in Rajapur that we draw upon here, with support from Katie Oven and Shubheksha Rana who designed and led the overall study.
13 Unless otherwise stated, the quotations used in the chapter are paraphrased based on the detailed summary notes prepared on each interview and FGD.
14 A tole is a small settlement or community within a ward which shares local resources, for example, a temple, water tap or community forest. They are a social rather than a mandatory administrative structure.
15 The *Tharu* population was a target group for the Maoists due to their history of dispossession and they recruited heavily in Bardiya District (for details see Adhikari 2014).
16 At the time of the research, there was one *Badghar* per tole. They were elected by the community and changed every year.
17 It is important to note that in the 2017 local elections in Nepal, 14,000 [40%] locally elected representatives were women (Baruah and Reyes 2017). This was the highest number of women ever elected to public office in Nepal. As noted by Baruah and Reyes (2017) "Certainly, these are significant developments toward building inclusive participatory spaces for women. [. . .] But initial observations and voices from the field are concerning, and reflect that women elected leaders are still confronted by deep-seated gender expectations, questions on capacities, and often divisive political ideologies, caste, ethnicity, religion, and class".

References

Adger, W. Neil. 2000. "Social and Ecological Resilience: Are They Related?" *Progress in Human Geography* 24, no. 3: 347–64.

Adhikari, Aditya. 2014. *The Bullet and the Ballot Box. The Story of Nepal's Maoist Revolution*. New Delhi: Aleph Book Company.

Allen, Katrina M. 2006. "Community-Based Disaster Preparedness and Climate Adaptation: Local Capacity-Building in the Philippines." *Disasters* 30, no. 1: 81–101.

Allendorf, Keera. 2007. "Do Women's Land Rights Promote Empowerment and Child Health in Nepal?" *World Development* 35, no. 11: 1975–88.

Baruah, Nandita, and Jerryll Reyes. 2017. "Nepal Elections: More Women Have a Seat at the Table, but Will They Have a Voice?" *Asia Foundation* (Blog), December 13. https://asiafoundation.org/2017/12/13/nepal-elections-women-seat-table-will-voice/.

Bohler-Muller, Narnia. 2002. "Really Listening? Women's Voices and Ethnic of Care in Post-Colonial Africa." *Agenda: Empowering Women for Gender Equity* 54: 86–91.

Bradshaw, Sarah. 2014. "Engendering Development and Disasters." *Disasters* 39, no. S1: S54–S75.

Cook, Bill, and Uma Kothari. 2001. *Participation: The New Tyranny*. London: Zed Books.

Cornwall, Andrea. 2003. "Whose Voices? Whose Choices? Reflections on Gender and Participatory Development." *World Development* 31, no. 8: 1325–42.

Crutzen, Paul J. 2002. "Geology of Mankind." *Nature* 415, no. 6867: 23.

Cupples, Julie. 2007. "Gender and Hurricane Mitch: Reconstructing Subjectivities after Disaster." *Disasters* 31, no. 2: 155–75.

Delica-Willison, Zenaida, and Jean-Christophe Gaillard. 2012. "Community Action and Disaster." In *Handbook of Hazards and Disaster Risk Reduction*, edited by Ben Wisner, Jean-Christophe Gaillard, and Ilan Kelman, 711–22. London: Routledge.

Edwards, Michael, and David Hulme. 1996. "Too Close for Comfort? The Impact of Official Aid on Non-Governmental Organizations." *World Development* 24, no. 6: 961–73.

Falk, Monica Lindberg. 2012. "Gender, Buddhism and Social Resilience in the Aftermath of the Tsunami in Thailand." *South East Asia Research* 20, no. 2: 175–90.

Flagship 4. 2013. *Flagship 4 Handbook. Nepal's 9 Minimum Characteristics of a Disaster Resilient Community*. Report. Kathmandu: Nepal Risk Reduction Consortium. http://flagship4.nrrc.org.np/sites/default/files/documents/NRRC%20-%20Flagship%204%20Handbook_14%20Nov%2013.pdf.

Folbre, Nancy. 1986. "Cleaning House: New Perspectives on Household and Economic Development." *Journal of Development Economics* 22, no. 1: 5–40.

Fordham, Maureen. 1998. "Making Women Visible in Disasters: Problematizing the Private Domain." *Disasters* 22, no. 2: 126–43.

Gender Equality. 2017. Nepal Flood Response 2017. Gender Equality Update No. 12. Report. https://reliefweb.int/sites/reliefweb.int/files/resources/Gender%20Equality%20Update%20No%2012%20%2831%20August%202017%29%20Final%20version.pdf.

Gladfelter, Sierra. 2018. "The Politics of Participation in Community-Based Early Warning Systems: Building Resilience or Precarity through Local Roles in Disseminating Disaster Information?" *International Journal of Disaster Risk Reduction* 30 (Part A): 120–31.

Gurung, Dibya Devi, and Suman Bisht. 2014. "Women's Empowerment at the Frontline of Adaptation: Emerging Issues, Adaptive Practices, and Priorities in Nepal." ICIMOD Working Paper 2014/3. Kathmandu: International Centre for Integrated Mountain Development. www.cabdirect.org/cabdirect/FullTextPDF/2015/20153232932.pdf.

Hickey, Samuel, and Giles Mohan. 2006. *Participation – From Tyranny to Transformation?: Exploring New Approaches to Participation in Development*. London: Zed Books.

Jones, Samantha, Komal Aryal, and Andrew Collins. 2013. "Local-Level Governance of Risk and Resilience in Nepal." *Disasters* 373: 442–67.

Jones, Samantha, Katie J. Oven, Bernard Manyena, and Komal Aryal. 2014. "Governance Struggles and Policy Processes in Disaster Risk Reduction: A Case Study from Nepal." *Geoforum* 57: 78–90.

Kleinman, Michael. 2017. "Development Is Not a Science and Cannot Be Measured. This Is Not a Bad Thing." *The Guardian*, June 1. www.theguardian.com/global-development-professionals-network/2017/jun/01/development-is-not-a-science-and-cannot-be-measured-that-is-not-a-bad-thing.

Li, Tania Murray. 2007. *The Will to Improve. Governmentality, Development and the Practice of Politics*. London: Duke University Press.

Li, Tania Murray. 2016. "Governing Rural Indonesia: Convergence on the Project System." *Critical Policy Studies* 10, no. 1: 79–94.

Mansuri, Ghazala, and Vijayendra Rao. 2013. "Localizing Development: Does Participation Work?" Policy Research Report. Washington, DC: World Bank. https://openknowledge.worldbank.org/handle/10986/11859.

Manuel, Marcus, Francesca Grandi, Stephanie Manea, Amy Kirbyshire, and Emma Lovell. 2018. "'Leave No One Behind' Index 2018." ODI Briefing Note. London: ODI. www.odi.org/sites/odi.org.uk/files/resource-documents/12304.pdf.

Maskrey, Andrew. 2011. "Revisiting Community-Based Disaster Risk Management." *Environmental Hazards* 10, no. 1: 42–52.

Ministry of Home Affairs. 2015. *National Progress Report on the Implementation of the Hyogo Framework for Action (2013–2015)*. Report, June 7. Ministry of Home Affairs, Government of Nepal. www.preventionweb.net/files/41755_NPL_NationalHFAprogress_2013-15.pdf.

Ministry of Local Development. 2011. "Local Disaster Risk Management Planning (LDRMP) Guideline, 2011." *Policy Report*, Unofficial Translation. Kathmandu, Nepal: Government of Nepal, Ministry of Local Development.

Moser, Caroline O. N. 1989. "Gender Planning in the Third World: Meeting Practical and Strategic Gender Needs." *World Development* 17, no. 11: 1799–825.

Mueller-Hirth, Natascha. 2012. "If You Don't Count, You Don't Count: Monitoring and Evaluation in South African NGOs." *Development and Change* 43, no. 3: 649–70.

Muni, S. D. 2015. "Nepal's New Constitution. Towards Progress or Chaos?" *Economic and Political Weekly* 50, no. 40: 15–19. www.epw.in/journal/2015/40/commentary/nepals-new-constitution.html.

Nagoda, Sigrid. 2015. "New Discourses but Same Old Development Approaches? Climate Change Adaptation Policies, Chronic Food Insecurity and Development Interventions in Northwestern Nepal." *Global Environmental Change* 35: 570–79.

Nagoda, Sigrid, and Siri Erikson. 2015. "The Role of Local Power Relations in Household Vulnerability to Climate Change in Humla, Nepal." In *Climate Change, Adaptation and Development: Transforming Paradigms and Practices*, edited by Tor Håkon Inderberg, Siri Erikson, Karen O'Brien, and Linda Sygna, 200–18. London: Routledge.

Nagoda, Sigrid, and Andrea Nightingale. 2017. "Participation and Power in Climate Change Adaptation Policies: Vulnerability in Food Security Programs in Nepal." *World Development* 100: 85–93.

National Planning Commission. 2017. *Nepal Flood 2017. Post Flood Recovery Needs Assessment*. Evaluation Report. National Planning Commission, Government of Nepal. www.npc.gov.np/images/category/PFRNA_Report_Final.pdf.

National Strategy for Disaster Risk Management in Nepal (NSDRM). 2008. Final Draft, March 2008. National Society for Earthquake Technology, UNDP-Nepal, and European Commission for Humanitarian Aid. http://www.nrcs.org/sites/default/files/pro-doc/NSDRM%20Nepal.pdf

Nightingale, Andrea J. 2002. "Participating or Just Sitting in? The Dynamics of Gender and Caste in Community Forestry." *Journal of Forest and Livelihood* 2, no. 1: 17–24.

Nightingale, Andrea J. 2005. "'The Experts Taught Us All We Know': Professionalisation and Knowledge in Nepalese Community Forestry." *Antipode* 37, no. 3: 581–604.

Nightingale, Andrea. 2015a. "A Socionature Approach to Adaptation: Political Transition, Intersectionality, and Climate Change Programmes in Nepal." In *Climate Change, Adaptation and Development: Transforming Paradigms and Practices*, edited by Tor Håkon Inderberg, Siri Erikson, Karen O'Brien, and Linda Sygna, 219–34. London: Routledge.

Nightingale, Andrea. 2015b. "Challenging the Romance with Resilience: Communities, Scale and Climate Change." In *Practising Feminist Political Ecologies. Moving Beyond the "Green Economy"*, edited by Wendy Harcourt and Ingrid L. Nelson, 182–208. London: Zed Books.

Nightingale, Andrea J. and Katharine Rankin. 2014. "Political Transformations: Collaborative Feminist Scholarship in Nepal." *Himalaya* 34, no. 1, Article 15: 105–17.

O'Brien, Karen, Siri Eriksen, Lynn P. Nygaard, and Ane Schjolden. 2007. "Why Different Interpretations of Vulnerability Matter in Climate Change Discourses." *Climate Policy* 7, no. 1: 73–88.

Ojha, Hemant R., Sharad Ghimire, Adam Pain, Andrea Nightingale, Dil B. Khatri, and Hari Dhungana. 2016. "Policy Without Politics: Technocratic Control of Climate Change Adaptation Policy Making in Nepal." *Climate Policy* 16, no. 4: 415–33.

Oven, Katie J., Shailendra Sigdel, Shubheksha Rana, Ben Wisner, Ajoy Datta, Samantha Jones, and Alex Densmore. 2017. "Review of the Nine Minimum Characteristics of a Disaster Resilient Community in Nepal." DFID Research Report. Research Report for the Ministry of Federal Affairs and Local Development, Government of Nepal, and the Nepal Risk Reduction Consortium. Durham University. www.gov.uk/dfid-research-outputs/review-of-the-nine-minimum-characteristics-of-a-disaster-resilient-community-in-nepal-research-report.

Oven, K. (2019). Natural hazards governance in Nepal. In *Oxford Research Encyclopedia of Natural Hazard Science*. Oxford University Press. DOI: http://dx.doi.org/10.1093/acrefore/9780199389407.013.312

Oxfam. 2005. "The Tsunami's Impacts on Women." Briefing Note. Oxford: Oxfam. https://policy-practice.oxfam.org.uk/publications/the-tsunamis-impact-on-women-115038.

Pant, Bijan, and Kay Standing. (2011). "Citizenship Rights and Women's Roles in Development in Post-Conflict Nepal." *Gender and Development* 19, no. 3: 409–21.

Panthi, Kamal. 2018. "Sonahas Want to Be Categorised Under Indigenous or Janajati Group." *Kathmandu Post*, February 9. http://kathmandupost.ekantipur.com/printedition/news/2018-02-09/sonahas-want-to-be-categorised-under-indigenous-or-janajati-group.html.

Pelling, Mark. 2007. "Learning from Others: The Scope and Challenges for Participatory Disaster Risk Assessment." *Disasters* 31, no. 4: 373–85.

Petal, Marla, Sushil Baral, Santosh Giri, Sumedha Rajbanshi, Subash Gajurel, Rebekah Paci-Green, Bishnu Pandey, and Kimberley Shoaf. 2017. *Causes of Death and Injuries in the 2015 Gorkha (Nepal) Earthquake*. Research Report. Kathmandu: Save the Children. https://resourcecentre.savethechildren.net/sites/default/files/documents/causes_of_deaths_and_injuries_nepal_earthquake_report_eng_2017.pdf.

Pradhan, Ajit, Pradeep Poudel, Deborah Thomas, and Sarah Barnett. 2011. *A Review of the Evidence: Suicide Among Women in Nepal*. Report. Options Consultancy Ltd.

www.medbox.org/countries/a-review-of-the-evidence-suicide-among-women-in-nepal/preview.

Rankin, Katharine N., Andrea J. Nightingale, Pushpa Hamal, and Tulasi S. Sigdel. 2018. "Roads of Change: Political Transition and State Formation in Nepal's Agrarian Districts." *The Journal of Peasant Studies* 45, no. 2: 280–99.

Ribot, Jesse. 2014. "Cause and Response: Vulnerability and Climate in the Anthropocene." *Journal of Peasant Studies* 41: 667–705.

Richardson, Diane, Nina Laurie, Meena Poudel, and Janet Townsend. 2016. "Women and Citizenship in Post-Trafficking: The Case of Nepal." *The Sociological Review* 64: 329–48.

Rigg, Jonathan, and Katie Oven. 2015. "Building Liberal Resilience? A Critical Review from Developing Rural Asia." *Global Environmental Change* 32: 175–86.

Rigg, Jonathan, Katie J. Oven, Gopi K. Basyal, and Richa Lamichhane. 2016. "Between a Rock and a Hard Place: Vulnerability and Precarity in Rural Nepal." *Geoforum* 76: 63–74.

Rigg, Jonathan, and Lisa R. Mason. 2018. "Five Dimensions of Climate Change Reductionism." *Nature Climate Change* 8: 1027–32.

Shneiderman, Sara. 2015. "Regionalism, Mobility, and "the Village" as a Set of Social Relations: Himalayan Reflections on a South Asian Theme." *Critique of Anthropology* 35 (3): 318–337.

Sugden, Fraser, Niki Maskey, Floriane Clement, Vidya Ramesh, Anil Philip, and Ashok Rai. 2014. "Agrarian Stress and Climate Change in the Eastern Gangetic Plains: Gendered Vulnerability on a Stratified Social Formation." *Global Environmental Change* 29: 258–69.

Sultana, Farhana. 2013. "Gendering Climate Change: Geographical Insights." *The Professional Geographer* 66, no. 3: 372–81.

Sunam, Ramesh. 2014. "Marginalised Dalits in International Labour Migration: Reconfiguring Economic and Social Relations in Nepal." *Journal of Ethnic and Migration Studies* 40, no. 12: 2013–48.

Tamang, Seira. "Democratic Deceits: Embedding the Constitution and Elections in Arrested Political and Social Change in Post-Earthquake Nepal." *Keynote Lecture*, 16th British – Nepal Academic Council Study Days, Durham University, April 16–17, 2018.

Thapaliya, Suresh, Pawan Sharma, and Kapil Upadhyaya. 2018. "Suicide and Self Harm in Nepal: A Scoping Review." *Asian Journal of Psychiatry* 32: 20–26.

Titz, Alexandra, Terry Cannon, and Fred Kruger. 2018. "Uncovering 'Community': Challenging an Elusive Concept in Development and Disaster Related Work." *Societies* 8, no. 3: 71–99.

Twigg, John. 2015. *Disaster Risk Reduction*. Good Practice Review 9. Evaluation Report. London: Humanitarian Policy Group, Overseas Development Institute (ODI). http://goodpracticereview.org/wp-content/uploads/2015/10/GPR-9-web-string-1.pdf.

United Nations Development Programme (UNDP). 2016. *Human Development Report 2016. Human Development for Everyone*. New York: UNDP. http://hdr.undp.org/sites/default/files/2016_human_development_report.pdf

United Nations Development Programme (UNDP). 2014. *Beyond Geography: Unlocking Human Potential*. Nepal Human Development Report 2014. Country Report. UNDP. www.hdr.undp.org/sites/default/files/nepal_nhdr_2014-final.pdf.

Wisner, Ben. 2006. "Self-Assessment of Coping Capacity: Participatory, Proactive and Qualitative Engagement of Communities in Their Own Risk Management." In *Measuring Vulnerability to Natural Hazards – Towards Disaster Resilient Societies*, edited by Jorn Birkmann, 316–28. Hong Kong: United Nations University Press.

Wisner, Ben, Greg Berger, and Jean-Christophe Gaillard. (2017) "We've Seen the Future, and It's Very Diverse: Beyond Gender and Disaster in West Hollywood, California." *Gender, Place and Culture* 24, no. 1: 27–36.

Yadav, Punam. 2016. *Social Transformation in Post-Conflict Nepal. A Gender Perspective*. Oxon: Routledge.

9 Gendered and ungendered bodies in the Tsunami

Experiences and ontological vulnerability in Southern Thailand

Claudia Merli

Introduction[1]

Focus on gender across academic disciplines and among agencies working on the ground in the field of disaster risk reduction, emergency intervention, risk management and humanitarian aid has gained momentum since the mid-1990s. International organisations, local and regional NGOs can tap into a copious literature tackling how managers, aid workers and emergency intervention planners should design and enact their projects within a frame, taking gender perspectives into account (see for example MDF-JRF[2] 2012; Oxfam[3] 2015; Wold Bank 2012a, 2012b, 2012c). This existing literature became the target of specific requests of information in the aftermath of the 2004 Indian Ocean tsunami when "NGOs, IGOs, and INGOs posted repeated e-mail requests for 'checklists and guidelines' for assisting women and children" (Enarson, Fothergill, and Peek 2007, 145).

Much of this long-awaited consideration addresses gendered violence and sexual violence against women and girls in groups and populations displaced by large-scale disasters (Enarson and Dhar Chakrabarti 2009; Felten-Biermann 2006). Gender is often read as synonymous to women, leaving much to desire in terms of deconstructing the homogeneity attributed to these categories into a matter of gender inequalities (Enarson, Fothergill, and Peek 2007, 141). Some interesting work focuses on patterns of land tenure and how post-disaster claims on land are affected by pre-existing gender disparities (Kusakabe, Shrestha, and Veena 2015), but also how land use is affected by disparities between and within nations and regions (Enarson, Fothergill, and Peek 2007, 143).

Women are still portrayed almost exclusively as victims of disasters (following the 'victim model'), attributing to them an inherent vulnerability with little scope for examining determinants such as poor maternal health (Enarson, Fothergill, and Peek 2007). In contrast to such simplistic approaches, women often play very active roles in grassroots level projects and through NGOs. They use these possibilities to activate their social networks in post-disaster situations, thus transforming disasters into possibilities of empowerment (Enarson, Fothergill, and Peek 2007, 138, 140, 142).

Gender features prominently as a category of analysis with regards to violence, trafficking and other forms of exploitation, targeting primarily children and women in times of crisis. However, the bodies and the embodied experiences of women and men in disasters, both those affected by the disaster and those involved in aid interventions, remain at best opaque if not completely invisible. This absence of bodies has been highlighted as a gap to be filled and an important direction for future disaster research, including the attention to aging bodies and bodies with disabilities (Enarson, Fothergill, and Peek 2007, 141).

This chapter focuses on the embodied experiences of women and men during and in the aftermath of the 2004 Indian Ocean tsunami. It draws on first-hand ethnographic material collected in Thailand's southernmost western province from December 2004 to March 2005 and on subsequent returns to the field to analyse how people lived the catastrophe through theirs and others' bodies.

Living through vulnerability

A tsunami is usually described as a "seismic-generated ocean surge that inundates coastal areas". As the waves become concentrated with a possibility of reaching a height of 30–40 metres, they especially ravage the ria coastlines with their "numerous estuaries along an embayed coast" (Sasaki and Yamakawa 2004, 170–1, 176). But tsunamis can also be triggered by a series of geological events, as in the recent Sunda Straits tsunami originated by a submarine landslide at the southern flank of Anak Krakatau volcano, on 23 December 2018. They can occur in alpine lakes, as in the Vajont dam disaster of 9 October 1963 in Northern Italy, triggered by the collapse of a large section of Mount Toc into the hydroelectric reservoir. The devastation to human life and the environment is similar, changing the physical geography as well as the social world of the survivors.

The megathrust earthquake (with the epicentre 30 km under the seabed off the west coast of Northern Sumatra in Banda Aceh, Indonesia), which on the early morning of 26 December 2004 registered a magnitude of 9.0,[4] wreaked devastation across the Indian Ocean with multiple tsunami waves. It has been considered one of the greatest disasters triggered by natural hazards, and concomitant humanitarian aid efforts, in contemporary history. The most affected region surrounding the Indian Ocean comprised Indonesia, Sri Lanka, India and Thailand. The estimates give an approximate number of at least 250,000 victims considered to have lost their lives or being presumed dead, including a large number of European visitors holidaying for the Christmas and New Year vacations, mainly in Thailand and Sri Lanka. Moreover, between 1.4 and 1.7 million people were displaced in 14 countries (Greenhough, Jazeel, and Massey 2005; Morgan et al. 2006; Telford, Cosgrave, and Houghton 2006; WHO 2014).[5]

The presence of many European tourists among the victims and the displaced brought forth, at least initially, the erasure of separate categories such as 'them' and 'us', 'Westerners' and 'locals' in a shared condition of devastation. The exposure to 'natural hazard' attributed to certain areas and others in the course of history, sanitation and colonialism, has been used to justify particular forms of relief operations in disaster-prone regions where social and economic conditions have made people vulnerable. This pattern continued into the 1970s and increasingly the 1990s (Bankoff 2003, 14–17; Gaillard 2019). Thus, a society's condition is attached to specific areas and geographic places as being more dangerous than others (Bankoff 2003, 11–12), where vulnerability has become a result of specific physical, human and psychological resources, often considered calculable and quantifiable (Handmer 2004, 88; Kelman 2018, 284). This accepted vulnerability paradigm of particular local contexts seemed to vanish in the 2004 tsunami's sudden commonality of experience, loss and grief, of local populations and Westerners alike. As Anthony Oliver-Smith and Susanna Hoffman have remarked, there is the spectre of a generalised vulnerability appearing when catastrophes happen (2002, 17). The 2004 tsunami put the Global North into a shared vulnerability in a natural disaster rarely envisioned in a technocratic society. The discourse of vulnerability implies that there are measures a society can take to counter the effects of natural physical phenomena, like "improving scientific prediction" (Bankoff 2003, 11).

As on other occasions of crisis, examples of discrimination in distribution of assistance (reproducing pre-existing divisions of inequality along ethnic, class, or gender lines), and stories of human rights violations across the Indian Ocean countries, emerged (cf. Kälin 2005, 10). Local epistemologies are often obliterated in favour of a scientific and policy-oriented discourse or paradigm. To recover local epistemologies and understandings (including cosmologies and theodicies) following a disaster is instead a decolonising move to recapture the local context, the local capacity to make sense of catastrophic events, and overcome trauma (see Handmer 2004, 99; Gaillard 2019; Gaillard and Texier 2010; Merli 2010, 2012). This chapter attempts to recapture discourses about and through the bodies of those affected.

Note on methods and material

In December 2004, I was on a stint of fieldwork (November 2004 to March 2005) in Satun province, on Southern Thailand's western coast, bordering Malaysia. My anthropological research, which had started in 2003, focused on traditional midwifery, reproductive health and postpartum rituals, with a strong focus on bodily practices related to the life cycle.[6] I continued through 2007, with shorter sojourns from 2009 until today. The population in the area is mostly Thai speaking with reported 10 percent also speaking Malay, while around 70 percent of the population is Muslim (NSO 2001, 2010). Compared to other southern Thai provinces that were affected by the tsunami (Phuket, Krabi and Phangnga), the Satun province did not report many casualties (all six victims were in Tung Wa district in the northern part of the province), but coastal villages were damaged, fishing boats and equipment wrecked, cattle and poultry lost, agricultural land and fish farms devastated and the freshwater wells were contaminated by seawater and mud.

Several of my informants in the Tung Wa district had suffered losses to their properties, their means of livelihood, or both. Very soon after the tsunami, some visual material (printed and video) emerged on the market. This material carried religious interpretations, with references to Quranic descriptions of the *Qiyāmah* (the Resurrection, the end time), and projected the disaster as a punishment to sinners. A new dimension was thus added to the all-pervading subject of conversation, which impacted on people's perception of the event and of their own futures.

The material analysed in this chapter is based on data I collected in the Satun province from December 2004 to March 2005, and from March to June 2006, and again in 2009 and 2010–2011. It includes a prolonged involvement with the local community, participant observation, individual as well as focus group interviews. During the first three months following the tsunami, I recorded daily discussions and conversations with friends, informants and acquaintances. I also systematically collected news published in both English and Thai, and video material circulating on the local market, in the form of VCDs. I also collected 44 semi-structured interviews with local fishermen and people residing in coastal villages directly affected by the tsunami, with prominent Muslim and Buddhist religious representatives, and with government officials at both provincial and district levels. I personally conducted all the interviews, aided by my long-term assistant for translations. Theoretically, my analysis is complemented with scholarly articles based on research in other areas of Thailand and the Indian Ocean.

The bodies of others

On 25 and 26 December, I was supposed to have been on Koh Bulon Island camping with local friends, but because of the distance from the mainland, the long ferry connections and a previously scheduled interview with an informant 150 km away from Satun province on the 27 December, I decided not to travel to the island. Later, I learned that my friends, and other tourists camping or staying at resorts on Koh Bulon, had been running up the hills to save their lives when the tsunami came. For several days, communication with the islands off Satun was difficult; mobile telephone connections were silent and provincial authorities scrambled to evacuate people from the islands as news stories of other possible incoming waves started to emerge.[7] In the following days, information circulated of the many evacuees and injured transferred to different hospitals in the south and researchers and scholars in the region kept calling each other. One of these colleagues alerted me to the fact that medical/public health authorities were looking for people who could speak Swedish, Norwegian and other languages beyond English to reach hospitals where tourists had been transported.[8]

On 29 and 30 December I was at Songkhla Nakarin Hospital in Hat Yai where some injured tourists, mainly from the Khao Lak area (in Phangnga province), were being treated. I remember, as if it were today, the body of a young Scandinavian man who was caught by the wave while snorkelling and tumbled over the coral reefs; his body was covered with infected cuts, the skin was swollen and yellowed because of the iodine tincture, the coral cuts were black and brown with red edges, with the white bedsheet only partially covering his body to let the lesions air. I wondered what would have happened to him, had he not been an athletic strong man. His wife remained uninjured, as she had stayed in the hotel that morning, and was now in the hospital, looking after him. She had no clothes with her (we were told by the nurses), wearing only a pyjama or cotton shorts and shirt and a pair of rubber sandals on her feet and no personal belongings. Many of those of us who had come to the hospital to provide translation assistance went to the nearest shopping outlet to buy toothpaste, toothbrushes, antiperspirants, soap bars, combs, razors, shaving cream and something to read in English for the evacuees.

Theodicies and religious discourses advanced by Islamic religious authorities in the region (extending beyond southern Thailand to Indonesia and Malaysia for example) responded to different interpretations of the event: as divine signs, as punishment of individuals and collectives, as a wake-up call for the whole of humanity, as a vengeful response by a devastated nature. Buddhist interpretations also resorted to some of these views albeit with a specific reference to, for example, *karma* (see Merli 2010 for a deeper analysis of these discourses). Often such sins were associated with immoral behaviour, sexual sin, tourism and the entertainment industry and human bodies. An important connection was drawn between fish and humans, their relations, the dependence of fishermen on fish, the subversion of natural marine life, and their bodies.

One prominent body was attributed to a mythical creature. A local Buddhist abbot told me a story that he had heard from the elders when he was a child, explaining the existence of a mythical creature living in the oceanic abysses, lying still on the seabed; a gigantic fish called Anon (อานนท์), whose sudden movements and turning around would set the surrounding waters in motion, causing colossal waves and even earthquakes. In popular Buddhism, Anon is one of seven giant fish of the Himmaphan (Himavanta or Himavan forest) who carry the weight of the world and encircle Mount Meru. Anon is identified as having a part in the end of the world, following a timeline when humans do not follow or do not know the precepts for conducting a moral life, and are therefore sinful (Ruenruthai 2014).

A different and more intimate relation between fish and human bodies turned out to impact the aftermath of the tsunami via a tangible fear of unintentional anthropophagy. Since fish and seafood are the Muslim local population's main sources of animal protein and fishing its most important economic activity for the coastal communities, the perceived vengeance of Nature or God undermined the basic conditions for survival in the area, and the local daily fresh fish market. People stopped eating fish for fear of indirectly ingesting other 'people's bodies' and, as a consequence, the local fish and seafood industry collapsed. Despite a fall in prices, customers tried to outsource other products. Small catch and fish that customers could ensure were from fish farms as well as freshwater fish were still popular, but larger fish from the sea were left unsold on the stalls. I followed the local approach to the problem and suspended eating fish and seafood and was thus in a sense joining a "community of abstinence" (see Fausto 2007, 502–3, and his reference to the debate on this concept). As my female friend Nong[9] told me in Malay: "People used to eat fish, but now fish eat people";[10] the eaten had transformed into man-eaters. The situation was experienced as a reversal of both the natural and cultural orders. The same understanding extended to squids; stories surfaced of corpses retrieved from the sea with squids tenaciously attached to them, with beaks deeply rooted in the human flesh, especially the ears. Fishermen would confirm that these stories were not myths by recalling previous occasions on which they had witnessed the retrieval of bodies at sea with squids partially eating them.[11] Potentially, eating fish became an act of indirect cannibalism or anthropophagy.[12]

The suspicion towards fish extended beyond the coastal locations, which the Department of Fisheries and the Public Health Ministry took to the newspapers with an article in *The Bangkok Post* on 8 February entitled 'Fancy fish?'. Officials were particularly concerned with the economic outcome of this aversion to fish and tried to persuade people that fish would not feast on human corpses. A campaign was designed as the first of a series, "Eating Fish and Seafood" to educate people that the gruesome pictures of rotten bodies broadcasted on TV should not be connected to fish. A spokesman from Department of Fisheries declared:

> Let me confirm that fish won't eat rotten meat. They eat plankton. Besides, corpses were found on the shore, not in the sea. We have tried to clarify this with the public. As well, the Public Health Ministry has continuously done random testing and found that fish is free of disease and safe to eat.
>
> (Valarak Chaiyatap 2005)

The food crisis was aggravated by news about several outbreaks of bird flu in different locations in Thailand, making poultry an additional consumption entailing risk. In Thailand, environmental encroachment – which is understood as a form of cosmological invasion – and overfishing, as well as stories hinting at certain natural and cultural subversions at the narrative spectre of anthropophagy, testify to the necessity of re-establishing respect for the boundaries, if not between categorical nature and culture, then between the modalities of relations between humans and non-humans.[13] In her analysis of eating and feeding, Marilyn Strathern reviews a number of studies on eating[14] (several by scholars of Amazonian perspectivism) and the differentiation made between 'persons' (that do not eat one another) and other 'agents' (who do eat one another) which in turn is related to the delineation between cannibalism (eating others as subjects) and anthropophagy (eating others as objects) (Strathern 2012, 11). Strathern refers to Carlos Fausto's definition of anthropophagy as eating that requires "desubjectifying the prey" (Fausto 2007, 504, also cited in Strathern 2012, 4). Although regionally and thematically distant, I consider these perspectivistic analyses very significant for reflecting on the endangered definition of subjectivity, objectivity and personhood of the post-tsunami's potential fish eaters; of eating not only as an action but as a perspective, and something that opens up to the risk of being fed (Strathern 2012, 6).

The negative perception associated with fish did not pertain only to the nutritional and cosmological understanding of locals but affected the livelihood of all those whose primary occupation and income were related to fishing. I had the possibility to interview several fishermen in the Tung Wa district, among these the elderly indigenous midwife Hodya and her family.[15] They told me of how fishing had become an anxious activity performed in a hurry before going back home, but without an alternative source of income it was not possible to renounce it. Similarly to what was happening on the provincial capital's market, Hodya's family also selected the fish to be eaten (from their own fishing and from the local market); small catch was considered safe because of its closeness to the shore,[16] while squids were abhorred. Hodya's daughter, Prani, explained her predicament of having to sell the daily catch.

> There is a lot of fish now in the sea. When I go fishing I make a big catch but cannot sell it, [so] I dry it with salt and send it to Tung Wa. There are no customers for fish. After the giant wave came it took fish from another sea here, but you cannot sell it or eat it. It is not like in the past when it was difficult to find enough [fish], now there is plenty of it. Before the giant wave I could earn 200–300 Bahts[17] per day, now I catch fish for 1,000 or more but cannot sell it. It would be good to sell now . . . but one cannot. That is not good because in my heart I only think about the giant wave coming and I am scared. When I go to sea I work quickly, fish and come back home quickly, I don't want to stay on the beach.

This sudden abundance of fish, that can be neither sold nor consumed, was interpreted differently by various fishermen in Satun, but often with a common understanding that it was originating from the moving of sea areas – the arrival of water from 'elsewhere'. Later reports revealed that the force of the water had shattered local fish schools and fish farms' cages, setting the fish free. After an initial period of abundance, fishermen in Langu district could not find any fish for a year. The description of a changed relation with a disordered environment and a hurried dealing with it, also found expression via Prani's husband's testimony, Mohammad who also was a fisherman. During my visit, he described how he since the tsunami had not been able to take his boat out at sea. He had attempted a few times but whenever he reached the beach to start working, his body would turn cold at the view of the water and would start trembling. His mother-in-law asked me if I knew of any job opportunities for him in Satun town stressing; "he would adapt to any job, he is a hard-working man but is not going fishing anymore".

Other bodies were put in direct relation with the triggering of the disaster. Among local Muslims there was a general belief that Allah had sent the *tsunami* to clean up places considered 'dirty' in both a physical and a moral sense, removing sinners and rubbish. At Tan Yon Lanai, a coastal location in the Tung Wa district, people described the wave as being so high that it was impossible to see the islands on the horizon. A small seaside platform restaurant had been recently built after the tsunami. The restaurant's owner described how the tsunami reached Tan Yon Lanai:

> The wave reached the street, it cleaned the rubbish from the beach, all the dirt was carried to the street and now the beach is beautiful, the water is crystal clear. I had the idea of building a small resort with bungalows here but now I think that it would encourage *zina* ['extra-marital intercourse']. And then it could be that Allah would want to destroy everything just as in Khao Lak and Phuket.

The fact that the *tsunami*, which was seen as God's punishment, struck hard at Khao Lak and Phuket, confirmed the general conviction that these places were 'dirty', i.e., places where people committed sin or morally polluted. Local people associated sexual sin with mass tourism reaching locations such as Phuket, Krabi and Phi Phi, which Westerners (especially men) are thought to visit for the purpose of satisfying their sexual lusts, often by accessing local provisions of sexual services.[18] The owner of the restaurant in Tan Yon Lanai was afraid that he would be similarly punished had he provided facilities that could encourage sexual sins. Tan Yon Lanai is not an international tourist resort, but the locals certainly appreciate the beautiful scenery of the place. Basic tourist facilities such as tables, chairs, public toilets and showers were provided on nearby Rawai Beach. The problem for the locals was to take full economic advantage of this natural beauty by expanding local tourism without letting the situation get out of hand and encouraging licentiousness.

The subject of corrupted morality was central to many interpretations given by local interviewees. The theme of 'cleansing dirty places' recurred in many interviews and conversations. It is a discourse on moral pollution that can be compared to other ideas of physical pollution in Islam, and to the means used to regain purity. "In brief, the purity of the soul is derived from physical purity which, if defiled, must be purified by 'absolute water'" (Khuri 2001, 30). If this elaboration applies to the case of tsunami, the enormous amount of absolute water released over Sri Lanka, Southeast Asia and the Andaman coast, interpreted by local people in Satun as 'cleaning dirty places', underlies the twofold dimension of this purification. A rubber plantation owner from the hilly Tung Maprang area nearer the provincial capital was one of those who was convinced that the catastrophe was the result of people's misconduct; "Allah wants to do something, to clean these areas, in Islam it is called *balaq*,[19] and even the good [people] die. But dirty areas were hit, like Aceh". Local discourses, resorting to what I called 'context-bound Islamic theodicies' (Merli 2010), for explaining the tsunami were complex and implied mutual reflections on innocence of those who perished in the disaster. A large number of readily available video material at the local market stalls explored these religious interpretations. While some of these videos were locally produced, others consisted of videos of destructions across the Indian Ocean with commentaries in Arabic.

Perceptive, inscribed and absent bodies

The aftermath of the tsunami was also marked by the presence of deeply inscribed human and non-human bodies as well as the very absence of bodies. The body's condition of openness to the world found expression in stories concerning enhanced bodily perceptions of the surrounding environment. People made sense of what they had noticed in terms of unusual bodily perceptions in the days and hours preceding the tsunami, as well as after the disaster. They also recognised signs of inscriptions in the dead bodies and in the landscape, reconnecting the signs to the activity of 'unnatural' or supernatural agency.

On the day preceding the tsunami, fishermen had noticed several events that they could not find an explanation for. Their stories specified animals' peculiar behaviours, debris floating in an unnatural vertical position in the sea, snakes swimming away from islands and the presence of large numbers of insects by the shores. All these signs went undetected before the tsunami, and were mainly thought of as not belonging to their usual experience of nature. The sea felt threatening to many and fishermen did not trust their own capacity to read its signs and appearance. Ordinary sounds and views had lost their familiarity and the course of interaction with nature had been altered. Many local people interpreted the changes as expressions of supernatural agency. The inscription of this agency in the body of water in the Indian Ocean was exemplified by a satellite image that started circulating immediately after the tsunami, picturing the receding tsunami wave in Kalutara (south-western Sri Lanka). The shape of the wave was readily interpreted as replicating the Arabic for Allah in the billows. The picture and the associated script were often seen side by side in Satun's cafes and other public places after the tsunami. Several Muslim communities in Thailand, Malaysia, Indonesia and elsewhere considered this picture as a specific warning and/or punishment meant for the Muslims (Sugirtharajah 2007, 126). The retreating and menacing tsunami's circular waves in the Phangnga province were equally considered unnatural, as shown in a number of videos. Likewise, fishermen described how the direction and the rhythm of the waves in the Tung Wa district in Satun differed from their usual forms, and how they travelled from a different direction. The Malaysian newspaper *Berita Harian*'s supplement *Dimensi* published on 9 October 2005 a series of colourful pictures of natural phenomena on which the name of Allah was inscribed. Among these the most noteworthy were an eggshell, the inner part of a tomato, the tsunami waves in Kalutara, clouds, a fish and a beehive. The authors of the article explain these phenomena as referring to several surah, among these Ar-Rahmān (55, 3–6), which indicates that all Creation adores Allah.

One of the characteristics noted by many after the tsunami was the presence of a constant and intense wind blowing for weeks. When talking to a group of men (mostly fishermen and the local Imam) during a return visit to a coastal fishing community in Satun in 2011, one fisherman said that the destruction, the damaged coral reefs, was a 'human effect' that will need about seven years to recover. As if it was a body of a sick patient, he said that the waves and the wind will treat (รักษา) the sea and make it clean again: "it is like with people, if the doctor does not treat them they might die". Another term these Malay-speaking fishermen used with reference to the sea was '*nyanyok*',[20] a term usually employed to describe erratic behaviour in a very old person, due to senility. The local imam explained how a *nyanyok* person "says he has not eaten whereas he has, and wants to eat again because he forgot he has already eaten. This is the time of the end of the sea". I asked Tok Imam if he meant that the sea can die. He replied affirmatively "it is too old now, older than anything else". A local fisherman, who was present at the meeting, added that the sea is like a deep beautiful woman, in the sense that all the water in the world is one person, "one woman which the Buddhist call Mae

Thorani", Mother Earth (แม่ธรณี). He continued, saying that fishing boats with their gears injure the sea; for this reason, "God took away the soul [วิญญาณ] of this woman so that the sea des not suffer pain anymore". The local Imam intervened to amend this (unorthodox) view and insisted that the tsunami was a punishment (*balaq*, annihilation) and put it in relation to the increasing occurrence of serious floods in Hat Yai, which he then associated with the rise in prostitution. He referred to women working in the sex industry as *ying kali* (หญิงกาลี), a term that is very different from the one used locally to indicate a prostitute, *ying borikan* (หญิงบริการ, service woman). Whereas the latter denotes a woman who engages in prostitution out of economic necessity, to support her family, *ying kali* is used to indicate a woman who has an evil side, whose heart is dark. The term also refers to the Hindu goddess Kali, Phra Mae Kali (พระแม่กาลี). Women's bodies are repository and givers of lives, like the sea-Mother Earth, now a senile elderly woman who does not know how to behave anymore, or represent negative sexual bodies of evil sinners.

If these two antithetical representations of women were provided by a group of local men, local women's bodies were telling other stories. An elderly woman said that on the day of the earthquake and the *tsunami*, and for several days afterwards, she could feel the trembling in her limbs, as if her body could register the earthquake like a seismograph. Other people told me of similar bodily sensations. A midwife described how on the day of the *tsunami*, she felt a tremor while she was praying *asri* (the prayer between 15.40 and 18.00) later in the afternoon, "It was nearly six o'clock and I was praying again when Amin arrived home and said 'Mak, Mak must go! The big wave is coming again' and I ran away". The same thing happened to her a few days later when, according to news reports, there were a series of earthquakes in Iran that were (supposedly) totally unrelated to the events in Indonesia. In her narrative, her body became an echoing chamber for perceiving and registering geological events very remote from her place of residence. Other bodies remained invisible, or not taken into consideration, in the days and weeks following the tsunami, when several local and national NGOs operated to bring relief to shattered communities.

Some of the interventions aimed at supporting recovering communities in the Satun area after the tsunami highlighted the implementation of commendable ideas without, in some cases, much foresight. Two of these attempts were evident in the coastal area of the Tung Wa district to which I returned for fieldwork in 2006, where Tok Cik (then 84-year-old), one of the oldest midwives I had met and interviewed regularly in 2004, was living. The tsunami had badly damaged her traditional one-room house on stilts, with a floor of irregular wooden planks and walls made of either woven pandanus leaves or corrugated iron sheets, built just a few metres from the shore. The house was still standing but the force of the water had bent the entire house and the stilts were angling inland, making it an even more precarious construction. She had been living in this house together with her son, a man in his forties, born without legs. I had met them shortly after the tsunami, in February 2005, and they were still occupying the damaged wooden house. But they had eventually moved into a nearby construction at the ground level, built with a shallow foundation and concrete bricks, constituting one room with a simple latrine at the back. The post-tsunami intervention of reconstruction included plans for economic support in the form of a fishpond, simply excavated next to the new construction. The dug-out pond did not have any water supply and required continuous filling by way of a bucket as well as cleaning of the proliferating algae. The small house did not have any water supply either. Considering the limited mobility of the elderly midwife and her son, the project was doomed to be a failure. Shortly after its inception, the fish pond was a squared murky hole filled with rotting fish, covered in a green plastic tarpaulin net. Tok Cik complained about the hard work it took, and the cleaning afterwards. When I returned again in December 2010, only a couple of poles remained standing of the old wooden house. Tok Cik had passed away at the beginning of the same year and her son was living alone in the small concrete house, in precarious hygienic conditions. We again talked about the absurdity of that fish pond.

Other bodies were inscribed indelibly by the tsunami. Human bodies of victims recovered after a few days were all equally disfigured, with features marked by blackness or extreme whiteness, swollenness and faces with protruding eyes and tongues. Ethnic characteristics were erased from the bodies' surface especially when the putrefaction process had turned into liquefaction. Only approximate age and growth together with sexual traits could be used as primary signs of visual differentiation. Images broadcasted in Thai news or available in other visual materials (photos, VCDs) compelled the horror of a homogeneous disfiguring death. Apart from the biological process of decomposition and chemical changes, leading to the cells' loss of 'structural integrity' (Hart and Timmermans 2012, 231), the lack of structural integrity extended to the social order by way of a disordered but inescapably visible death onto which it was extremely difficult to impose a semiotic order in culturally appropriate ways.

Identifying bodies

The treatment of human corpses in the aftermath of the 2004 tsunami was on a scale not previously seen for any single environmental disaster. One of the greatest challenges pertained to the formal identification of victims and, when possible, the performance of appropriate funerary rituals. Because of the magnitude of the disaster and the large presence of Western tourists in the affected region, especially in Thailand, global aid and humanitarian intervention, as well as diplomatic and logistic support for the repatriation of the victims, were promptly initiated. In Thailand there were 5,935 reported deaths, of which approximately half were foreign tourists (Tun et al. 2005). The international scrutiny and the deployment of technical, as well as scientific expertise and support to the Thai forensic teams, also made the event into a complex bureaucratic and medical international collaboration. Whereas affected countries generally hit by disasters take care of the resulting dead themselves, 30 international forensic teams joined the effort (Tun et al. 2005; Scanlon 2008).

The sheer number of victims made individual inhumation an impossible task. Pictures reveal the implementation of mass burials in countries which did not receive the same attention by international forensic teams.[21] Just a few years after the Indian Ocean tsunami, when cyclone Nargis brought devastation to Myanmar in May 2008, the military used explosives to destroy corpses as a way to deal with the huge number of cadavers. Collective rituals replaced individual funerals (Brac de la Perrière 2010). While in Thailand great efforts were made to identify all victims, the practical impossibility of retrieving all bodies led to the organisation of funerals being conducted without bodies, a condition that Falk's Buddhist research participants described as experiencing an 'ambiguous loss' (Falk 2010). Although Falk explains in great detail how locals arranged to deal with loss via a combination of a 'counterfeit funeral' ending in a 'counterfeit cremation', the two rites were not counterfeit but were carried out 'as if' a body was present. This sometimes resulted in a substitute mock body being placed in its stead. The rituals were performed and carried out to their full meaning of separating the dead from the living. The risks of not effecting the separation is well illustrated by the stories of wandering spirits in the areas hit by the tsunami, where 'ghosts' or 'sightings' were reported in the Thai news.

In Myanmar and Thailand, the presence of Buddhist monks did not dispel the fear of the presence of malevolent ghosts. In Myanmar after the cyclone Nargis, people thought that these spirits (*kyap*) could not reincarnate (Brac de la Perrière 2010). In Thailand survivors 'made merit' for the deceased by being ordered as monks (women as *mae chii*).[22] Monks played an essential role in post-tsunami Thailand. They communicated between the two worlds and compared themselves to post offices for sending off a person's *winjan* (spirit or soul), thus aiding the recovery process (Falk 2010, 2015). In Myanmar, in comparison, people would shout the name of the popular organisation supporting the present Junta, 'Djankaïnyé', in order to frighten the *kyap* taking human semblances. Westerners who brought aid were also identified as ghosts to be appeased (Brac de la Perrière 2010). Ghosts were not the only unruly agencies of the post-tsunami context in Thailand. If Buddhist monks were appeasing spirits to leave the locations, Islamic clerics led prayers to avert further destructions, while forensic experts were carrying out their own struggles to make bodies stay, to govern the decomposition process. Thus:

> We can also read the desperate attempt to halt that destruction when it is the body that takes it further, as an expression of necropolitics. This post-mortem unruly agency can be construed as cadaveric counterconduct.
>
> (Merli and Buck 2015, 12)

The landscape of the tsunami's aftermath was governed by the processes of death and decomposition, a processual reality onto which procedures were arranged that reframed mass death as eminently bureaucratic and in which sophisticated forensic techniques seemed to reassure people with 'virtues of science' (Merli and Buck 2015, 12). The post-tsunami situation can be read as an illustration of 'thanatopolitics', a government of death, as essential to the body politic.

> In a sense, the central problematic of the body politic is this ontomedical revenant, the body that remains. In other words, the primary concern of the body politic is neither a theology of spirit not a physiology of organism, not a physics of mechanism, but rather a necrology of the corpse.
>
> (Thacker 2011, 151 cited in Merli and Buck 2015, 14)

Beyond the immense effort to reaffirm the cogency of individual bodies and individual identities, coherent and incoherent decomposed bodies were met with the attempt to establish a coherence of the body politic, a nationality of these corpses. This was done by re-drawing national boundaries into bureaucratic forms that needed to be filed and into the corpses themselves. Methods to contain the dissolving bodies, freezing them, analysing them and storing them in body bags, also speak of a desperate attempt to distinguish their respective body politic (Merli and Buck 2015, 16).

I have discussed earlier how the common disfigurement of victims' bodies was considered a sign of divine intervention. A stark visual opposition between the pre-tsunami life and this homogenous death can be exemplified by the display of victims' pictures and missing or presumed dead on walls set up at different locations where bodies were collected, as well as at sites where relatives would converge to find out information from the unflagging Thai and international volunteers managing databases and collecting information. The lined-up pictures of the missing with details of personal identity and eventual indication of particular signs can be contrasted with pictures of corpses, only a few steps away, blackened, with swollen bodies and faces, with no identification other than a label indicating the sex of the body in Thai language: '*phu chai*' (ผู้ชาย, man), '*phu ying*' (ผู้หญิง, woman). But the level of disfiguring in the latter group of pictures, which made relatives' task or recognising somebody basically impossible, also led to the inability of forensic experts to distinguish and separate Asians from Europeans in a reliable manner (Black 2009; Cohen 2009, 191). This inability resulted in considerable international acrimony that was sufficiently vociferous to actually threaten the continuation of the whole international DVI programme (Black 2009). The attempt to separate bodies in terms of 'race'/ethnicity to reflect the nationality of the victims, was itself problematic as several countries affected by the tsunami have mixed demographic populations as well as citizens of Asian descent (Australia for example).

The distinction between ethnicity and nationality in this context was a large biopolitical problem for the differential treatment, at least initially, afforded to these bodies. The bodies of 700–900 'Asian' victims were given temporary burials (a solution that would slow down their decomposition), with the intention to retrieve them later for identification purposes (Cohen 2009). However, accusations surfaced that these bodies were buried without correctly identifying them

(Black 2009). They were eventually all exhumed after a few days because of the possibility that Western bodies could have been erroneously classified and buried (Black 2009; Cohen 2009, 192).

This situation testified to the fact that whereas a common human vulnerability of the body was initially assumed as the hallmark of an unprecedented disaster equally affecting 'us' and 'them', it was not long before inequalities emerged. The possibility to access dental records for Western victims by forwarding them electronically was, for instance, a step that was largely unavailable for local people as dental records could not be obtained or went missing in the debris. These discrepancies in medical pre-mortem information were associated with a parallel 'hierarchy of grief', meaning that some bodies and their stories became more visible (certainly as accounted for in global media reporting), while others were omitted and in a certain sense even de-realised (Butler 2004, 32–4).

Amid the multiple evidential regimes and legal requirements, forensic experts needed to negotiate and respond to competing demands of the living and the dead in addition to the scientific demands pertaining to forensic identification proper (Crossland 2013). The legal requirement for identification took the place of the actual naming of a victim. One prominent example of this complexity was the fate of 83 identified bodies of Burmese citizens who could not be repatriated due to the lack of cooperation of Burmese authorities, with the embassy in Thailand refusing to certify that the bodies were in fact Burmese (Cohen 2009, 195).

In a mass disaster event that involves individuals of diverse nationalities, the management of dead bodies can become a process fraught with political and diplomatic complexities and filled with multiple kinds of loss: of personal and national identity, which can lead to the impossibility of repatriation.[23] Still months after the tsunami, bodies were retrieved in such a state of deterioration that it was initially difficult to establish if they belonged to humans or animals (Cohen 2009, 190). The decomposition process created a homogenous death population as a spectral new mass identity, where visual identification was at best unreliable or nearly impossible without the support of other forensic and medical means; it also removed the multiple and distinct identities of lived individuals.

The dissolution of bodies was accompanied by a parallel dissolution of social identities, despite an encumbering materiality of the corpses not allowing for social and ethnic differences to be recognisable. The term identification goes beyond the description of a technical forensic procedure to encompass also the process of identity-making itself (Jenkins 2004). As objects and practices after death inform the socialisation of the dead (Hockey and Draper 2005, 47), the post-tsunami process of identification also became a massive generator of sociality and relations. The stories about and of bodies can be read as expressions of a generalised ontological vulnerability.

Conclusions

By giving centrality to the body we can capture important dimensions of women's and men's experiences of living through a disaster. In the context of the 2004

Indian Ocean tsunami, understanding the place, agency and precarity attributed to human bodies in relation to other non-human bodies reveal different levels of vulnerability. Bodies appear in local epistemologies as both gendered and ungendered. In the prevalent fishing communities, where the ethnographic material mentioned in this chapter was collected, the relation of fishermen with their catch was reversed, revealing a subversion of the food chain that made humans the (potential) fodder for fish, casting the dread of anthropophagy as a primary reason to suspend consuming fish. The body of a gigantic fish mentioned in the stories of old myths was, at the same time, the origin and cause of earthquakes and tsunamis.

Gendered bodies as sexually active bodies were mentioned as a remote cause of the tsunami in religious interpretations that attributed the triggering of the disaster to unbridled sexuality and moral corruption. A further distinction was made between sex workers forced to enter prostitution because of economic needs and others who would engage in it through intrinsic evilness – as in the specific term associating them with the Hindu goddess of death, Kali. Aging and disabled bodies were rendered invisible or were not given specific consideration in some recovery interventions, better suited for people with full mobility. The sea itself was described as an elderly person afflicted by senility and losing the capacity to behave appropriately, thus approaching its end.

The tsunami left behind around 250,000 dead bodies across the different countries affected by the waves. Thailand's coastal region was transformed into an immense morgue where 30 international forensic teams battled time and the decomposing process to identify 5,935 individuals. The forensic experts needed to negotiate different evidential regimes and demands, difficult accessibility to medical records, the relatives' trauma in the recognition process, and the bureaucratic process that would inscribe onto these bodies individual identities, ethnicities and nationality that had been erased by the uniformity of a mass death that, at least initially, had made tangible the ontological vulnerability residing in our bodies.

Notes

1 Sections of this chapter resemble passages from previous texts published and based on the same original ethnographic material (Merli 2005, 2010, 2012). This material is presented here with a novel analysis. The section on Identifying Bodies is based on the article I published together with forensic anthropologist Trudi Buck (Merli and Buck 2015). In this chapter, I do not consider technical details of forensic identification, limiting myself to my theoretical analysis of the political dimensions related to the disaster victim identification process.

2 Multi Donor Fund for Aceh and Nias (MDF) and the Java Reconstruction Fund (JRF).

3 Oxfam was recently at the centre of a scandal for reported sexual abuses, and sexual exploitation of local women by its male operatives in several settings.

4 The USGS calculated the magnitude to be 9.1. This was revised by other sources to be 9.0, although USGS maintains their first evaluation.

5 The exact number of victims remains uncertain as no definitive statistics are provided for Myanmar. Numbers provided in different publications vary between 217,000 to 283,000, to testify that the definition of these counts was part of a difficult assessment.

6 The research on reproductive health was carried out with grants received from Uppsala University, Svenska Sällskapet för Antropologi och Geografi (Sweden), and Donnerska Institutet, Åbo Akademi (Finland). A return to the field to continue investigating specifically the post-tsunami recovery and emerging epistemologies in 2006 was generously supported by the Margot and Rune Johansson Foundation.

7 Some sparse information on organisation of rescue were available, via www.mcot.org (MCOT TNA is at present an invalid domain) and posted and circulated on various websites www.lonelyplanet.com/thorntree/forums/asia-thailand/thailand/how-is-the-situation-on-ko-bulon-ko-lipe-and-ko-muk (last accessed 27 December 2018).

8 Beyond the medical care and the exceptional support provided by nurses, authorities were looking for listeners, as they thought the injured people may need to speak in their own language, to tell somebody what happened. This exposed the listeners to vivid descriptions of what the survivors had seen, heard and experienced, opening up to the possibility of what is called secondary traumatisation.

9 All the names of participants in this article are pseudonyms.

10 "Dahulu orang makan ikan, kini ikan makan orang".

11 These bodies did not have histories, but locals speculated that they could have been men forced to work in fishery, on big trawlers. They might have died of disease, killed because of disputes, or in any case thrown offboard.

12 An overview of the place of cannibalism and anthropophagy in anthropology (and in the Western imaginary), including the intradisciplinary controversies whenever these practices were either analysed or disputed is summarised by William Arens (1998).

13 Stories of fish feasting on people's bodies returned after cyclone Nargis in Myanmar in 2007 (see Brac de la Perrière 2010).

14 These studies build on ethnographic material from Amazonia, on eating animals and humans (Fausto 2007) and on funerary rites (Vilaça 2009).

15 All the interviews were conducted in Thai or Malay.

16 Closeness to the shore should have been considered more dangerous, since it is where the bodies were generally retrieved (at least according to the official declaration of the ministry official reported above).

17 In December 2004, the exchanged rate for 1 USD was approximately 40 THB.

18 This view does not seem to acknowledge the well-established Thai clientele, shadowed by the more prominent international outlook of places such as Pattaya.

19 *Balaq* is translated with 'to annihilate', 'to make waste of'.

20 Locally spelled *nyanyok*, it is in current Malay *nyanyuk*, senile.

21 In Banda Aceh, Indonesia, there were 14 mass graves, with the largest containing about 60,000–70,000 victims (Morgan et al. 2006, p. 196).

22 In popular Buddhism the performance of good deeds, acts of devotion or renunciation (including temporary ordination of men as monks or of women as nuns) are not only considered important to accumulate karmic 'merit' for oneself (as in orthodox textual Buddhism) but also to transfer merits to the deceased (see Keyes 1983; Obeyesekere 1968).

23 The response of the media over the treatment (or initially non-treatment) of the victims of the Malaysian Airline carrier shot down over the Ukraine on 17 July 2014 illustrates this process well and highlights the important political and socio-cultural implications of the need for identification.

References

Arens, William. 1998. "Rethinking Anthropophagy." In *Cannibalism and the Colonial World*, edited by Francis Barker, Peter Hulme, and Margaret Iversen, 39–62. Cambridge: Cambridge University Press.

Bankoff, Greg. 2003. *Cultures of Disaster. Society and Natural Hazard in the Philippines*. London and New York: RoutledgeCurzon.

Black, Sue. 2009. "Disaster Anthropology: The 2004 Asian Tsunami." In *Handbook of Forensic Anthropology and Archaeology*, edited by Soren Blau and Douglas H. Ubelaker, 397–406. Walnut Creek: Left Coast Press.

Brac de la Perrière, Bénédicte. 2010. "Le scrutin de Nargis: Le cyclone de 2008 en Birmanie [The Nargis Election: The 2008 Cyclone in Burma]." *Terrain* 54: 66–79.

Butler, Judith. 2004. *Precarious Life: The Powers of Mourning and Violence*. London and New York: Verso.

Cohen, Erik. 2009. "Death in Paradise: Tourist Fatalities in the Tsunami Disaster in Thailand." *Current Issues in Tourism* 12, no. 2: 183–99.

Crossland, Zoe. 2013. "Evidential Regimes of Forensic Archaeology." *Annual Review of Anthropology* 42: 121–37.

Enarson, Elaine, and P. G. Dhar Chakrabarti, eds. 2009. *Women, Gender and Disaster: Global Issues and Initiatives*. New Delhi: SAGE Publications India.

Enarson, Elaine, Alice Fothergill, and Lori Peek. 2007. "Gender and Disaster: Foundations and Directions." In *Handbook of Disaster Research*, edited by Havidan Rodriguez, Enrico L. Quarantelli, and Russell Dynes, 130–46. New York: Springer.

Falk, Monica Lindberg. 2010. "Recovery and Buddhist Practices in the Aftermath of the Tsunami in Southern Thailand." *Religion* 40, no. 2: 96–103.

Falk, Monica Lindberg. 2015. *Post-tsunami Recovery in Thailand: Sociocultural Responses*. London and New York: Routledge.

Fausto, Carlos. 2007. "Feasting on People: Eating Animals and Humans in Amazonia." *Current Anthropology* 48, no. 4: 497–530.

Felten-Biermann, Claudia. 2006. "Gender and Natural Disaster: Sexualized Violence and the Tsunami." *Development* 49, no. 3: 82–86.

Gaillard, Jean-Christophe. 2019. "Disaster Studies Inside Out." *Disasters* 43, no. S1: S7–S17.

Gaillard, Jean-Christophe, and Pauline Texier. 2010. "Religions, Natural Hazards, and Disasters: An Introduction." *Religion* 40, no. 2: 81–84.

Greenhough, Beth, Tariq Jazeel, and Doreen Massey. 2005. "The Indian Ocean Tsunami: Geographical Commentaries One Year On." *The Geographical Journal* 171, no. 4: 369–86.

Handmer, John. 2004. "Global Flooding." In *International Perspectives on Natural Disasters: Occurrence, Mitigation, and Consequences*, edited by Joseph P. Stoltman, John Lidstone, and Lisa M. Dechano, 87–106. Dordrecht, Boston and London: Kluwer Academic Publishers.

Hart, Lianna, and Stefan Timmermans. 2012. "Death Signals Life: A Semiotics of the Corpse." In *Routledge Handbook of Body Studies*, edited by Bryan S. Turner, 240–52. London and New York: Routledge.

Hockey, Jenny, and Janet Draper. 2005. "Beyond the Womb and the Tomb: Identity, (Dis)embodiment and the Life Course." *Body & Society* 11, no. 2: 41–57.

Jenkins, Richard. 2004. *Social Identity*. London: Routledge.

Kälin, Walter. 2005. "Natural Disasters and IDPs' Rights." *Forced Migration Review*. Special Issue on "Tsunami: Learning from the Humanitarian Response," July 2005: 10–11.

Kelman, Ilan. 2018. "Lost for Words Amongst Disaster Risk Science Vocabulary?" *International Journal of Disaster Risk Science* 9, no. 3: 281–91.

Keyes, Ch.F., 1983. "Merit-transference in the Kammic Theory of Popular Theravāda Buddhism." In *Karma: An Anthropological Inquiry*, edited by Charles F. Keyes and E. Valentine Daniel, 261–86. Berkeley, Los Angeles, London: University of California Press.

Khuri, Fuad Ishaq. 2001. *The Body in Islamic Culture*. London: Saqi Books.

Kusakabe, Kyoko, Rajendra Shrestha, and N. Veena, eds. 2015. *Gender and Land Tenure in the Context of Disaster in Asia*. Switzerland: Springer International Publishing.

MDF-JRF Secretariat. 2012. "More than Mainstreaming: Promoting Gender Equality and Empowering Women through Post-Disaster Reconstruction." MDF-JRF Working Paper Series No. 4. Jakarta: World Bank. https://openknowledge.worldbank.org/handle/10986/17633.

Merli, Claudia. 2005. "Religious Interpretations of Tsunami in Satun Province, Southern Thailand: Reflections on Ethnographic and Visual Materials." *Svensk Religionshistorisk Årsskrift* 14: 154–81.

Merli, Claudia. 2010. "Context-bound Islamic Theodicies: The Tsunami as Supernatural Retribution vs. Natural Catastrophe in Southern Thailand." *Religion* 40: 104–11.

Merli, Claudia. 2012. "Religion and Disaster in Anthropological Research." In *Critical Risk Research: Practices, Politics and Ethics*, edited by Matthew Kearnes, Francisco Klauser, and Stuart Lane, 43–58. Chichester: Wiley-Blackwell.

Merli, Claudia, and Trudi Buck. 2015. "Forensic Identification and Identity Politics in 2004 Post-tsunami Thailand: Negotiating Dissolving Boundaries." *Human Remains and Violence* 1, no. 1: 3–22.

Morgan Oliver W., Pongruk Sribanditmongkol, Clifford Perera, Yeddi Sulasmi, Dana Van Alphen, and Egbert Sondorp. 2006. "Mass Fatality Management Following the South Asian Tsunami Disaster: Case Studies in Thailand, Indonesia, and Sri Lanka." *PLoS Medicine* 3, no. 6: e195.

National Statistical Office of Thailand, Office of the Prime Minister (NSO). 2001. *The 2000 Population and Housing Census, Changwat Satun*. Bangkok: National Statistical Office.

National Statistical Office, Office of the Prime Minister (NSO). 2010. *The 2010 Population and Housing Census, Changwat Satun*. Bangkok: National Statistical Office.

Obeyesekere, Gananath. 1968. "Theodicy, Sin and Salvation in a Sociology of Buddhism." In *Dialectic in Practical Religion*, edited by Edmund R. Leach, 7–40. Cambridge: Cambridge University Press.

Oliver-Smith, Anthony, and Susanna M. Hoffman. 2002. "Introduction: Why Anthropologists Should Study Disasters." In *Catastrophe & Culture: The Anthropology of Disaster*, edited by Susana Hoffman and Anthony Oliver-Smith, 3–22. Oxford: James Currey.

Oxfam. 2015. "Gender, Disaster Risk Reduction, and Climate Change Adaptation: A Learning Companion." In *Green planet Blues: Critical Perspectives on Global Environmental Politics*, edited by Ken Cona and Geoffrey D. Dabelko, 333–46. New York, London: Routledge.

Ruenruthai Sujjapun. 2014. "อิทธิพล ของ คติ ไตรภูมิ ต่อ การ สร้างสรรค์ วรรณคดี [Influences of Trai Bhum Beliefs on Literary Creation]." *Journal of Humanities and Social Sciences* 4, no. 1: 1–34.

Sasaki, Hiroshi, and Shuji Yamakawa. 2004. "Natural Hazards in Japan." In *International Perspectives on Natural Disasters: Occurrence, Mitigation, and Consequences*, edited by Joseph P. Stoltman, John Lidstone, and Lisa M. Dechano, 163–80. Dordrecht, Boston, London: Kluwer Academic Publishers.

Scanlon, Joseph. 2008. "Identifying the Tsunami Dead in Thailand and Sri Lanka: Multi-National Emergent Organisations." *International Journal of Mass Emergencies and Disasters* 26, no. 1: 1–18.

Strathern, Marilyn. 2012. "Eating (and Feeding)." *Cambridge Anthropology* 30, no. 2: 1–14.

Sugirtharajah, R. S. 2007. "Tsunami, Text and Trauma: Hermeneutics After the Asian Tsunami." *Biblical Interpretation* 15, no. 2: 117–34.

Telford, John, John Cosgrave, and Rachel Houghton. 2006. *Joint Evaluation of the International Response to the Indian Ocean Tsunami: Synthesis Report*. London: Tsunami Evaluation Coalition.

Thacker, Eugene. 2011. "Necrologies; Or, the Death of the Body Politic." In *Beyond Biopolitics: Essays on the Governance of Life and Death*, edited by Patricia Ticineto Clough and Craig Willse, 139–62. Durham and London: Duke University Press.

Tun, Kan, Barbara Butcher, Pongruk Sribanditmongkol, Tom Brondolo, Theresa Caragine, Clifford Perera, and Karl Kent. 2005. "Panel 2.16: Forensic Aspects of Disaster Fatality Management." *Prehospital and Disaster Medicine* 20, no. 6: 455–58.

Valarak Chaiyatap. 2005. "Fancy Fish?" *Bangkok Post*, February 8.

Vilaça, Aparecida. 2009. "Bodies in Perspective: A Critique of the Embodiment Paradigm from the Point of View of Amazonian Ethnography." In *Social Bodies*, edited by Helen Lambert and Marion McDonald, 129–47. New York: Berghahn.

World Bank. 2012a. Gender-Sensitive Post-Disaster Assessments. East Asia and the Pacific *Region Sustainable Development*. Guidance Note No. 8. Washington, DC: World Bank. https://openknowledge.worldbank.org/handle/10986/17075.

World Bank. 2012b. Integrating Gender-Sensitive Disaster Risk Management into Community-Driven Development Programs. East Asia and the Pacific Region Sustainable Development. Guidance Note No. 6. Washington, DC: World Bank. https://open knowledge.worldbank.org/handle/10986/17078.

World Bank. 2012c. Making Women's Voices Count: Integrating Gender Issues in Disaster Risk Management – Overview and Resources for Guidance Notes. Washington, DC: World Bank. https://openknowledge.worldbank.org/handle/10986/26531.

World Health Organization (WHO). 2014. *A Year After the Tsunami of 26 December 2004*. Report. www.who.int/hac/crises/international/asia_tsunami/one_year_story/en/World Health Organization (WHO). 2014. *A Year After the Tsunami of 26 December 2004*. Report. www.who.int/hac/crises/international/asia_tsunami/one_year_story/en/.

Part III

10 Disasters and gendered violence in Pakistan

Religion, nationalism and masculinity

Sidsel Hansson and Catarina Kinnvall

Introduction

Pakistan is a country high on the climate vulnerability scale. Due to its geographical location, Pakistan has entered the heat surplus zone of the earth. It is expected that the precipitation level in Pakistan will change drastically and that such a change might alter or completely take away the occurrence of monsoons from the Indian-Pakistan region. Additional climate changes are likely to further deteriorate the conditions of the people living in the country and to make life more difficult as it did in 2010, 2011 and 2013 when Pakistan was hit by massive floods which affected around 18 million people. According to a UN Women report (2018), displacement due to climate catastrophes and flooding has increased the constancy of factors that make violence against women possible, such as ruptures of public and private spaces, loss of home structure, lack of water in close proximity to the household, which have exposed women to security check points and sexual abuse in search of clean water, as well as the camp experience itself. The report puts particular emphasis on the camp in which tension between various groups is being played out in relation to women's bodies.

While violence perpetrated by strangers can start immediately in crises when people situate themselves in mixed group sites, familial violence is more likely to set in after a time lapse of camp living as the post-humanitarian crisis make people feel compelled to find detrimental coping mechanisms. In addition, as government and civil society organisations (including Western organisations) have often been late in arriving to the devastation, Islamic groups have become involved in the relief efforts and have played a greater role in the reconstruction works (Gronewold 2010; Hansen 2006; Memon 2015; Rozan 2011). In the aftermath of the disaster of 2010, for instance, the Pakistan Talibans, the TTP, issued warnings to foreign workers and to the military to get involved in the relief efforts (ABC News 2010; Crilly 2010). The floods' aftermaths have further contributed to public perception of inefficiency and to political unrest, while the government's response has been complicated by insurgencies, growing urban sectarian discord, increasing suicide bombings against core institutions and unstable relations with India.

In this chapter we discuss the wider implications of the gendered effects of climate disasters in Pakistan by drawing on reports and interviews from across

Pakistan, as well as on fieldwork conducted in the Ahmadpur East *tehsil* of the Bahawalpur district in Southern Punjab. Drawing on this particular case, we look at situations in which reconstruction efforts, including displacement (where many are forced to live in make-shift tents and other 'camp-like' housing areas), have become recurrent and normalised due to reoccurring disasters. In particular we are concerned with how these processes shape and construct gender relations and gendered violence in relation to ongoing traumas of reconstructing livelihoods (often requiring short-term migration and/or going into debt), in the face of resurgence and militarism. The ways in which particular norms of masculinity are strengthened in areas struck by disaster and how these multiply through religious organisations and militaries involved in disaster relief are often at the heart of these experiences, influencing the situation and the agency of women and girls, and making their lives particularly difficult.

The chapter starts with a theoretical discussion of the relationship between gender, masculinity and crisis as this is played out in connection with climate disasters, where we pay especial attention to the role of the patriarchal order at times of crisis. Here Pakistan is a particular case in point as specific forms of masculinity have taken the shape of homogenous patriarchal structures and discourses associated with certain rituals, symbols and security-seeking behaviour linked to narrative masculinity. We then move on to look more closely at the existence of religious, military and other organisations to investigate how specific norms of masculinity have been strengthened in relation to climate disasters in Pakistan more generally, and how that have affected gendered violence and the situation and agency of women and girls. From there we turn to a more in-depth discussion of the disaster experience in gendered terms, drawing on official reports and our own fieldwork in Bahawalpur to analyse how these experiences have shaped and constructed gender relations and gendered violence in relation to ongoing traumas of reconstructing livelihoods. Finally, we discuss the Bahawalpur case in relation to the emergence of a *tripartite form of security-seeking masculinity* shaped by the co-existence of men's crisis; the consistent crises of religion, nationhood and the state; and the perpetuation of multiple large and small-scale crises in the country.

The disaster experience: crisis, gender and masculinity

One of the characteristic markers of the Anthropocene can be found in the fact that the vast majority of ecosystems on the planet now reflect the presence of people. There are, for instance, more trees on farms than in wild forests, and these anthropogenic biomes (i.e., the terrestrial biosphere in its contemporary, human-altered form) are spread throughout the planet in a way that previous 'pre-human' ecological arrangements were not. As discussed in the introduction: "[T]he Anthropocene implies that the human imprint on the global environment is now so large that the Earth has entered a new geological epoch; it is leaving the Holocene, the environment within which human societies themselves have developed" (Steffen et al. 2011, 4). The fossil record of the Anthropocene shows, for instance, a planetary ecosystem homogenised through domestication and human interference. As Lewis and Maslin (2018) point out, embracing the Anthropocene as an idea means the possibility of reversing this trend. It means treating humans not as insignificant observers of the natural world but as central to its workings, and elemental in their force.

Proceeding from an idea of the Anthropocene as a more profound realisation of the intertwinement of nature, culture and power structures, a number of studies have shown (e.g., Bradshaw 2013; Krüger et al. 2015; Memon 2014; UN Women 2016) how global environmental forces, such as floods, droughts, tsunamis and hurricanes affect women and men differently. As True (2012) has noted, the death rate of women after the 2004 Indian Ocean South Asian tsunami was at least three times higher than that of men in some communities. And among survivors of such disasters, a great deal of evidence (Akbar and Aldrich 2017; Aolain 2011; Fisher 2010; IFRC 2014) shows that women and girls face a higher risk of experiencing gender-based and sexual violence. Such evidence goes beyond the disaster experience itself, making the effects of a natural disaster anything but natural (True 2012). The preparedness for a disaster is obviously dependent on previous political decision-making processes as well as on economic interests. The risks associated with living in areas prone to storms, flooding and droughts (among others) are of course not unknown, but such knowledge does not prevent individuals and groups from habituating these areas. Often, as is the case of Pakistan, we can fit a map of the worst affected areas inflicted by flooding onto the socio-economically most marginalised communities (Memon 2014). Hence, a disaster affects groups and collectives differently depending on where they occur. The World Bank (2018) reports for instance that 95 percent of disaster-related deaths occur among the 66 percent of the world's population who live in the poorest countries, showing how disasters discriminate against groups with lesser capabilities, resources and opportunities. Due to socio-economic vulnerability, their resilience thresholds are very low and, in relation to malevolent floods in Pakistan, have hit the hardest against landless labourers. "In short, vulnerability to death and violence is highly differentiated; proximity disaster and the ability to anticipate, cope with, protect oneself, and recover in a disaster's aftermath (with support for evacuation through to insurance for rebuilding) are ultimately socially determined" (True 2012, 163, see also Jones and Murphy 2009).

Here a disaster is regarded as a crisis in terms of being related to an external threat, to the vulnerability of a community, and to the entrance into a state of uncertainty, what Kinnvall has referred to as being in a state of ontological insecurity – of not knowing what tomorrow holds (Kinnvall 2004, 2018; Kinnvall and Nesbitt-Larking 2011). However, as Walby (2015) has argued, it is important to move beyond a view of disasters as extreme events in order to analyse them as consequences of wider structural processes. There is, she argues, a tension between treating 'disaster' as if it were a politically constructed object and in treating it as if it were an already existing object that has political implications. As stated by Hay (1996, 255): "State power (the ability to impose a new trajectory upon the structure of the state) resides not only in the ability to *respond* to crises, but to *identify*, *define* and *constitute crisis* in the first place". As Squires and Hartman (2006, in True 2012) contend, there is no such thing as 'natural' or 'inevitable' about natural disasters, as past and present political decisions and economic interests shape every phase of a disaster, from preparedness and planning, to impact and responses. This also puts into focus contemporary notions of crisis as being a temporal experience related to traumatic events, such as violence, decease or bereavement, while for many people it is a continuous experience linked to the presence and possibility of conflict, poverty and disorder. As Koselleck (2002, 8; see also Kinnvall 2017; Vigh 2008) has argued, there is a tendency to think of crisis as a rupture in the order of things; "an intermediary moment of chaos where social and societal processes collapse upon themselves only to come to life after the crisis is overcome", which overlooks the fact that for many people, trauma is plural and the suffering arising from crisis the norm. Crisis, in this context, is chronic and forces people to make lives in volatile worlds rather than waiting for normalisation and reconfiguration (Vigh 2008). It is not, in other words, a short-term explosive situation or a decisive change, but a condition in which agency is not a question of capacity but of possibility. The question then is how people act in such a context and how they consume information?

Here it is important to note how beliefs and attitudes lead to particular ways of perceiving crisis and risks; values affect how people prioritise risks and how they relate to other people when dealing with risks with behaviours and outcomes being related to specific perceptions and values (IFRC 2014). Cultural values and beliefs often become (re)activated in relation to disasters, affecting how they are interpreted and responded to. For instance, it may be difficult to understand why people would rather face natural hazards in one location than the possibility of loss of reciprocity from neighbours and community, lack of employment or livelihood options, physical violence or crime in a new 'safe' place. "People who return feel that they can be more in control, that the variables they have to deal with are known to them and that they can make responses within existing framework of experiences" (IFRC 2014, 25). Particular cultures of violence are often at the heart of specific power relations in which certain attitudes and behaviours are legitimised and where certain groups (e.g., landless people, minorities, low class and low caste groups as well as females) may accept their position and perceive it as legitimate because it is regarded as cultural rather than exploitative.

All of this can strongly affect specific vulnerabilities and thus impact on the response and recovery process. Within such contexts religion and other beliefs play an important role because they can help to explain and sometimes justify why disasters occur and, in the case of Pakistan, are closely tied up with gendered norms of masculinity and femininity. For those living in dangerous places, a belief that a flooding can happen because of divine intervention can offer some kind of rationality and reason for why it happens. In circumstances where hazards may be frequent life experiences, local communities may have developed 'ways of normalising threats' (Krüger et al. 2015), expressed through practices that deal with the emotional and psychological requirements of living with uncertainty and which may influence normative values and religious beliefs. Environmental crises can thus be seen as an amplifier for intensified or stagnated gender inequalities in which traditional gender roles and hierarchies are often reinforced (Juran and Trivedi 2015; Sjoberg, Hudson, and Weber 2015)

Hence, it is the social experience of the disaster that works differently for men and women. This is closely related to the manifestation of a patriarchal order which, in the case of Pakistan, cannot be separated from ideas about crisis and masculinity. In all societies, girls and boys are taught to behave in certain stereotypical ways, affecting their treatments and behaviours as different messages are conveyed to them through the media, religious leaders, parents, school, peers and others. Through these prescribed patterns, young men typically learn that it is considered 'masculine' to be strong and dominant, sexually active, not to show emotions, and to exercise authority over women and children of their families (Aurat Foundation 2016; Connell 1995, 2005; Kimmel 2018). In societies like Pakistan, boys and men are also expected to support their parents financially throughout their lives, expectations that have consequences for both females and males and which work to sustain gender inequalities and the perpetuation of harmful masculine norms that govern gender roles (Rozan 2011).

Masculinity as a 'dominant' and 'superior' gender position is thus produced through customary laws, family, religion, norms and sanctions, popular culture and media, as well as through state regulations, and has specific consequences for women (Srivastava n.d.). Hence, it is the society that makes us masculine and feminine through the process of gendered socialisation where dominant or hegemonic masculinity tend to be those masculinities by which men measure themselves and other masculinities, as discussed in the introduction (Connell 2005). As such, they often cause the social pressures and societal expectations faced by boys and men. Domination, aggressiveness, competitiveness, athletic prowess, stoicism and control characterise hegemonic masculinity, which means that love, affection, pain and grief are then viewed as improper displays of emotion. "Any male [or women trying to be hegemonically masculine] who fails to qualify in any one of these ways is likely to view himself [or herself] as unworthy, incomplete, and inferior" (Goffman 1963, 128). Hegemonic masculinities usually rest upon and generate patriarchy – a system of power inequalities and imbalances between women and men – which uses violence as a means to ensure that these power imbalances stay in place. Gender-based violence becomes a means to acquire resources and deny access of these resources to others. It also becomes a means to solidify relations of domination that uphold this "structure of inequality [. . .] involving a massive dispossession of social resources" (Connell 2005, 83).

This has led to a focus in much gender work on discourses, narratives and subjectivities, but also to the realisation that the structures of gender relations are formed and transformed over time in which multiple forms of masculinities arise. In the aftermath of climate disasters, it is thus important to investigate those socio-cultural, economic and political structures and masculinities that allow for everyday abuse and violence of women and girls (Horton and Rydstrom 2011). In the next section we discuss how such structures and masculinities have come together in a patriarchal order as manifest in the Pakistani society and how they have impacted on particular gendered disaster experiences of women and girls.

Climate disasters and the manifestation of a patriarchal order in Pakistan

Three root causes have been named as particularly relevant in terms of physical and environmental conditions when discussing climate disasters in Pakistan. These are Pakistan's mountainous topography, triggering flash floods in the Kyber Pakhtunkhwa Province (KPK) in Northern Pakistan; anthropogenic climate change, linked to weathering anomalies such as the monsoon rainfalls in Punjab; and certain natural climate variability, such as the ENSO (El Niño-Southern Oscillation) (Mustafa and Wrathall 2011; Witting 2012). Deforestation has also had a major bearing on the flood impact, partly due to lack of regular flood flows and partly due to avaricious elements in politics and bureaucracy aggravating the flood impact (Memon 2015).

However, equally important are root causes related to the political environment, like the instable security situation and post-civil conflict impact, especially in the Northern Provinces, in which military offensives against anti-government elements have destroyed key infrastructure and forced people to migrate. Failed perceptions of the national government in terms of asking for assistance from the international community at times of flooding have also been important, partly due to the National Disaster Management Agency (NDMA) conferring its operational mandate to the Provincial Disaster Management Agency (PDMA). Inadequate government commitment to disaster risk reduction (RDM) and a lack of commitment to development, poverty and malnutrition, linked to the absence of a legal framework and insufficient law enforcement for operating and monitoring responsibilities, have also played a major role. Limited governmental resources and capacities have further influenced forecasting and early-warning performances. Additionally, traditional ways of farming can be viewed as a root cause, in which farmers tend to settle in high-exposure areas, such as floodplains in order to benefit from fertile soils. High levels of corruption and feudal landownership and power structures have further worsened social inequalities and vulnerability conditions, while lack of knowledge and education is linked to missing building codes and inadequate urban and land use planning (Memon 2015; Witting 2012). Finally, religious norms and gender roles have strongly influenced the susceptibility of women. Due to their poor social integration, women often have a very low level of awareness of how to deal with certain threats (Bukhari and Rizvi 2015; Maheen and Hoban 2017). Additionally, rescue activities for unaccompanied women have been particularly challenging during emergency relief phases.

The gendered dimension of flood disasters in Pakistan has been described in detail by Memon (2012, 2015) in reports by the SPO, a non-governmental organisation focused on Strengthening Participatory Organization. In particular he highlights the lack of proper data for the numbers of affected women and children, disaggregated by sex, age, geographic location and occupation/livelihood. Also, the government does not recognise women-headed households, preferring to register such households under the name of the family patriarch or the nearest male sibling (SPDC 2015; UN Women 2016). Equity and equality in compensation and the feminisation of poverty also go unrecognised in compensation efforts and women in rural areas often lack national identity cards that can give them access to bank accounts (as well as giving them a voice in elections). To this can be added the need to focus on shelter and infrastructure, especially on latrines/toilets, kitchens, water sources, hygiene, sanitation and sewerage in the shelter component which particularly affect women. Health issues in relation to pregnancies, treatment of water-borne gastric diseases, where mostly boys and men receive treatment as cultural constraints often prevent women from seeing male doctors and paramedics, are also important factors together with an inadequate focus on girls' and women's vaccination and education (Memon 2012, 2015; see also Bukhari and Rizvi 2015; Neumayer and Plumber 2007). The loss of family, community, shelter, livelihoods, incomes and the onset of disease, disability and displacement have also caused, in both women and men, a widespread loss of coping mechanisms, self-respect, dignity and self-confidence, leading to anger, insecurity and trauma, which need both short- and long-term interventions (Memon 2015).

Violence against women and issues of protection have been listed as one of the most important dimensions of the flood disasters in Pakistan, with unverified reports of criminal elements kidnapping girls during the initial emergency and rescue phases, under the guise of priority evacuation, food and shelter provision. Also, trafficking, sexual abuse, forced prostitution, child labour, divorce, desertion, abandonment of women and senior citizens, sale or coercive 'adoption', have been reported but not received enough attention due to poverty-stricken families being too scared to file FIRs (first information reports) (Memon 2015; Najam-u-din 2010). To this can be added how many flood-affected areas have high poverty and low literacy rates, especially among girls, and how the loss of teachers has further affected their situation. Women are also mainly responsible for gathering wood and fuel for domestic consumption and fodder for the livestock, and massive landslides, soil erosion and unprecedented flooding (due to pre-flood deforestation), have made this task very difficult. Reports also point to increased gendered violence as women and girls have to travel further in search of wood, fodder and water (Najam-u-din 2010; SPDC 2015).

This patriarchal order is not limited to Pakistan, but takes particular forms in the Pakistani society. Here it should be noted that even under normal circumstances, many females in the rural areas affected by flooding, such as Bahawalpur, have no outdoor exposure and have to stay at home around the clock. They are often not allowed to receive medical assistance even under critical conditions from male doctors. Nor are they allowed to join private or public sector jobs, with the Gender Development Index of the 2010 Human Development Report stating that the female participation in the labour force was 21.80 percent in comparison to 86.70 percent male participation rate, while the ratio of secondary education rate amongst Pakistani women is just 23 percent as compared to 46 percent for men (UNDP 2011). Moreover, in Pakistan 50 percent of females are married before the age of 20, a figure that is much higher in rural areas, and the rates of teenage marriages have increased significantly during flooding, leaving many women pregnant without adequate healthcare. Fearful for their safety, parents of young girls prefer marrying them off to protect them and to avoid the stigma attached to rape and harassment (Bukhari and Rizvi 2015; UN Women 2016).

Patriarchy is an established informal system, with clear hierarchies of power and authority that is transferred from one generation to another with its roots deeply embedded in cultural settings and in the fabrics of society. In this sense it is not limited to the actions of individuals but a consequence of societal conditions and expectations. As Filemoni-Tofaeno and Johnson (2006, 11) have argued:

> It is therefore necessary to dispel the myth that acts of violence against women are simply unfortunate instances of individual men momentarily losing control. On the contrary, such acts are most often the end of a mindset congruent with the rules of patriarchal power, in which the perpetrator views the woman not as a person [. . .] but as an object.

A patriarchal order thus gives value and privileges to men over women and permit or encourage men's domination, oppression and exploitation over women, in which women are perceived as an 'object' rather than a 'subject' and are given low status in society. It implies that deep-rooted ideas about women's low status and male supremacy enable men to exercise unlimited power over women and legitimise that power and domination. As an order it is often internalised, where both men and women regard many instances of abusive conduct as normal and part of life and in which male violence is exercised at will; whereas women often do not perceive themselves as abused unless they have experienced severe physical violence (Hadi 2018; Dobash and Dobash 1979).

There are certain rituals and cultural practices that strengthen the particular form of patriarchy existing in Pakistan and the rest of the subcontinent. Patriarchal control over women are exercised through institutionalised codes of behaviour, gender segregation and an ideology which associates family honour with female virtue. Customary practices which aim to preserve the subjugation of women are often defended and sanctified as cultural traditions and given religious associations. In terms of gendered violence, such traditional practices include honour killings, rape and sexual assault, sexual harassment, being burned, kidnapping, domestic violence, dowry murder, forced marriages, custodial abuse and torture. When women speak up against physical or sexual violence, they are often regarded with contempt and considered to have lost their and their families' dignity, resulting in many cases of rape and abuse going unreported. Domestic violence is also commonly treated as private matters and many men still regard it as their basic right to threaten or be physically violent to their wives to correct disobedient behaviour (Hadi 2018; Rozan 2012). As one of Abi Nobil Ahmed's (2008, 135) interviewees explained:

> Pointing to his shoe, he told me that in his village, a woman's worth is no greater than the height of a man's foot; that a brother shoots his own sister if she dares even suggest a preference for a marriage partner to her male relatives.

Joint and extended families tend to live together in which husband, wife, husband's brothers and their families (spouses and children), parents of husband, and unmarried sisters of husbands live together under one roof. In this arrangement, the grandfather is the ultimate authority as he owns all the property and wealth of the family. Unpacking this family structure, a clear hierarchy can be distinguished, which indicates how power and authority lies within the husband's household which specifies the elder male who has more authority over both younger men and all sisters as well as how the wife experiences relations once she is married, where her own brothers and sisters are less important than her husband's (Saeed 2015). The extended family system likewise means that also in-laws perpetrate violence on issues related to dowry and petty family disputes. Both the Human Rights Watch (1999) and the Aurat Foundation (2012) have estimated that 70 to 90 percent of Pakistani women are subjected to intimate partner violence, while also acknowledging that the incidences of intimate partner violence are grossly under-reported (Hadi 2018).

Structural discrimination of women and girls thus means that they often (already) lack the agency to report violence to the police and obtain proper healthcare (Aolain 2011). At times of disasters, reports also show how women have been sexually abused by authorities, such as the police and the military, which constitutes an important reason for not reporting such violence. The patriarchal order, in addition, points to how matters of violence are often supposed to be taken care of in the private in which the woman's responsibilities lie with the family and where reporting assaults might bring dishonour to the family and prove that she is not foremost concerned with her family's welfare (Fisher 2010; Khan et al. 2013). In Pakistan, this patriarchal structure and culture are linked to particular notions of masculinity as played out in religious, nationalist and military terms.

Religion, nationalism and the military: masculinity in practice

As a discourse, nationalism relies on an essentialised historical narrative of the *nation-as-this* and the *people-as-one*, which are supposed to guide social and political action in the name of a particular *ethnos* (being Pakistani for instance) and a certain imagined national space (Pakistan as the locus of Pakistanness) (Kinnvall 2006; Torfing 1999;). Such discourses on national identity have often been linked to religious justifications and legitimisations and have, in the case of Pakistan, resulted in the two-nation theory professed by M. A. Jinnah and the Muslim League. This theory, which found realisation in the establishment of Pakistan in 1947 (see Svensson 2013), maintain that Muslims and Hindus constitute separate nations, thus ascribing Muslims and Hindus nationality on the basis of membership in a socio-religious community.

Despite intricate linkages between religion and nationalism, the religious element in the study of nationalist movements is often neglected or dismissed. One reason for this can be found in the modernist narrative and its wish to depart from the 'primitive Sacred' to the 'modern Secular' (Casanova 2008). This narrative equates religion with irrationality and superstition, often in opposition to the progress of science, secularism and the inherent rationality of the enlightenment (Beyer 1994; Haynes 1999). Religion, compared to the nation, is also commonly viewed as being de-territorialised – as a transnational space or movement – thus obscuring how both national and religious identity make claims to a monolithic and abstract identity, i.e., to *one* stable identity that answers to the need to securitise self and subjectivity (Kinnvall 2004, 2006). As a clearly defined historical body religion often becomes a stabilising anchor in an otherwise chaotic and changing world, linking the past and the present to future action. In Pakistan, the dominant institution, the military, has recurrently employed religion to secure its own position in society and in the political sphere. It might, consequently, be suggested that religion has functioned, and still functions, as an inclusive, unifying vessel and foundation for Pakistani identity, thus becoming the locus of attempts to secure and stabilise, i.e., to govern, the Pakistani nation-state. The entwinement of national belonging with notions of survival, and with a need to be vigilant towards state failure and disintegration, turns religion into an unwarranted and undesirable focal point for securitising practices. As a governing process religion is more than a psychological experience based on the individual mind; it has ideological functions that serve oppressive interests. Nomisation, or the order of meaning in society, is never neutral or altogether positive but is mediated through social relations (Haynes 1997; Vanaik 1997).

As traditional discourses of religion (in this case specifically Islam) represent significant components of Pakistani identity in societal power struggles, women have as a consequence faced multiple levels of opposition in both the family and the state. Colonial structures have historically been buttressed by cultural and religious obstacles to women's rights, thus reinforcing gender hierarchies, while the postcolonial world has seen new nationalisms leaving patriarchal social structures largely unchallenged in state formations (Standish 2014). In Pakistan, generations of rulers have turned to religion in order to unite a divided and composite country. Through processes of Islamisation of the Pakistani state and society, the political elites have often used the narratives provided by religious leaders to divert attention away from other issues at the same time as more resources have been devoted to religious control. The number of religious seminars, so-called madrasas, have also grown significantly with Amir Rana at the Pakistan Institute of Peace Studies estimating a growth from 576 in 1979 to around 20.000 in 2008 (interview 9 October 2008). Most of these are jihadist or sectarian in their beliefs, working actively against the democratic system. Within these struggles over space and minds, religion has been invoked repeatedly as the state has used Islamic symbolism to galvanise support against the alleged division from the country's diverse ethnic groups, as well as against the perceived threat from India (Khan 2017; Kinnvall 2010). This support for militants Islamists has had its costs in interethnic relations, but has also affected gender relations and has been detrimental to women and girls in areas affected by militancy as well as by different forms of natural disasters.

In most disaster zones in Pakistan, different relief organisations have been active, with religious organisations and parties, as well as the military, often taking an active role in the relief efforts. The most prominent 'Islamist' welfare group active in the floods has been Al Khidmat Trust, which is linked to the Jamaat Islami, a lawful political party in Pakistan. As an organisation, it has been well organised and been able to deploy volunteers from the earliest stages by establishing relief camps and offering medical relief, which is consistent with the ways in which the Jamaat Islami has traditionally been involved in grassroots politics (Semple 2011, 68–9). There are mixed reports in terms of how religiously motivated militant organisations, such as the Taliban or Sipah-e-Sahaba, a Sunni Deobandi movement with an anti-Shite ideology, have been able to capitalise on the floods in terms of recruitment from or mobilisation of flood-affected communities. However, as Arai (2012, 58) has argued: "the perceived ineffectiveness and injustice of government facilitated resource distribution and discriminatory social structures could offer these militant networks greater scope to mobilise widespread popular resentment". This resentment of the government is something repeatedly referred to in our interviews, as explored below. In addition, in response to government sponsored schools being destroyed by the floods, many development practitioners in Punjab and Khyber-Pakhtunkhwa have argued that madrassas, or Islamic seminaries, are often able to offer more attractive and affordable alternatives to government-sponsored schools. "In short, class tensions strained by the floods, if left untransformed, might be compounded by religious militancy over time, with implications for deepening public discontent and undermining the legitimacy of civilian governance" (Arai 2012, 58). In his study of the consequences of the earthquake disaster that struck Pakistan in 2005, David Hansen shows, for instance, how religious extremist leaders were telling those affected by the earthquake that they only had themselves to blame for the disaster, using a surah (verse) from the Holy Koran called 'The Quaking', encouraging the victims to show remorse by displaying a deeper dedication to Islam.

> This thing that happened, the earthquakes, it is really our own fault. There is a story about it in the holy Koran. It [the story] is called the Earthquake [sic]. They say it is our own fault [. . .] the religious leaders.
>
> (Hansen 2006, 11)

As pointed out in many of our interviews with civil society organisations working with disasters: religious organisations and the military were mostly the first to respond to natural disasters. And while the Pakistani military has been criticised for their passive presence in the recovery and reconstruction efforts (Aurat Foundation 2016), religious organisations have often been praised for their involvement in the day-to-day relief efforts. This has frequently involved justifying the causes and consequences of disasters on a religious basis. Thus, it is important to emphasise how religious ideology, especially as it has become institutionalised in the state structure of Pakistan, is frequently used to justify particular kinds of hegemonic masculinity associated with sexual superiority and possessiveness.

"Men have been documented to believe that they are entitled to 'sexual acts' any time they feel like. There is religious interpretation also attached with this. Men believe that a woman who refuses to have sexual intercourse with her husband will be punished as per the instructions of the Quran" (Aurat Foundation 2016, 14). In Pakistan, sexual virility and the ability to produce male off-springs is another dimension of masculinity, in which the production of a male child and being sexually powerful are some of the expectations of being a real man. Many local accounts suggest however (Rozan 2012; Bhanbhro et al. 2013), that gendered violence and honour killings are not solely driven by religion, customs and traditions, but are also propagated by the local gender system in their conceptions of manhood and the complicit role of state institutions, the military and law enforcement agencies. Hence, documented incidents of honour killings show how feudal lords and state institutions, particularly members of the police force, are often listed as perpetrators.

Below we discuss the gendered ramifications of living with recurrent floods using evidence from a field study carried out in one of the most flood prone locations in the Indus valley river basin in Southern Punjab. This study, we argue, goes some way to illustrate the gendered ramifications for the most vulnerable populations living with climate change induced disasters in Pakistan. It brings together the disaster experience with masculinity, a patriarchal order and gendered violence, in a setting where religious extremism and militancy are always at its doorstep although not explicitly referred to in direct interviews. The interviews also show how, among both men and women, the loss of family, community, shelter, livelihoods, incomes and displacement have caused a deep sense of loss, often in terms of self-confidence, dignity and pride, which in turn has resulted in anger, insecurity and trauma, as discussed previously. These are both physical and psychological aspects that need short- and long-term interventions. What the study also shows is how resilience to the consequences of devastation is regularly emphasised as a coping mechanism after the disaster, especially in the recovery and rehabilitation phases. In gendered terms, however, such resilience is often associated with strict gender complementary roles and arrangements within the household unit. Here we elaborate how the different phases, the recovery and the rehabilitation phase as well as the emergency phase, have resulted in different gendered dynamics and, as discussed in the final concluding section, how these cannot be viewed apart from the overall traumatic experiences of living in a society defined by gendered existential crisis played out in religious and nationalist terms.

Gender, masculinity and disaster in Bahawalpur

This field study was conducted in the Ahmedpur East *tehsil* of the Bahawalpur district in Southern Punjab, a district surrounded by three canals, separated from each other by 5 kilometres and connected to the river Punjnad Head which is formed by the merger of the key rivers of Punjab in Pakistan. While the area was greatly affected by floods in 2010 and 2014, almost every year the area is engulfed by water during the rainy season as the rivers swell and raise the level in the canals, thus affecting Ahmadpur East, which is at a lower level. Pakistan and India are both signatories of the 1960 Indus Water Treaty, which gives Pakistan three western rivers and India, three eastern rivers, all three of which ultimately goes through Bahawalpur – directly or indirectly. Interestingly, the river bed of some of these is dry; and yet at other times, there is flood. Some political parties, especially parties on the right and in Punjab, blame both occurrences on India, arguing that it is the release of water (or not) in India that is behind both droughts and floods. Interestingly, one of the banned organisations, Jamaat-ud-Dawa (JuD), claims that the main reason for it being targeted is for talking about the 'water terrorism' of India, while its members argue that they are merely a social welfare organisation undertaking relief activities for those affected by floods and other disasters (India Today 2014).[1] Their relief organisation, Falah-i-Insaniat Foundation, as well as Jamaat-e-Islami's charity section, Al-Khidmat Organization, have both been involved in relief activities in the area and the area is known to be a hub of extremist activities. As pointed out by Ali (2009), of the 363 madrassahs in Ahmedpur, 80 percent of Deobandi, 70 percent of Shia, 25 percent of Barelvi and 14 percent of Ahl-e-Hadith madrassahs were involved in sectarian violence. Since Ahmedpur is considered a hotbed of sectarian violence, these numbers may be much higher than an average area in Pakistan, but the case still offers an informative window into the interaction between customs, tradition, religion and politics at a micro level.

In the Indus valley river basin, the worst affected populations tend to be located in rural pockets marred by very low development and excessively high social inequality. In Southern Punjab, economic and political power is more closely linked to landownership than anywhere else in rural Pakistan (Saleem 2009; Martin 2014). In the particular area studied here, 96 percent of the population live on the very edge of subsistence, and only 8 percent of the women are literate. Hence, the setting of this study represents one of the most flood-prone and socioeconomically marginalised areas in the country (Memon 2014). Minor floods that landlords regard as mainly natural hazards often become devastating disasters for the local subsistence farmers and farm workers (Raza 2017). The participants in this study, young women and men in the age group of 18–35 belong to the worst affected population living on low lying lands within a defunct irrigation system zone near Panjnab, a tributary to the Indus River in the Bahawalpur district of Southern Punjab (Mohsin 2013). Depending on their age, these young people recall having experienced from three to four, to eight to nine devastating floods in their life time.

Here we are interested in their experiences of living with what has become a recurrent cycle of natural disasters. Our focus is on their individual accounts of the disaster cycle and its different phases of recovery and rehabilitation on the one hand, and the emergency phase on the other, with attention paid to what the participants perceived to be the risks involved, the forms of resilience needed, and the norms and values that guided their actions. As shown, a set of social and gendered risks, and the women and men's ability to mitigate these, came to the forefront as the most crucial concerns in the emergency phase. The relief camps were especially associated with precarious conditions, excruciating hardships and conflicts over scarce government provisions, as well as intense contestations over women's and children's bodies (Bukhari and Rizvi 2015).

Under these tense conditions, men were expected to demonstrate the ability to exercise control and mitigate a wide range of social and gendered risks. Risks of sexual harassment and abuse as well as of child abduction – some of the risks generally associated with displacement and rescue camps in Pakistan – were also present here. Attention is also given to the ways in which the enactment and imaginary of social conflicts unfold in these make-shift camps at the very periphery of Pakistani society, and how these are informed by and feed into the dominant symbolic narrative of gendered body politics in the country. As discussed earlier, and observed elsewhere, the recovery and rehabilitation phases of the disaster cycle were mainly associated with family violence, while the emergency phase were more characterised by violence from strangers.

The cycle of disasters: the recovery and rehabilitation phases

When invited to share their experiences of the floods, the participants invariably steered the conversation away from the question of their individual emotional responses and traumas, and instead pointed to how they over time had developed the necessary foresight, preparation and skills to confront the devastation and hardships of the disaster situations:

> We actually feel stressed during floods, due to the heavy loss of our crops and the monetary [costs]. But still, we have to handle things, not only for our survival, but for our future too. We were already living a hand-to-mouth existence, so in all situations we have to keep [the flood] in mind, before it comes and hits our lives badly. We have to stay prepared at all times.
>
> (Woman, age 31, farmer)

Studies show that elsewhere in Pakistan disaster experiences often result in post-traumatic stress disorders and depression, and that these ailments in turn are associated with increased gendered violence (see, e.g., Aslam and Kamal 2016; Mubeen et al. 2013; Riaz et al. 2015). The participants of this study, however, tended to downplay issues of emotional stress and human vulnerabilities, and instead, more readily articulate their fears about the inevitable damages to their homes and livelihoods, and the more long-term consequences of being ensnared in what many described as a continuous loop of devastation, loss and need for reconstruction. These were anxieties about their livelihoods, their children's education and future prospects, and the lack of infrastructure and development in the area. When considering the recurrent disaster cycles in a long-term perspective, the participants expressed uncertainty about their capacity to sustain the necessary human resilience and ability of economic recovery over time. "[I]t takes a lot of time to restore the losses, and during the process of restoration another flood comes" (Man, age 18, farmer). Some felt that they were already caught in a downward spiral towards economic destitution, "We used to have an income of one *lakh*, and now we are poor and helpless" (Man, age 29, farmer).

Crucial to their capacity to restore livelihoods are the households' social networks, and in the case of Pakistan's poorest these tend to be very limited. Nevertheless, the network patrons reportedly mediated the households' access to the government's disaster compensation schemes and the local formal and informal loan markets. The participants either complained that they had no access to the government's compensation scheme, or they had, but received far less compensation than the amounts they were entitled to. Although loans and debts are sensitive issues and difficult to probe into, it was clear that the local loan markets also had social and gendered ramifications related to the disaster. A few participants offered some insights, however, without going into details about the fallout of having amassed substantial, non-repayable debts to a landowner:

> It is hard to rebuild our life after the floods, because all our savings have been swept away. We lease land from a contractor. Once after a flood, we went into huge debt. When we had nothing in hand, how could we repay it? But the contractor couldn't understand this.
>
> (Woman, age 28, tenant farmer)

In this particular case, the elders of the community reportedly came in to negotiate an acceptable solution. However, and more generally in Pakistan, the predicament described here often amounts to a first step towards servitude or bonded labour for the most vulnerable households, such as tenant farmers and day labourers, and also involves the risk of the household's children being commodified. Although our data here is ambiguous, what seemed to be at play, other than the rather well-known power dynamics of the feudal order, was the more underlying tensions of what Arai (2012, 58) has referred to as "[t]he widening gap between the lowest income households, or the 'have-nots', and the 'super have-nots'", in the worst affected disaster areas in Pakistan. This will be discussed further in the next section on the emergency phase.

Here we are more concerned with what the participants perceived as the emotional stress points during the recovery and rehabilitation phases. The participants, and especially the men, avoided going into details about their emotional experiences, with the exception for what they saw as high-tension situations of periodical food scarcity and the assessment of damages and loss. A few men related how they found the latter to be a particularly anxiety-laden experience. Some women observed that these situations were associated with risks to women and children:

> I never get into fights with anyone because of this situation [the aftermath of the flood]. One is upset. It is wearisome, since we poor are facing great losses. It is true that the number of fights between men and women goes up after the floods, because the male takes out his loss on the wife and the children. That is not good. But, by the grace of God, my husband does not fight with me.
>
> (Woman, age unknown, farmer)

While some female participants, like this farmer, seemed confident about their husbands' capacity to demonstrate self-restraint, others maintained that it was actually the women who bore responsibility for avoiding confrontations that could escalate into conflict and violence in these situations. As is the case elsewhere in Pakistan, the entire disaster cycle was associated with resource scarcity and food insecurity. In this regard, both women and men readily shared their experiences, with emphasis on the anguish of having to let the elders, the sickly and especially, the children go hungry. This occurred, as they pointed out, in extremely strenuous situations when it was vital to maintain one's physical stamina. Some reported that periodical food shortage resulted in conflicts within the households: "When it [the crops] is destroyed, we have nothing to eat, and then fights between family members happen" (Man, age 26, farmer). And, food shortage was also associated with men's under performance, and their manhood and masculinity being questioned:

> During the floods, when the situation got bad and there was nothing to eat, my wife did not [any more] give me any importance [. . .], and then we fought. Afterwards, we resolved it, and I tried my best to fulfil all the needs of my wife.
>
> (Man, age unknown, farmer)

At the same time, and irrespective of their individual propensities to share their experiences of conflicts and violence, the participants commonly considered physical violence unacceptable and incompatible with the norms and values of their families and community. What they instead endorsed were the norms of self-discipline, mutual support and social unity, in other words, the norms and values that were core to their coping strategies and resilience in disaster situations:

> During disasters we come under a lot of pressure [. . .] we know that for our survival and for the future of our children we have to work harder than what we usually do. [. . .] What can we do in such a critical situation, other than being patient and controlling our nerves? That actually keeps us motivated, too.
>
> (Woman, age 20, handicraft worker)

> It makes relationships even more solid. [. . .] People do not fight, but stick together. We are facing very difficult times, and this [unity] gives us the strength to resume our lives.
>
> (Man, age 29, farmer)

In the interviews, the participants largely portrayed the recovery and rehabilitations phases as a resilience building process. The common characteristics of this was personal self-discipline and endurance, familial cohesion and household self-sufficiency. Familial cohesion was closely associated with strict gender complementary roles and arrangements within the household unit, with no or little dependency on outside resources, other than perhaps, the households' immediate network. In broader terms, as discussed earlier, these phases are generally associated with familial violence, and perhaps, as our data indicate, can partly be seen as triggered by the male sense of underperformance when having to affront the challenges of resource scarcity, and material damage and loss. Interestingly, while the participants largely agreed on the qualities that went into the making of human resilience at times of a disaster, they had highly divergent views on how far these qualities played out and characterised social interactions, especially at the community level during the emergency phase.

The cycle of disasters: the emergency phase

In the emergency phase the participants became more acutely dependent on external assistance, and had to engage with a wider, complex web of social dynamics and socially dominant norms and values. When accounting for this phase, the participants tended to emphasise the social and gendered risks involved, and the challenges of mitigating these risks while being in positions of acute vulnerability. As for the external assistance provided during the emergency phase, the participants largely agreed that the early-warning system set up in the area was reliable, with information being issued well beforehand through the mosques, and by visiting army and government staff. However, they considered the assistance interventions, especially from the government, to be unreliable and insufficient. The grievances that the participants felt free to articulate were the ways in which they became more or less acutely vulnerable due to the shortcomings of the assistance provided by the provincial government. These shortcomings were largely seen as a matter of gross neglect of the population's rights and entitlements. For some it was also a matter of the unfair distribution of whatever assistance was provided. As many pointed out, the performance of the provincial government could be contrasted with that of the army whose interventions tended to cater to the needs of all, and being more impartial in its distribution of supplies. This runs counter to the observations made by Hansen (2006) in relation to the 2005 earthquake, in which the army seemed to have been more preoccupied by keeping terrorists at bay than with relief efforts.

In Bahawalpur, in contrast, the socio-political context was a complex web of feudal power with political aspirations that, at least at the local level, fed on and into a periphery of criminal gangs and religious militants. The dynamics of these power relations and how it played out in the emergency phase were largely deemphasised by the participants. As noted elsewhere (Memon 2015), the assistance from the provincial government was probably mediated by the feudal power in ways that strengthened its local patronage network. Hence, while some participants found the local landlords catering to their needs, others found them indifferent or entirely inattentive, or highly preferential in their propositions for how the government assistance should be distributed. As seen elsewhere in Pakistan (Akbar and Adrich 2017), the perceived unreliability and unfairness of the external assistance probably contributed to a social and institutional distrust that may also have aggravated the aforementioned tensions between the 'have-nots' and 'super have-nots'. This however, was a contested point among the participants. While some maintained that the adversity of recurrent disasters had made the community come together to stand more unified than ever before, others were more uncertain, or felt that the community had degenerated into a situation of 'every man for himself' (itself a gendered perception) and was in a process of social disintegration.

When discussing their experiences of the emergency phase, the participants yet again tended to highlight their acquired resilience. At the same time, however, it became apparent that household resilience thresholds varied between those who were able to stay at home, those who could move in with relatives located in more flood-averse locations, and those who had no other option than to be transferred to the make-shift camps during the peak period of the floods. Over the years some had managed to build up their resilience threshold by moving into more flood-averse homes at higher altitudes, or shifted from mud-clay to concrete dwellings. Irrespective of how safe they considered their current dwellings to be, the participants' preferred option invariably was to remain at home, even when this involved having to spend days or weeks on the rooftops of their dwellings and become dependent on highly unreliable food supplies and other forms of external assistance. A less attractive option was to move in with relatives located in less flood-prone areas. Apart from safeguarding human lives, livestock and property, they were intensely concerned about going through the emergency phase without compromising the household's family honour:

> The army men would tell us to go [to the camp], but how could I flaunt my honour so publicly! To us honour is more important than our lives.
> (Woman, age 33, tenant farmer)

> We can stay in our house without food, but we cannot go to the camps or someone else's house where [we are at risk of] losing our integrity.
> (Man, age unknown, farmer)

What was at stake here was the risk of women's and children's bodies being exposed to violations, with the concurrent symbolic loss of manhood for the men and social marginalisation of the household. The common opinion articulated by the participants was that only the most vulnerable, those with no other survival options, could accept the social risks of staying in the improvised, open air make-shift relief camps at higher altitudes in the surrounding areas. A young woman with personal experience of being displaced to the camps, explained:

> The dignity of women and children is not secured. This is why only the helpless are willing go to the camps. Everyone else prefers staying at the rooftop of their house. Why would anyone go, when the government cannot ensure proper security in the camps? The government comes for a day or two, and

then it does not care anymore. And, the food provided is not sufficient. [. . .] There were fights over the distribution of flour packets, [. . .] there are lots of fights in the camps.

(Woman, age 24, tenant farmer)

The precarious conditions of the relief camps in Pakistan, and the kinds of risks that particularly women, children and other vulnerable groups are exposed to have been discussed earlier (see, e.g., Bukhari and Rizvi 2015; Maheen and Hoban 2017; UNIFEM 2011). In the make-shift camps organised in this area, which typically came with the lack of shelter, basic facilities and security arrangements, the risks for the most vulnerable groups became intensified. However, the participants who had stayed in the camps still emphasised their determination to cope with the social tensions and physical hardships involved by employing resilience skills. A few men and women even found their camp stays to have been an empowering experience, "Those conditions taught me much. Who helps you in hard times, and who abandons you? And, how to take control in tough times. How to gain confidence" (Man, age 30, factory worker). Still, the social boundary transgressions that frequently occurred in this setting, with power contests among men over women's and children's bodies, were a much more dreaded and anxiety-triggering experience that crucially challenged men's performance of their masculinity and manhood:

During the floods the main task is to take our family members and livestock to safe places. We do not feel safe in those places [the make-shift camps], because mostly these are open air places. Like, in our area, women prefer to live in privacy [*pardah*], but during the floods they [must] sit with outsiders. Providing protection to women and children is another matter. Everyone sits with their family in camps that are without proper walls and with little privacy. Because it is a difficult time, we somehow manage to live under such conditions. During the day we somehow manage. But during the night some of the men stay awake for the sake of security. Only then can women and children go to sleep.

(Man, age 19, farmer)

The male participants tended to be more restrained in their descriptions, while the female participants more readily detailed the conflicts, the violence and the stakeholders:

People from different areas were staying there, including some ill-mannered men. These men would cat-call. It was especially difficult at night. Some stole luggage. Once, a man tried to rape a girl. They beat this culprit really badly. I had to stay up all night to look after my daughters. My husband and sons had to stay up all night. Women had no toilet facility or washroom. They had to go to the fields. We were just surviving.

(Woman, age unknown, tenant farmer)

The men who harassed women by doing immoral gestures had their own ladies in the camp, and yet they dared! Once we heard about an incident where two girls had been abducted. It later was resolved by the tribal leaders, and the girls were released. But once women have lost their dignity, it is lost forever. A young girl's dignity is vulnerable. Once also, a baby boy was raped by a man.

(Woman, age 24, tenant farmer)

As these accounts suggest, and as observed elsewhere in Pakistan (Ajaz and Majeed 2018, 323), the make-shift camps were experienced as a site for sexual and gendered violence perpetrated by strangers. In the context of recurrent disasters, this site, for some of the most vulnerable participants, became a repeatedly visited crisis scene. By employing the dominant cultural and religious script of honour, the violence against women and children became meaningful as ritualised enactments of social humiliation with symbolic loss of manhood in power contests among men. As such, these contests were not only meaningful in the local context at the periphery of the society, but potentially also, as part of the broader social crisis scene in Pakistan.

Conclusion: disaster as a tripartite crisis of masculinity

Although our interviews only briefly touched upon the role of the military and militant groups in the area (and mainly in passing), the growth of religious madrasas and sectarian violence did not go completely unnoticed. Some of our interview subjects also spoke about religion in relation to the floods, but mostly in terms of how disasters could be viewed as a punishment:

There is something good in Allah's every work, these floods are kind of lesson for us. When Allah will not be satisfied with us then he will punish us, people do injustice, tell lies, don't think about poor and because of it Allah will be unhappy.

(man, age 19, farmer)

In his study of the growth of religious militancy in the area, Ali (2009) finds that the key explanatory variable for sectarian violence in Ahmedpur is class conflict, as religious-political parties become vehicles for peasants to challenge the political elite. This is an area in which feudal landowners monopolise political and economic power and where 96 percent of Ahmedpur's population has less than five acres of land. Additionally, Ali maintains that the state has involved madrassahs in regional conflicts and domestic politics. Within this pattern of development, it is difficult not to notice how the state's policy has been to involve madrassahs in regional conflicts and domestic politics, while granting them astonishing room for independent manoeuvre. Parliament's unwillingness to repeal or even reform discriminatory laws, the absence of a national domestic violence law and a gender-insensitive, dysfunctional criminal justice system are thus significant factors in the state's failure to protect women from endemic violence.

The extent to which rights violations against women go unpunished may not be as alarming as it is in the open conflict zones of FATA and Kyber Pakhtunkhwa (KPK), where women have been subjected to state-sanctioned discrimination, militant violence, religious extremism and sexual violence. In these regions, militants target women's rights activists, political leaders and development workers without consequences (ICC 2015). However, the prevalence of informal justice mechanisms in many parts of Pakistan, including Bahawalpur, are highly discriminatory towards women; and the government's indiscriminate military operations, which have displaced millions, have further aggravated the challenges they face in the conflict zones.

An influential factor here, we argue, is the emergence of a particular *tripartite form of security-seeking masculinity* in Pakistan in general and in Bahwalpur in particular. This masculinity is shaped by the co-existence of men's crisis; the consistent crises of religion, nationhood and the state; and the perpetuation of multiple large and small-scale crises in the country, where climate disasters take a particular form. Most of the participants in our study did not openly talk about the radicalisation in the area and how this had played out during the disaster, which agrees with a study by AWAAZ (2016, 6), who conducted focus groups on radicalisation in Bahawalpur, noting:

> At an FGD [focus group discussion] held at Mari Sheikh Shajra, Goth Laal, Bahawalpur, women showed little interest in the questions and they were not too eager to respond. Sitting on charpoys, most of them stared blankly at the interviewer, not understanding the necessity to respond to questions specifically on women's mobility and their public and private spheres.

However, other women in the disaster zones spoke about the negative effects of the militants (AWAAZ 2016, 11):

> Our culture and tradition have been deliberately destroyed. I blame the state and religious extremists for whom there is no law. Extremism is at its peak, preventing girls from attending schools. Women teachers are also forced to give up their jobs. Everybody is against women! exclaimed Ms Afzal.

The jihadist expansion in Punjab, previously known for its tolerance and syncretic form of Islam, has become possible through proxies, financial support from foreign countries, particularly Saudi and Gulf countries, and the fact that the state borders insurgency-hit and lawless regions of the country, while also sharing a border with India. With state sponsorship and a pervasive climate of impunity enhancing jihadist groups recruitment potential, the risks of joining are far lower than potential gains that include employment and other financial rewards, social status and a sense of purpose (ICC 2016). For those affected by economic hardship and trauma, as well as struggles against oppressive landlords, the disaster experience can become a motivating factor for joining. In out interviews with the head of Potohar Mental Health Association (PHMA), an organisation that works with men's psychological needs at times of disasters, Zulqurnain Asghar stressed the strong association between mental health, gender-based violence and sectarian violence. He especially underlined how in disaster situations men may kill their women because of honour, as their own survival becomes most important, thus arguing that men respond to disaster situations in one of two ways: "either they become more humanised, or they become violent":

> In traumatic disaster situations we see that men feel that they have lost their power, because they have lost the control of their family, and especially the men who have become homeless, often become depressed. In disaster situations, rape occurs more frequently. The landlords think that the women are their property, they are seen as being entitled to rape, and even gang-rape them whenever they want, regardless of the women's age, whether 8 or 65. You find this pattern in Southern Punjab. 98 percent of rape cases goes unreported, it is a matter of honor. The men tend to protest initially, by saying, no, this is our women. If rape turns out to be unavoidable, then what bothers them is not the fact that the women have been raped, but that they as a consequence, have lost their manhood. The deeper, reconciliatory reasoning among the affected men is that the women have been sacrificed, so that they themselves at least, can stay safe. When men are unable to control their women and their property, they primarily understand this as not being in control of their women, and consequently feel that they are not in control at all. They feel powerless, and start wondering, what kind of men are we?

The fact that the women (and men) we interviewed were reluctant to speak about either the trauma of the disaster situation or the militants in the area and their role in the post-disaster setting, can be related to women's general confinement to the home and to gender specific norms. Because of entrenched patriarchy women are the last ones to have their opinions heard, particularly in times of dire circumstances, and are not acculturated to speak out. Natural disasters exacerbate these challenges; where women face particular difficulties in speaking about political events as well as gendered violence. As discussed earlier, the patriarchal order is normalised into patriarchal control exercised through institutionalised codes of behaviour, gender segregation and an ideology which associates family honour with female virtue. Hence, traditional and religious practices which aim to preserve the subjugation of women are often defended and sanctified as cultural traditions and given religious associations. The current crisis in Pakistani society can thus not be separated from the crisis in manhood as disasters strike, which in turn must be connected to particular identity boundaries that emanate from the preoccupation with religious and customary practices of masculinity.

Note

1 According to the UN, JuD is the reincarnation of Lashkar-e-Tayyaba, which carried out the Mumbai attacks in 2008.

References

ABC News. 2010. "Taliban Urges Pakistan to Reject Foreign Flood Aid." August 11. www. abc.net.au/news/2010-08-11/taliban-urges-pakistan-to-reject-foreign-flood-aid/940748.

Ahmed, Ali Nobil. 2008. "The Romantic Appeal of Illegal Migration: Gender, Masculinity and Human Smuggling from Pakistan." In *Illegal Migration and Gender in a Global and Historical Perspective*, edited by Marlou Schrover, Joanne van der Leun, Leo Lucassen, and Chris Quispel, 127–50. Amsterdam: Imiscoe University Press.

Ajaz, Tahseen and Muhammad Tariq Majeed. 2018. "Changing Climate Patterns and Women Health: An Empirical Analysis of District Rawalpindi, Pakistan." *Global Social Sciences Review* 3, no. 4: 320–42. https://gssrjournal.com/article/Changing-Climate-Patterns-and-Women-Health:-An-Empirical-Analysis-of-District-Rawalpindi-Pakistan

Akbar, Muhammad Siddique, and Daniel P. Aldrich. 2017. "Determinants of Post-flood Social and Institutional Trust Among Disaster Victims." *Journal of Contingencies and Crisis Management* 25, no. 4: 279–88.

Aolain, Fionnuala Ni. 2011. "Women, Vulnerability, and Humanitarian Emergencies." *Michigan Journal of Gender and Law* 18, no. 1: 1–23.

Arai, Tatsushi. 2012. "Rebuilding Pakistan in the Aftermath of the Floods: Disaster Relief as Conflict Prevention." *Journal of Peacebuilding & Development* 7, no. 1: 51–65.

Aslam, Naeem, and Anila Kamal. 2016. "Stress, Anxiety, Depression, and Posttraumatic Stress Disorder among General Population Affected by Floods in Pakistan." *Pakistan Journal of Medical Research* 55, no. 1: 29–32.

Aurat Foundation. 2012. Pakistan: NGO Alternative Report on CEDAW 2012. https:// www.af.org.pk/pub_files/1358697993.pdf [accessed 15 August 2018].

Aurat Foundation. 2016. "Masculinity in Pakistan: A Formative Research Study." http:// af.org.pk/gep/images/GEP%20Gender%20Studies/Masculinity%20in%20Pakistan.pdf.

AWAAZ 2016. "Radicalization and Engendered Space: Women, Voice and Visibility." AWAAZ Programme 2016.

Beyer, Peter. 1994. *Religion and Globalization*. London: Sage.

Bhanbhro, Sadiq, Raque Wassan, Muhbat Ali Shah, Ashfaq A. Talpur, and Aijaz Ali Wassan. 2013. "Karo Kari – The Murder of Honour in Sindh Pakistan: An Ethnographic Study." *International Journal of Asian Social Science* 3, no. 7: 1467–84.

Bradshaw, Sarah. 2013. *Gender, Development and Disasters*. Cheltenham and Northhampton: Edward Elgar.

Bukhari, Syed Iazaz Ahmad, and Shahid Hassan Rizvi. 2015. "Impact of Floods on Women: With Special Reference to Flooding Experience of 2010 Flood in Pakistan." *Journal of Geography & Natural Disasters* 5, no. 2.

Casanova, José. 2008. "Secular Imaginaries: Introduction." *International Journal of Politics, Culture, and Society* 21, no. 1/4: 1–4.

Connell, Raewyn W. 1995. *Masculinities*. Cambridge: Polity.

Connell, Raewyn W. 2005. "Hegemonic Masculinity: Rethinking the Concept." *Gender & Society* 19, no. 6: 829–59.

Crilly, Rob. 2010. "Pakistan Flood: Taliban Vows to Kidnap Foreign Aid Workers." *The Daily Telegraph*, August 26. www.telegraph.co.uk/news/worldnews/asia/pakistan/7965241/Pakistan-floods-Taliban-vows-to-kidnap-foreign-aid-workers.html.

Dobash, R. P., and Dobash, R. E. (1979). *Violence Against Wives: A Case Against the Patriarchy*. New York, NY: Free Press.

Filemoni-Tofaeono, Joan, and Lydia Johnson. 2006. *Reweaving the Relational Map (Religion and Violence)*. London: Routledge.

Fisher, Sarah. 2010. "Violence Against Women and Natural Disasters: Findings From Post-Tsunami Sri Lanka." *Violence Against Women* 16, no. 8: 902–18.

Goffman, Erving. 1963. *Stigma: Notes on the Management of Spoiled Identity*. New York: Simon & Schuster Inc.

Gronewold, Nathan. 2010. "Is the Flooding in Pakistan a Climate Change Disaster? Devastating Flooding in Pakistan May Foreshadow Extreme Weather to Come as a Result of Global Warming." *Scientific American*, Climate Wire, August 18, 2010.

Hadi, Abdul. 2018. "Intimate Partner Violence and It's Under-Reporting in Pakistan." *European Journal of Social Sciences Education and Research* 12, no. 1: 254–60.

Hansen, David. 2006. "The 10/8 Pakistan Earthquakes: Potential for (Further) Conflict." Paper Delivered at the Nordic Institute of Asian Studies' (NIAS), Intensive Researcher Training Course. May 2006.

Hay, Colin. 1996. "Narrating Crisis: The Discursive Construction of the 'Winter of Discontent'." *Sociology* 30, no. 2: 253–77.

Haynes, Jeff. 1997. 'Religion, Secularisation and Politics: A Postmodern Conspectus', *Third World Quarterly* 18, no. 4. pp. 705–28.

Haynes, Jeff. ed. 1999. *Religion, Globalization and Political Culture in the Third World*. Great Britain: Macmillan Press.

Horton, Paul, and Helle Rydstrom. 2011. "Heterosexual Masculinity in Contemporary Vietnam: Privileges, Pleasures, Protests." *Men and Masculinities* 14, no. 5: 542–64.

Human Rights Watch. 1999. "Crime or Custom? Violence Against Women in Pakistan". October 19, 1999. https://www.hrw.org/report/1999/10/19/crime-or-custom/violence-against-women-pakistan [accessed 25 December 2018].

India Today. 2014. "India Behind Pakistan Flood, Tweet Hafiz Saed." September 10. www.indiatoday.in/world/pakistan/story/hafiz-saeed-jamat-ud-dawah-india-jammu-and-kashmir-floods-208086-2014-09-10.

International Crisis Group (ICC). 2015. "Women, Violence and Conflict in Pakistan." Asia Report No. 265, April 8, 2015.

International Crisis Group (ICC). 2016. "Pakistan's Jihadist Heartland: Southern Punjab." Asia Report No. 279, May 30, 2016.

International Federation of Red Cross and Red Crescent Societies (IFRC). 2014. World Disaster Report: Focus on Culture and Risk. Report. www.ifrc.org/world-disasters-report-2014.

Jones, Erik C., and Arthur D. Murphy, eds. 2009. *The Political Economy of Hazards and Disasters*. Lanham, MD: AltaMira Press.

Juran, Luke, and Jennifer Trivedi. 2015. "Women, Gender Norms, and Natural Disasters in Bangladesh." *Geographical Review* 105, no. 4: 601–11.

Khan, Hazif Muhammad, Riaz Hussain, Khan Sindher Ather, and Irshad Hussain. 2013. "Studying the Role of Education in Eliminating Violence against Women." *Pakistan Journal of Commerce and Social Sciences* 7, no. 2: 405–16.

Khan, Ismail M. 2017. "Religion, Ethnicity and Violence in Pakistan." In *Routledge Handbook of Contemporary Pakistan*, edited by Aparna Pande. London: Routledge.

Kinnvall, Catarina. 2004. "Globalization and Religious Nationalism: Self, Identity and the Search for Ontological Security." *Political Psychology* 25, no. 4: 741–67.

Kinnvall, Catarina. 2006. *Globalization and Religious Nationalism in India: The Search for Ontological Security*. London: Routledge.

Kinnvall, Catarina. 2010. "Pakistan: Inside and Outside Threats." *Journal of Islamic State Practices in International Law* 6, no. 2: 3–29.

Kinnvall, Catarina. 2017. "Feeling Ontologically (In)Secure: States, Traumas and the Governing of Gendered Space." *Cooperation and Conflict* 52, no. 1: 90–108.

Kinnvall, Catarina. 2018. "Ontological Insecurities and Postcolonial Imaginaries: The Emotional Appeal of Populism." *Humanity and Society* 42, no. 4: 1–21.

Kinnvall, Catarina, and Paul Nesbitt-Larking. 2011. *The Political Psychology of Globalization: Muslims in the West*. Oxford and New York: Oxford University Press.

Kimmel, Michael. 2018. *Healing From Hate: How Young Men get Into and Out of Violent Extremism*. California: University of California Press.

Koselleck, Reinhart. 2002. *The Practice of Conceptual History: Timing History, Spacing Concepts*. Stanford: Stanford University Press.

Krüger, Fred, Greg Bankoff, Terry Cannon, Benedict Orlowski, and E. Lisa F. Shipper. 2015. *Cultures and Disasters: Understanding Cultural Framings in Disaster Risk Reduction*. London: Routledge.

Lewis, Simon L., and Mark A. Maslin. 2018. "Welcome to the Anthropocene." *IPPR Progressive Review* 25, no. 2.

Maheen, Humaira, and Elizabeth Hoban. 2017. "Rural Women's Experience of Living and Giving Birth in Relief Camps in Pakistan." *PLoS Currents*, January, 202.

Martin, Nicolas. 2014. "The Dark Side of Political Society: Patronage and the Reproduction of Social Inequality." *Journal of Agrarian Change* 14, no. 3: 419–34.

Memon, Naseer. 2012. Malevolent Floods of Pakistan. Report by SPO (Strengthening Participatory Organization), Pakistan.

Memon, Naseer. 2015. Climate Change and Natural Disasters in Pakistan. Report by SPO (Strengthening Participatory Organization), Pakistan.

Mohsin, Muhammad. 2013. "Evaluation of 2013 Flood Damages in Pakistan: A Case Study of Ahmedpur East, Bahawalpur, Pakistan." *International Journal of Asian Social Science* 3, no. 10: 2204–20.

Mubeen, Syed Muhammed, Seema Nigah-e-Mumtaz, and Saqib Gul. 2013. "Prevalence of Post-Traumatic Stress Disorder and Depression Among Flood Affected Individuals of Sindh, Pakistan: A Cross-Sectional Survey in Camps Five Months after the Flood." *Pakistan Journal of Medical Research* 52, no. 4: 111–15.

Mustafa, Daanish, and Wrathall, David. 2011. "Lessons from the Flood." *China Dialogue*, March 11. www.chinadialogue.net/article/show/single/en/4155-Lessons-from-the-flood-1.

Najam, U. Din. 2010. *Internal Displacement in Pakistan: Contemporary Challenges*. Report. Human Rights Commission of Pakistan.

Neumayer, Eric, and Thomas Plümper. 2007. "The Gendered Nature of Natural Disasters: The Impact of Catastrophic Events on the Gender Gap in Life Expectancy, 1981–2002." *Annals of the Association of American Geographers* 97, no. 3: 551–66.

Raza, Hassan. 2017. "Using a Mixed Method Approach to Discuss the Intersectionalities of Class, Education, and Gender in Natural Disasters for Rural Vulnerable Communities in Pakistan." *Journal of Rural & Community Development* 12, no. 1: 128–48.

Riaz, Muhammad Naveed, Sadia Malik, Sehrish Nawaz, Muhammad Akram Riaz, Naila Batool, and Jawwad Muhammad Shujaat. 2015. "Well-Being and Post-Traumatic Stress Disorder Due to Natural and Man-Made Disasters on Adults." *Pakistan Journal of Medical Research* 54, no. 1: 25–28.

Rozan. 2011. "Engaging with Boys and Young Men to Address Gender Based Violence and Masculinities." *Training Module*. Rozan, Islamabad, Pakistan. http://menengage.org/resources/engaging-boys-men-address-gbv-masculinities-training-module/.

Rozan. 2012. "Will the Real Men Please Stand Up? Stories of five Men and their Affirmative Action against Violence 2012." Rozan, Islamabad, Pakistan. www.partners4prevention.org/resource/will-real-men-please-stand-stories-five-men-and-their-affirmative-action-against-sexual-vio.

Saeed, Rizwan. 2015. "Patriarchy and Masculinity in the Culture and Language of Pakistan & India." Published on Rozan. www.rozan.org.

Saleem H. Ali. 2009. *Islam and Education: Conflict and Conformity in Pakistan's Madrassas* Oxford: Oxford University Press.

Semple, Michael. 2011. "Breach of Trust: People's Experiences of the Pakistan Floods and their Aftermath, July 2010–July 2011." Pattan Development Organisation, Islamabad.

Standish, Katerina. 2014. "Understanding Cultural Violence and Gender: Honour Killings; Dowry Murder; the Zina Ordinance and Blood-feuds." *Journal of Gender Studies* 23, no. 2: 111–24.

Steffen, Will, Åsa Persson, Lisa Deutsch, Jan Zalasiewicz, Mark Williams, Katherine Richardson, Carole Crumley, Paul Crutzen, Carl Folke, Line Gordon, Mario Molina, Veerabhadran Ramanathan, Johan Rockström, Marten Sheffer, Hans Joachim Schellnhuber, and Uno Svedin. 2011. "The Anthropocene: From Global Change to Planetary Stewardship." *Ambio* 40, no. 7: 739–61.

Sjoberg, Laura, Heidi Hudson, and Cynthia Weber. 2015. "Gender and Crisis in Global Politics: Introduction." *International Feminist Journal of Politics* 17, no. 4: 529–35.

Social Policy and Development Centre (SPDC). 2015. "Gender and Social Vulnerability to Climate Change: A Study of Disaster Prone Areas in Sindh." Social Policy and Development Centre, Karachi.

Squires, Gregory D., and Chester Hartman, eds. 2006. *There Is No Such Thing as a Natural Disaster: Race, Class, and Hurricane Katrina.* New York: Routledge.

Srivastava, Sanjay. n.d. "Masculinity and its Role in Gender-Based Violence in Public Spaces." *Cequin.* http://cequinindia.org/wp-content/uploads/2018/01/MASCULINITY-AND-ITS-ROLE-IN-GBV-IN-PUBLIC-PLACES-Sanjay-Srivastav.pdf.

Svensson, Ted. 2013. *Production of Postcolonial India and Pakistan: Meanings of Partition.* London: Routledge.

Torfing, Jacob. 1999. *New Theories of Discourse: Laclau, Mouffe and Zizek.* Malden, MA: Blackwell.

True Jacqui. 2012. *The Political Economy of Violence Against Women.* Oxford: Oxford University Press.

UNIFEM Report. 2011. "Facts and Figures: Violence against Women." Unifem.org.

United Nations Development Programme (UNDP). 2011. Human Development Report: Sustainability and Equity: A Better Future for All. Report. http://hdr.undp.org/sites/default/files/reports/271/hdr_2011_en_complete.pdf.UN Women. 2016. Women's Economic Participations and Empowerment in Pakistan: Status Report 2016. Report. Center of Gender and Policy Studies, Islamabad, Pakistan.

UN Women. 2018. "Facts and Figures: Ending Violence against Women: Status Report 2018." www.unwomen.org/en/what-we-do/ending-violence-against-women/facts-and-figures.

Vanaik, Achin. 1997. *The Furies of Indian Communalism: Religion, Modernity and Secularization.* London: Verso.

Vigh, Henrik. 2008. "Crisis and Chronicity: Anthropological perspectives on Continuous Conflict and Decline." *Ethnos* 73, nos. 1–2: 136–48.

Walby, Sylvia. 2015. *Crisis.* Cambridge: Cambridge University Press.

Witting, Maximilian. 2012. "Detecting disaster root causes: A framework and an analytical tool for practitioners.". German Committee for Disaster Reduction (DKKV). Schloemer Gruppe, Düren.

World Bank. 2018. "Disaster Risk Management." www.worldbank.org/en/topic/disasterrisk management/overview#2.

11 Crises, ruination and slow harm

Masculinised livelihoods and gendered ramifications of storms in Vietnam

Helle Rydstrom

Introduction

A gender-blind perspective has predominated in disaster studies (MacGregor 2010) and only recently have gender and masculinity been included into climate disaster research (Bradshaw 2013; Enarson and Pease 2016). With some exceptions (e.g., Godfrey and Torres 2016; True 2013), these studies primarily apply a macro structural approach which tends to overlook ethnographic experiences from the ground (e.g., Enarson and Chakrabarti 2009; Fordham 2011). A gender specific crisis perspective, I would suggest, offers an alternative analytical entry point to the study of climate disasters in the era of the Anthropocene. Such a perspective provides a prism for an analysis of the entanglements between a crisis of emergency and a broad spectrum of contextual crises antecedents spurred by gender specific lifeworlds, livelihoods, hierarchies, powers and privileges (Denton 2002; Ginige, Amaratunga, and Haigh 2014; True 2013).

A sudden incident like a typhoon is ruining as an immediate crisis which brings "great and irretrievable disaster upon" people and things (Stoler 2013, 9).[1] An emergency situation caused by a disaster inflicts instant ruination but might even foster additional crises, or fortify various types of crises, which were already in place prior to the catastrophe. Yet a crisis of emergency tends to be studied as an abrupt incident; as a bracketing of daily life which renders people vulnerable and demands resilience in order for survivors to recover and return to how things used to be (Adger 2006; Fordham 2011; Hewitt 1983; Wisner 2016). Pre-disaster normalcy thus is expected to be re-established within a not too-distant future even though this might not always be possible (Bradshaw 2013; MacGregor 2017; UNADPC 2010).

In this chapter, I focus on climate disasters in central coastal Vietnam by exploring gender specific and differentiated types of crises, hazards and ramifications inflicted by storms upon people and communities. Ethnographic data on the ways in which local configurations of masculinities and femininities define hierarchies, privileges and powers and interact with various types of crises, I would argue, enhance the study of climate disasters in the era of the Anthropocene (cf. Hoffman and Oliver-Smith 2002; Oliver-Smith 1996, 1999). My analysis draws on material collected through the conduction of fieldwork in patrilineally organised Long

Lanh, which is located in central coastal Vietnam. The community is defined by the livelihood of fishing, which is associated with a certain kind of masculinity, social order and male sociality. Fishing, as I will highlight, not only informs daily life but even the ways in which various types of crises unfold prior to, during and in the aftermath of storms.[2]

Crises: emergency, *catharsis* and chronicity

As a decisive moment, which holds power to differentiate, select and separate, a crisis can be understood as "conditions that make outcomes unpredictable" (Habermas 1992 [1976], 1). The word *crisis* is a Latinised form of the Greek *krisis*, which refers to "a turning point in a disease" (Etymology Dictionary; Merriam-Webster). Such a turning point could be for better or for worse, as the crisis could lead to recovery or death. A crisis is in this view defined by its temporality and embeddedness in the course of history and thus supposed to be demarcated by a beginning and an end regardless of whether it is personal, political, economic or socio-cultural. Even though a crisis could result in a negative outcome, a crisis usually is taken as a promise for betterment and improvement. When suffering has come to an end, a process of recuperation is assumed to take place and strengthen those who went through the crisis. A crisis thus tends to be associated with a figurative purification (e.g., in literature), *catharsis*, in terms of providing an avenue to a new beginning (Endres and Six-Hohenbalken 2014; Vigh 2008; Walby 2015).

Though, a crisis might not necessarily offer a path to renewal but might as well be a road fraught with difficulties and long-lasting challenges (Etymology Dictionary; Merriam-Webster). In disrupting coherency and augmenting uncertainty for future prospects (Habermas 1992 [1976]), a crisis could alter from a sudden 'state of emergency' (Benjamin 1999) into a condition of 'ahistorical permanence' (Bhabha 1994). Such a crisis might even morph into what Henrik Vigh (2008, 9) refers to as a crisis of chronicity understood as a space with a blurred horizon wherein which, as Anthony Gramsci (1971, 275–6) explains, "the crisis consists precisely in the fact that the old is dying and the new cannot be born; in this interregnum a great variety of morbid symptoms appear".

As Giorgio Agamben (1998, 2005) and scholars of the study of conflict, war and disaster have shown, a crisis of emergency could alter into overall encompassing socio-economic and political conditions that jeopardise basic human rights and security (cf. Butler 2010; Das 2000; Nordstrom 2004; Rydstrom 2012; Stoler 2013). Thus, instead of being a pivotal moment imbued with promises of a better future, a crisis might transmute into a new normalcy; into a context that persistently ruins and thereby imposes slow harm upon lifeworlds, livelihoods and environments (Rydstrom 2019; see also Arendt 1970; Stoler 2013; Vigh 2008).

Spaces of conundrum and lines of entanglement

Considering crisis as chronicity, Janet Roitman (2014) argues, means to turn a crisis into a condition and experience which is stretching across time, as if beyond history and freed "from its temporal confines" (Vigh 2008, 9). A crisis is inherently temporal and hence part of the progression of history, according to Roitman (2014), who notes that conceptualising crisis as chronicity is oxymoronic. However, I would approach temporality from another angle; from a non-linear and contextual perspective which makes a crisis of chronicity appear as less of an oxymoron and more as anchored in the contingencies not of one history but of specific histories.

Conceptualising temporality as shaped by lines of entanglements and spaces of conundrum, not unlike a rhizome as introduced by Gilles Deleuze and Félix Guattari (2000), means to analyse time as configured by differentiated energies, dynamics and rhymes within a complex web of imbroglios nodes. A rhizomatic understanding of temporality, I would argue, captures the intricate ways in which temporality is composed of multiple tempo-spatial zones rather than one. In each of these zones, a crisis can instantiate differently due to its specific intensities and modalities. Thus, time is not just time; in some tempo-spatial pockets, a crisis might pass quickly while it in others might compartmentalise as a new order; as chronicity.

My findings suggest that we see climate disaster related crisis as disordered order, which does not necessarily come to an end in the not too distant future. An abruptly inflicted crisis ruins the ordinary order; when a typhoon has slowed down or propelled off to other sites, a wave train of a tsunami has come to an end and an earthquake gone quiet, people and places are left with a disordered order (cf. Stoler 2013; Roitman 2014; Vigh 2008). In this space, gender and masculinity informed societal asymmetries can lead to the production of new types of precariousness.

The specific temporalities, intensities and modalities of a crisis make each crisis different and the ways in which it might render the vulnerable even more vulnerable by encroaching unequal ramifications upon people, livelihoods and environments. Ben Wisner and colleagues (2003, 11) identify disaster vulnerability as "the characteristics of a person or group and their situation that influence their capacity to anticipate, cope with, resist and recover from the impact of a natural hazard" (see also Introduction).

Gender specific vulnerability is indicated by the ways in which women and girls "always tend to suffer most from the impact of disasters" (UNADPC 2010, 8). Elaine Enarson (2000) thus pinpoints more than 20 indicators including low-income status, men's violence and the need for pre-natal care which render female populations particularly precarious and foster gender specific crises conditions prior to a climate disaster, during the catastrophe and in its aftermath (see also Fordham 2011; MacGregor 2017; True 2013).

Recognising how vulnerability is an integrated dimension of climate disaster both means to take the non-technical impact of a climate disaster into account (Bradshaw 2013) and to be vigilant to how the notion of vulnerability might imply passivity in disaster survivors (Fordham et al. 2013). A resilience perspective, on the other hand, emphasises local capabilities to adjust to extraordinary and socially askew conditions. Yet, a resilience approach might be too optimistic, or even appear as a dictum, as regards how local communities can cope with the consequences of a catastrophe (Oliver-Smith 1996, 1999; Rigg et al. 2008; Steffen, Crutzen, and McNeill 2007; see also introduction).

Vietnam in the era of the Anthropocene

The Anthropocene has provoked new levels of precariousness, especially in the Global South, as indicated by a conspicuous imbalance in the extent to which the Global South vis-à-vis the Global North is confronted with climate disasters and their negative impacts (Enarson, Fothergill, and Peek 2007; MacGregor 2017; Momtaz and Asaduzzaman 2018). While risks and insecure conditions follow in the maelstrom of a climate disaster, these should not distract from socio-economic and political inequalities already defining local communities regardless of the Anthropocene (Hewitt 1983; Rydstrom 2012; Sternberg 2019).

In the era of the Anthropocene, human interaction with environment is assumed to force the Earth into planetary *terra incognita* (Crutzen and Stoermer 2000; Galaz 2014; Steffen, Crutzen, and McNeill 2007).[3] The Earth is becoming warmer and less biologically diverse with less forests and weather, which is both wetter and stormier (Steffen, Crutzen, and McNeill 2007, 614).[4] While the Anthropocene appears as a determining global reality (Haraway et al. 2015), it is also uneven (Momtaz and Asaduzzaman 2018; Wisner et al. 2003). As a result of climate change, some places are more disposed to ecological ruination, or slow environmental harm, and some lifeworlds and livelihoods are more susceptible than others to the negative consequences of climate change (Galaz 2014; Oliver-Smith 1996, 1999; Wisner, Gaillard, and Kelman 2012).[5]

The impacts of climate change in terms of frequent storms, as faced by Vietnamese society, epitomise the way in which the Asia-Pacific region is particularly susceptible to recurrent storms and water related disasters in the époque of the Anthropocene (UNESCAP, ADB, and UNDP 2010). In Vietnam, the average temperature has increased by 7°C and sea levels have risen by 20 centimetres, over the last 50 years (*Reliefweb*, 5 October 2011; see also IPCC 2018). These changes are officially recognised in Vietnam as consequences of climate change and the Vietnamese Government has signed the United Nations Framework Convention on Climate Change and the Kyoto Protocol.

Hence, Vietnam's Prime Minister Nguyen Tan Dung approved in November 2007 a national strategy on natural disaster control to 2020 and, moreover, ratified a national climate disaster coping programme on 2 December 2008 (Government of Vietnam, Decision 158/2008/QD-TTg; *Reliefweb* 5 October 2011). The national climate change coping and management programme is run by the Vietnam Institute of Meteorology, Hydrology and Environment (IMHEN), which is under the Ministry of Natural Resources and Environment and provides the guidelines for local disaster mitigation and coping strategies (Government of Vietnam, Decision 158/2008/QD-TTg; *Reliefweb* 5 October 2011).

According to the Long Lanh People's Committee and Climate Mitigation Group, climate change has resulted in more frequent and serious storms in coastal Vietnam. Virtually every fall, the central coastal area is confronted with the realities of the Anthropocene as exemplified by the storms brought to the fore in this chapter (see, e.g., *The Sydney Morning Herald*, 26 December 2017). These storms provide a backdrop for local memory and create an underpinning insecurity regarding the ravaging forces of potential future disasters. The storm Durian, for instance, tore into central and southern Vietnam in November 2006 to kill at least 46 people and sink hundreds of fishing boats (*The New York Times*, 5 November 2006) while the devastating storm Linda slammed into southern Vietnam on 2 November 1997 to kill almost 500 persons, injured 800–900 people, disappear more than 3,200 persons, sink more than 3,000 boats, and destroy more than 200,000 houses (*Reliefweb*, 14 November 1997).

More recently, Nari, also known as storm number 11 (*Bão số 11*) came in from the Philippines to hit central coastal Vietnam in October 2013. The storm had killed 13 people and displaced more than 43,000 persons in the Philippines (*Reliefweb* 2013; *Weather Underground*, 10 November 2013). Soon thereafter, in November 2013, super typhoon Haiyan (Yolanda),[6] introduced in the opening chapter of this volume and discussed in several chapters, headed for Vietnam from the Philippines. Incoming reports about the terrifying havoc wrecked by Haiyan in the Philippines led the Vietnamese government to carry out a comprehensive mitigation strategy, which included evacuation of almost 900,000 people from exposed coastal areas. When reaching Vietnam, Haiyan was downgraded from a Category 5 to a Category 1 storm, yet it killed 13 persons, injured 81 and caused severe material damage in the Vietnamese context (Red Cross and Red Crescent Societies, 12 November 2013; *Weather Underground*, 10 November 2013).

Associating a surge in the number of serious storms with climate change, local authorities are dedicated, they emphasized, to pepetually improve disaster mitigation plans for instance by demanding that all wards develop and implement storm warning systems and coping strategies. As a gender specific measure during storms single female-headed households, which are thought to be in ample need of manpower to prepare their houses, are offered support by male volunteers or the army. When a storm is approaching, Long Lanh inhabitants told, evacuation centres are set up and patrolled by police officers and firearm forces.

The sea, the boats and the patriliny

Analysing the interactional dynamics of various types of crises means to avoid conflating the disruption caused by a crisis of emergency with other socioeconomic antecedents, such as the patriliny. In the Vietnamese setting, the patriliny not only frames daily life but also allows for the fostering of particular types of gendered crises in the household and society at large (see also King and Stone 2010; Sandgren 2009). Gender provides an analytical lens for critical assessments of how socio-cultural logics stimulate the fabrication of particular images, narratives and practices in regard to men and women, yet my data encourage a focus on the ways in which masculinity in a patrilineally organised world assigns men with privileges and powers that shape daily life (cf. Butler 2004; Petersson and Runyan 2013).

I thus explore how "men must signify possession of a masculine self to maintain a superior position in the gender hierarchy" (Schrock and Schwalbe 2009, 280) and thereby contribute to the perpetuation of gendered inequality. I do so to identify how masculinity, male powers and privileges in Long Lanh, first, interact with a crisis of emergency caused by a storm, and, second, incite or augment various types of gendered crises in the local context (cf. Hearn 2015; Louie and Low 2003; Trinh 2008).

Even though not all men adhere to a pervasive masculinity, they are under pressure to confirm to influential ideas about men, maleness and masculinity as are women to assumptions about women, femaleness and femininity. Whenever engaging with pervasive images and narratives about maleness and masculinity, men can be seen as facilitating the (re-)production of a hegemonic masculinity understood as "a social ascendancy achieved in a play of social forces that extends beyond contests of brute power into the organization of private life and cultural processes" (Connell 1987, 184).

The patrilineal and heterosexual family remains common in Vietnamese society despite an increase in same-sex partnership, single-headed households and cohabitation prior to marriage (Braemer 2014; Horton 2014; Newton 2012; Nguyen 2015; Rydstrom 2016). The 'Pillar of the House' (*trụ cột gia đình;* literally the pillar of the family) is a role which the most senior male of a household is supposed to hold. The continuation of the patrilineage is considered critical and male progeny is thus appreciated. As 'inside lineage' (*họ nội*), sons enjoy privileges and powers which daughters do not. Descent traced along the male line means that daughters come to stand in a position of exteriority to the patrilineage as 'outside lineage' (*họ ngoai*) (Rydstrom 2003a, 2003b; Sandgren 2009). The special status of sons was described by Van, who at the same time also highlighted how masculinity and the patriliny are intimately intertwined with the livelihood of fishing:

My father does not allow me to work as a fisherman, because I am the only son in my family. Fishermen are people who live by the sea and who even might die by the sea. So, fishermen could face a sudden death [at sea]. When they die, their body might never be found and the family will have to bury an empty coffin.

The patriliny intersects with the occupation of fishing, which mostly is held by men, and cultivates a masculinity defined by a livelihood at sea, a vast ocean and the sociality shared by men on the boats and even on land. While loving the ocean and their boats, as Long Lanh fishermen told, they are also painfully aware that the sea is powerful and occasionally takes lives. Liem, a local fisherman, thus recalled how his uncle never returned after going fishing because his boat had been caught in an off-coast storm which he sadly could not escape.

The boats leave early in the morning or late in the evening, and while some of the larger vessels might be at sea for weeks or even for months, most boats return within 12 hours to sell their catch. Men are the owners of the fishing vessels or roundabout boats, or they rent boats. Teams, usually consisting of men only, work on the vessels while a roundabout boat is an individual task. Working in teams (i.e., 4–10 men) on the vessels, means "to stick together", as Duc, a Long Lanh fisherman underscored. The teams, he explained, resemble military platoons and the brotherhood found in the army in the sense that men work together, exchange experiences and join forces to fight storms' (cf. Rydstrom 2012). When returning with their boats, fishermen gather in bars and cafés to drink, gamble and maybe even buy sex (cf. Horton and Rydstrom 2011; Nguyen-vo 2008).

Gendered harmony, ontologies and dichotomies

In Long Lanh and in Vietnam more broadly, men and women usually are rendered meaningful in a dichotomic and ontological way with men being associated with masculinity and women with femininity. Recognised as the result of the merging of 'physiology' (*sinh lý học*), 'psychology' (*tâm lý*) and a person's 'character' (*tính cách*), male bodies are connected with the forces of *Dương* (Yang in Chinese) and female bodies with the forces of *Âm* (Yin in Chinese). Due to the forces of *Dương*, a man is said to be 'hot' (*nóng;* also meaning bad tempered), while a female body is thought to be 'cold' (*lạnh*) due to the forces of *Âm* (see also Sandgren 2009). 'Hot' bodies are associated with active and centripetal energies, or masculinity, and 'cool' bodies with passive and centrifugal energies, or femininity. Ideally, these two forces would complement one another and foster harmony in the household and by extension in society (Jamieson 1993; Leshkowich 2008; Ngo T.N.B. 2004).

Women's inferior position is expected to be balanced through a sociality of femininity called *tình cảm* (sentiments/emotions/feelings) and the qualities with which it is associated including showing 'respect' (*kính*), 'self-denial' (*nhường*), 'endurance' (*chịu*) and 'holding back oneself' (*nhịn*). Even men have *tình cảm*, but living with *tình cảm* is critical for girls and women as a social capacity by which their 'good morality' (*đạo đức tốt*) can be demonstrated and the asymmetrical reciprocity intrinsic to the patriliny coped with.[7] The femininity associated with *tình cảm* is assumed to stimulate the fostering of a 'Happy and Harmonious Family Life' (*gia đình hạnh phúc hòa thuận*) (Rydstrom 2003b, 2017). Women are expected to remain calm by practicing *tình cảm* and in doing so prevent the rice from 'boiling' (*sôi*); or rather a husband from being 'enraged' (*nổi khùng*) and maybe even violent. A patrilineal organisation, and the ways in which it facilitates the production of masculinity and femininity, creates "openings towards violence [and] towards misogyny" (Connell 1987, 185–6). A Long Lanh woman, however, is expected to facilitate peace in the household and avoid conflicts and violence (see Kwiatkowski 2008, 2011; Rydstrom 2017).

Storms, crises and emergency

An incoming storm brings about serious hazards and ramifications. Together with her family, Mai lives near one of the recently constructed roads in a newly built white washed two-story house. Thanks to their solid house, the family is able to cope with the recurrent storms as was the case in November 2013, when super typhoon Haiyan hit southern and central Vietnam. Vietnamese authorities, Mai recalled, had predicted "a very horrible" storm and inhabitants had hurried to the stores to buy equipment to strengthen their houses before the storm came in. Mai and her family took hide in their safe house by which they were protected while waiting for the storm to pass. The instant crisis brought by the storm became a bracketing of daily life; an incident after which life could continue as usual for Mai and her family.

When a storm is forecasted, as Mai explained, "men do the heavy tasks", which would include preparing the house and saving the boats. Women, on the other hand, are in charge of the household. This common gender specific division of labour in times of disasters fortifies men's proximity to the sea. While the generations of fathers and grandfathers are well aware of the dangers involved in saving the boats during a storm, and therefore take appropriate precautions such as saving the boats in advance of an incoming storm, younger men show what local authorities deem as poor judgement.

Young men usually rush to the shore to save their boats during a typhoon while items are whirling around and the sea is running wild. When the storm Durian ruptured in Long Lanh in 2006, Hoa's eldest son, for instance stayed put on the shore to protect his boat. He experienced the dangers of a catastrophe first hand when knocked unconscious by a hovering door but luckily survived the incident. Tu, another local fisherman, remembered how his father had struggled with the disasters and finally decided to get rid of his boats due to the perils involved in keeping these safe whenever a storm came in:

> He [i.e., father] used to have his own boats, but he sold all of them because of the storms. If we own the boats, we must take care of the boats to keep them safe during a storm. If we [are hired to] work on other people's boats, we don't need to [save boats]. We can even take days off if we want to and if we are sick, we don't need to worry about the boats.

Men are eager to protect the valuable boats and prevent them from drifting off shore during a storm. Losing one's boat means losing one's livelihood, something which would have devastating consequences for a fisherman and his family.

Storms, crises and renewal

Being the 'Pillar of the House' is associated with masculine responsibilities, which take new shape when a disaster breaks out. Quang, a fisherman from Long Lanh, explained how men typically would be in charge of protecting house and home during a storm:

> It is necessary that one person stays at home [when a storm is coming in]. The local authorities also told us to assign one person [in the family] to not

leave the home to confront and fight the storm. We [men] are more flexible than women and children so in some families, one or two men would stay [during the storm].

As already mentioned, about a month prior to Haiyan, the storm Nari was forecasted. Luan, a retired soldier who even has worked as a welder and in a state-owned firm, decided to evacuate his family as the family's house was unlikely to stand the storm. After his family had left for safety at a relative's place, Luan and his father tried to solidify the shutters and doors with used steel wire hoping the house would stay somewhat intact during the storm. When Luan's family returned to their house after the storm had passed, the roof was missing and the ruination of the home a reality. In the aftermath of a disaster repairing and rebuilding their home become critical as an effort to bring life back to how it used to be prior to the disaster. Luan and his family could rely on their savings and therefore both renovate and improve their house after the storm (cf. Godfrey and Torres 2016).

For Quang's household, Nari was a different experience due to the family's limited savings. Yet, the crisis of emergency eventually meant that also Quang and his family could improve their house after the storm. For a fisherman married to a small trader, Quang noted, a household's income is not only unstable but even unpredictable, because, as he explained, "some days I earn money and other days I don't earn any money. If I don't get any money, my wife will also grumble". Building an economic buffer for times of disasters has been a challenge for Quang's household, not least because the couple is devoted to invest any economic surplus in their 18-year-old daughter's education. (Her 6-year-old brother is in public school, which is covered by the government.)

When Nari was approaching the coast of Longh Lanh, Quang used ropes to tie the corrugated iron roof of the family's small one-level house. In spite of the preparations, the roof was ripped off by the powerful winds and the front of the house damaged. In the wake of the storm, the local government would provide economic support to survivors given a family had no economic resources and their house had been totally destroyed, Quang explained. While Quang's household used to be classified as 'poor' (*nhà nghèo*), the family has in spite of economic hardship been able to increase its income and distance itself from the category of impoverishment. The damages of the house and the family's improved economic status meant that it did not qualify for public renovation support. Hence, Quang and his wife had to use their few savings and in addition take loan from neighbours to repair and renew their house.

Storms, crises and chronicity

Women who do not act with *tình cảm* tend to be seen as unable to balance a 'hot' partner by 'cooling' him down. Such assumed feminine 'shortcomings' are frequently taken as justification for a man's beating of his wife by men, women and even by official representatives (Rydstrom 2003b, 2017; see also Kwiatkowski 2008). Thus, a woman who 'grumbles/complains' (*càm ràm*) would not uncommonly be condemned in the Vietnamese context for not stimulating harmony in the household through the demonstration of *tình cảm* (Rydstrom 2003a, 2019; see also Nguyen and Rydstrom 2018; Yount et al. 2015).

This is the backdrop against which testimonies of male-to-female violence should be understood and the ways in which such experiences might be minimised by those involved. Oanh from Long Lanh, for instance, recalled the abuse to which she had been subjected by her husband while at the same time downplaying the incident. She thus told: "I was beaten once by my husband because I 'complained'. It was not a big thing. He was drunk and hot-tempered and then he beat me". Giang, who also lives in Long Lanh, told about the violent marriage in which she stayed for years and how her husband repeatedly abused her. Resembling Oanh's experiences about intimate partner violence, Giang recalled how she, as she said, had begun to 'grumble/complain' about her husband's extramarital affairs and how he had responded with violence. To escape her husband's abuse, Giang would often flee to her parent's place together with the couple's children, yet she would always return to her abusive husband.

Violence is prohibited by Vietnamese laws but as elsewhere legislation aimed at protecting women and girls (and anybody else) from violence might not always be sufficiently implemented.[8] A domestic space in which male violence is inflicted upon females with impunity, takes shape as a 'Zone of Exception' (Agamben 1998; Mbembe 2003), in which the laws do not apply and a beaten woman is caught in her home in a crisis which might be experienced as permanent (Rydstrom 2017). After a particularly horrifying and brutal incident of abuse, Giang fled from her violent husband with her children to her parents' house but this time they stayed for good. However, the expanded household struggles to survive regardless of Giang's devotedness to increase the family's income by taking various jobs. Even today, the family is living hand-to-mouth and the household has been classified as 'poor' by the authorities.

For Giang and her family, the storm Nari became a socio-economic tilting point by exhausting Giang, a violence survivor, and by worsening the family's constrained financial situation (cf. Peluso and Watts 2001). The house in which Giang and her family live usually is flooded during storms because it is located below sea level. When Nari swept over Long Lanh, Giang's father refused to leave the house unguarded and the family therefore stayed in their modest and fragile house while witnessing how the roof was ripped off and a wall collapsed.

In the aftermath of the storm, the family came to realise that the damages on their house did not qualify for any governmental renovation funding, only for minor emergency support. Giang and her family could not afford the materials needed for the rebuilding of their house, so they bought corrugated iron sheets, which had been whirled off their neighbours' houses during the storm. Thanks to the second-hand material, Giang's father, brother and son could repair the family's house, however, not sufficiently and even today the house has not yet been fully reconstructed. A ruining storm engaged with gender specific predisaster crises conditions thus turned a crisis of emergency into a crisis marked by chronicity.

Conclusions

The recurrent storms in coastal Vietnam highlight the ways in which a climate disaster produces and facilitates varied types of crises. While differing in temporality, intensity and modality, existing crises experiences, realities and conditions engage with a crisis of emergency caused by a climate disaster. Depending on gender, age, sexuality, ethnicity, class and bodyableness, a crisis brought by a climate disaster imposes gender-specific hazards and ramifications.

A crisis of emergency caused by a climate disaster might be a parenthesis of daily life, a path to renewal, or an incident which transmutes into a more permanent crisis condition. When impairing socio-economic conditions engage with a climate disaster, the horizon of a crisis of emergency is rendered opaque, as are the prospects of its closure. A crisis can become as a new normalcy in terms of a context of disordered order, which brings ruination and inflicts slow harm upon lifeworlds, livelihoods and environments in the era of the Anthropocene.

At the same time, however, a crisis is volatile as a phenomenon, as an experience and as a political construction because of the ways in which it perpetually oscillates between various crisis characteristics including a breakdown of the order of daily life, an event that can be overcome and result in positive change, or a hampering context (Vigh 2008; Walby 2015). Though, even when a crisis seems chronic, resilience, resistance and agency point towards changes which go beyond the difficulties of a present framed by gender-specific disaster provoked hazards and ramifications.

Acknowledgements

The chapter is part of a current research project on "Climate Disasters and Gendered Violence in Asia: A Study on the Vulnerability and (In)Security of Women and Girls in the Aftermath of Recent Catastrophes in Pakistan, the Philippines, and Vietnam" funded by the Swedish Research Council (*Vetenskapsrådet*). Project participants include Helle Rydstrom (PI), Department of Gender Studies, Lund University, Sweden, Catarina Kinnvall, Department of Political Science, Lund University and Huong Nguyen, Department of Gender Studies, Lund University and Department of Anthropology, Hanoi University, Vietnam. I wish to recognise my collaboration with the Vietnamese Institute of Gender and Family Studies; Vietnam National Institute of Educational Sciences; and the Department of Anthropology, Hanoi University, Vietnam. I am grateful to Jonathan Rigg and the Asia Research Institute at the National University of Singapore for generously hosting me as a visiting scholar in the winter of 2017–2018. I appreciate the very kind and invaluable fieldwork related support of Dinh Hoai, Ho Ha, Ngo Huong, Nguyen Hanh, Nguyen Mai, Nguyen Huu Minh, Nguyen Huong, Pham Hoan, Thanh Xuan, Tran Nhung and Tran Thuan. Communication with Don Kulick, Nguyen-vo Thu-huong and the Pufendorf Institute for Advanced Studies Theme on CRISIS has been stimulating. Minor portions of this chapter connote an article, which I have published in *Ethnos* under the title "Disaster, Ruins, and Crises: Masculinity and Ramifications of Storms in Vietnam" (2019).

Notes

1 See Introduction for a definition of disaster. See also the United Nation's International Strategy for Disaster Reduction (UNISDR) and Bradshaw (2013).
2 In this chapter, I am drawing on ethnographic data gathered during various periods of anthropological fieldwork conducted in Vietnam. I carried out fieldwork from 1994 to 1995 in a northern rural commune, which I call Thinh Tri, to study gender socialisation (e.g., Rydstrom 2003a, 2009). From 2000 to 2001, I conducted yet another fieldwork in Thinh Tri now with a focus on violence and sexuality studied in an inter-generational perspective (e.g., Rydstrom 2012). In 2004, 2006, 2012–2013, and 2016, I conducted periods of fieldwork in the larger region of Hanoi and in a semi-rural area (i.e., called Quang Vinh) to study gender. These research projects have been funded by the Swedish International Development Coorporation Agency; the Swedish Foundation for Humanities and Social Sciences (*Riksbankens Jubileumsfond*); and the Swedish Research Council (*Vetenskapsrådet*). This chapter in particular refers to the gender and disaster study funded by the Swedish Research Council. In 2015, I collected data in central coastal Vietnam in the community, which I call Long Lanh. The material includes 35 in-depth interviews with Long Lanh men and women (ages 20–65) as well as same-sex focus group interviews with local inhabitants and interviews with representatives of the Province and Long Lanh Women's Union; Long Lanh Farmer Union; Long Lanh Ministry of Labour, Invalids, and Social Affairs (MOLISA); Province and Long Lanh People's Committee; and the Province Teacher Training College. In addition to the qualitative data, a questionnaire was distributed to about 50 men (ages 20–65) and 50 women (ages 20–65) on disaster, gender, crisis and violence. Furthermore, the data include observations in the community and communication with Vietnamese scholars and organisations. A follow-up fieldwork was carried out in Long Lanh in 2017 of which Huong T. Nguyen was in charge while a Post-Doctoral Research Fellow in the 'Climate Disasters and Gendered Violence in Asia' project. This fieldwork offers a project point of reference with 12 in-depth interviews with men (ages 20–65) on climate disasters, livelihood and masculinity as well as observations and communication with Vietnamese activists and organisations. The Long Lanh fieldwork material also benefits from previous studies in the region, in which I have been involved, as well as secondary sources, official reports and other types of data such as legislation and media debates. The possibility to compare project data from Vietnam and the Philippines also has inspired the analysis (see Nguyen and Rydstrom 2018). For anonymity reasons, all names of persons and places referred to in this chapter are pseudonyms.
3 I use the notion of the Anthropocene as an indication of registered and emergent climate changes manifested as ruining storms. For a more comprehensive discussion of the Anthropocene, see the introduction to this volume but see also Donna Haraway et al. (2015) and Bruno Latour (2014).
4 For details on climate change, see Victor Galaz (2014, 3).
5 For a definition of climate change, see introduction and UNISDR (2009).
6 See *Reliefweb* (2013) for an overview of the damages caused by Haiyan.
7 On gender and asymmetrical reciprocity, see Iris M. Young (1997).
8 See Law on Domestic Violence Prevention and Control, 2007; Law on Marriage and Family of 2000; and Penal Code No. 15/1999/QH10 (Art. 104).

References

Adger, Neil W. 2006. "Vulnerability." *Global Environmental Change* 16: 28–281.
Agamben, Giorgio. 1998. *Homo Sacer: Sovereign Power and Bare Life*. Stanford: Stanford University Press.
Agamben, Giorgio. 2005. *State of Exception*. Chicago: The University of Chicago Press.

Arendt, Hannah. 1970. *On Violence*. London and New York: Harcourt Brace and Company.

Benjamin, Walter. 1999. *Illuminations* (2nd ed.). London: Pimlico Press.

Bhabha, Homi K. 1994. *The Location of Culture*. London and New York: Routledge.

Bradshaw, Sarah. 2013. *Gender, Development and Disasters*. Cheltenham and Northampton: Edward Elgar.

Braemer, Marie. 2014. *Love Matters: Dilemmas of Desire in Transcultural Relationships in Hanoi*. Ph.D. Dissertation. Aarhus University, Aarhus.

Butler, Judith. 2004. *Undoing Gender*. London and NY: Routledge.

Butler, Judith. 2010. *Frames of War*. London and New York: Verso.

Connell, Raewyn W. 1987. *Gender and Power: Society, the Person, and Sexual Politics*. Stanford: Stanford University Press.

Crutzen, Paul, and Eugene F. Stoermer. 2000. "The 'Anthropocene.'" *Global Change Newsletter* 41: 1–17.

Das, Veena. 2000. "The Act of Witnessing: Violence, Poisonous Knowledge, and Subjectivity." In *Violence and Subjectivity*, edited by Veena Das, Arthur Kleinman, Mamphela Ramphele and Pamela Reynolds, 205–26. Berkeley: University of California Press.

Deleuze, Gilles, and Félix Guattari. 2000. *A Thousand Plateaus*. London and New York: Continuum.

Denton, Fatme. 2002. "Climate Change Vulnerability, Impacts, and Adaptation: Why Does Gender Matter?" *Gender & Development* 10, no. 2: 10–20.

Enarson, Elaine. 2000. "Gender and Natural Disasters." In *Focus Program on Crisis Response and Reconstruction*. Geneva: International Labor Organization (ILO).

Enarson, Elaine, and Dhar Chakrabarti. 2009. *Women, Gender and Disaster*. New York: Springer.

Enarson, Elaine, Alice Fothergill, and Lori Peek. 2007. "Gender and Disaster: Foundations and Directions." In *Handbook of Disaster Research*, edited by Havidán Rodríguez, Enrico L. Quarantelli, and Russell R. Dynes, 130–46. New York: Springer.

Enarson, Elaine, and Bob Pease. 2016. "The Gendered Terrain of Disaster: Thinking about Men and Masculinities." In *Men, Masculinities, and Disasters*, edited by Elaine Enarson and Bob Pease, 3–19. London and New York: Routledge.Endres, Kirsten, and Maria Six-Hohenbalken. 2014. "Introduction to 'Risks, Ruptures and Uncertainties: Dealing with Crisis in Asia's Emerging Economies." *The Cambridge Journal of Anthropology* 32, no. 2: 42–48.

Etymology Dictionary "Crisis" https://www.etymonline.com/word/crisis [accessed 28 March 2019].

Fordham, Maureen, ed. 2011. *Gender, Sexuality, and Disaster: The Routledge Handbook of Hazards and Risk Reduction*. London, New York: Routledge.

Fordham, Maureen, William E. Lovekamp, Deborah S. K. Thomas, and Brenda D. Phillips, eds. 2013. "Understanding Social Vulnerability." In *Social Vulnerability to Disasters*, edited by Deborah S. K. Thomas, Brenda D. Phillips, William E. Lovekamp, and Alice Fothergill, 1–32. London and New York: CRC Press.

Galaz, Victor. 2014. "Planetary terra incognita." In *Global Environmental Governance, Technology and Politics*, edited by Victor Galaz, 1–20. Cheltenham and Northampton: Edward Elgar Publishing Limited.

Ginige, Kanchana, Dilanthi Amaratunga, and Richard Haigh. 2014. "Tackling Women's Vulnerabilities through Integrating a Gender Perspective into Disaster Risk Reduction in the Built Environment." *Procedia: Economics and Finance* 18: 327–35.

Godfrey, Phoebe, and Denise Torres. 2016. "World Turning: Worlds Colliding?" In *Systematic Crises of Global Climate Change: Intersections of Race, Class and Gender*, edited by Phoebe Godfrey and Denise Torres, 1–17. London and New York: Routledge.

Government of Vietnam. 2008. "Approving the National Target for Climate Change." Decision No. 158/2008/QD-TTg. December 2. http://hethongphapluatvietnam.com/decision-no-158-2008-qd-ttg-of-december-2-2008-approving-the-national-target-program-on-response-to-climate-change.html.

Gramsci, Antonio. 1971. *Selections from the Prison Notebooks*. London: Lawrence and Wishart.

Habermas, Jürgen. 1992 [1976]. *Legitimation Crisis* (2nd ed.). Cambridge: Cambridge University Press.

Haraway, Donna, Noboru Ishikawa, Scott F. Gilbert, Karen Olwig, Anna L. Tsing, and Nils Bubandt. 2015. "Anthropologists Are Talking – About the Anthropocene." *Ethnos* 31, no. 3: 535–64.

Hearn, Jeff. 2015. *Men of the World: Genders, Globalizations, Transnational Times*. London: Sage.

Hewitt, Kenneth. 1983. "The Idea of Calamity in a Technocratic Age." In *Interpretations of Calamity*, edited by Kenneth Hewitt, 3–32. Boston: Allen and Unwin.

Hoffman, Susanna M., and Antony Oliver-Smith, eds. 2002. *Catastrophe and Culture: The Anthropology of Disaster*. School of American Research Press, James Currey.

Horton, Paul. 2014. "'I Thought I was the Only One': The Misrecognition of LGBT Youth in Contemporary Vietnam." *Culture, Health & Sexuality* 16, no. 8: 960–73.

Horton, Paul, and Helle Rydstrom. 2011. "Heterosexual Masculinity in Contemporary Vietnam: Privileges, Pleasures, Protests." *Men and Masculinities* 14, no. 5: 542–64.

Intergovernmental Panel on Climate Change (IPCC). 2018. "Global Warming of 1.5 C°." www.ipcc.ch/report/sr15/.

International Federation of Red Cross and Red Crescent Societies (IFRC). 2013. "Emergency appeal Philippines: Typhoon Haiyan," November 12. www.ifrc.org/docs/Appeals/13/MDRPH014pea.pdf.

Jamieson, Neil L. 1993. *Understanding Vietnam*. Berkeley: University of California Press.

King, Diane, and Linda Stone. 2010. "Lineal Masculinity: Gendered Memory within Patriliny." *American Ethnologist* 37, no. 2: 323–36.

Kwiatkowski, Lynn. 2008. "Political Economy and the Health and Vulnerability of Battered Women in Northern Vietnam." *The Economics of Health and Wellness* 26: 199–226.

Kwiatkowski, Lynn. 2011. "Cultural Politics of a Global/Local Health Program for Battered Women in Vietnam." In *Anthropology at the Front Lines of Gender Based Violence*, edited by Wies, J. and H. J. Haldane, 139–65. Nashville: Vanderbilt University Press.

Latour, Bruno. 2014. "Anthropology at the Time of the Anthropocene. A Personal View of What is to Be Studied." Distinguished Lecture at the American Anthropologists Association Meeting in Washington, December 2014.

Law on Domestic Violence Prevention and Control. 2007. Law No. 02/2007/QH12.

Leshkowich, Ann Marie. 2008. "Entrepreneurial Families." *Education About Asia* 13, no. 1: 11–16.

Louie, Kam, and Morgan Low, eds. 2003. *Asian Masculinities: The Meaning and Practice of Manhood in China and Japan*. London, New York: Routledge.

MacGregor, Sherilyn. 2010. "A Stranger Silence Still: The Need for Feminist Social Research on Climate Change." *Sociological Review* 57: 124–40.

MacGregor, Sherilyn, ed. 2017. *Routledge Handbook of Gender and Environment*. London, New York: Routledge.

Marriage and Family Law. 2000. 22/2000/QH10, June 9, 2000. Official Gazette, no. 28. (31-7-2000). National Assembly of Vietnam.

Mbembe, Achille. 2003. "Necropolitics." *Public Culture* 15, no. 1: 11–40.

Merriam-Webster "Crisis." https://www.merriam-webster.com/dictionary/crisis [accessed 28 March 2019].

Momtaz, Salim, and Muhammad Asaduzzaman. 2018. *Climate Change Impacts and Women's Livelihood: Vulnerability in Developing Countries*. London and New York: Routledge.

Newton, Natalie. N. 2012. *A Queer Political Economy of "Community": Gender, Space, and the Transnational Politics of Community for Vietnamese Lesbians (les) in Saigon*. Ph.D. Dissertation, University of California, Irvine.

The New York Times. 2006. "Typhoon Durian Tears into Southern Vietnam." *The New York Times*, November 5. www.nytimes.com/2006/12/05/world/asia/05iht-storm.3784111.html.

Ngo, Thi Ngan Binh. 2004. "The Confucian Four Feminine Virtues (*Tu Duc*)." In *Gender Practices in Contemporary Vietnam*, edited by Lisa Drummond and Helle Rydstrom, 47–74. Singapore: Singapore University Press.

Nguyen, Huong T, and Helle Rydstrom 2018. "Climate Disaster, Gender, and Violence: Men's Infliction of Harm upon Women in the Philippines and Vietnam." *Women's Studies International Forum* 71: 56–62.

Nguyen, T. N. Minh. 2015. *Vietnam's Socialist Servants: Domesticity, Class, Gender, and Identity*. London and New York: Routledge.

Nguyen-vo, Thu-huong. 2008. *The Ironies of Freedom: Sex, Culture, and Neoliberal Governance in Vietnam*. Seattle: University of Washington Press.

Nordstorm, Carolyn. 2004. *Shadows of War*. Berkeley: University of California Press.

Oliver-Smith, Anthony. 1996. "Anthropological Research on Hazards and Disasters." *Annual Review of Anthropology* 25: 303–28.

Oliver-Smith, Anthony. 1999. "What is a Disaster? Anthropological Perspectives on a Persistent Question." In *The Angry Earth*, edited by Anthony Oliver-Smith and Susanna Hoffman, 18–34. London and New York: Routledge.

Peluso, Nancy Lee, and Michael Watts, eds. 2001. *Violent Environments*. Ithaca, London: Cornell University Press.

Penal Code. 1999. No: 15/1999/QH10. National Assembly of Vietnam.

Petersson, Anne S., and Spike V. Runyan. 2013. *Global Gender Issues in the New Millennium*. London: Westview Press.

Reliefweb. 1997. "Vietnam Typhoon Linda Situation Report No. 4." November 14. https://reliefweb.int/report/viet-nam/vietnam-typhoon-linda-situation-report-no4.

Reliefweb. 2011. "Vietnam Adopts Climate Change Mitigation Strategy," Government of Vietnam. October 5. https://reliefweb.int/report/viet-nam/vietnam-adopts-climate-change-mitigation-strategy.

Reliefweb. 2013. "Typhoon Haiyan – Nov 2013." https://reliefweb.int/disaster/tc-2013-000139-phl.

Rigg, Jonathan, Carl Grundy-Warr, Lisa Law, and May Tan-Mullins. 2008. "Grounding a Natural Disaster: Thailand and the 2004 Tsunami." *Asia Pacific Viewpoint* 49, no. 2: 137–54.

Roitman, Jane. 2014. *Anti-Crisis*. Durham: Duke University Press.

Rydstrom, Helle. 2003a. *Embodying Morality: Growing Up in Rural Northern Vietnam*. Honolulu: University of Hawai'i Press.

Rydstrom, Helle. 2003b. "Encountering 'Hot Anger': Domestic Violence in Contemporary Vietnam." *Violence Against Women*, 9, no. 6: 676–97.

Rydstrom, Helle. 2009. "Moralizing Sexuality: Young Women in Rural Vietnam." In *Rethinking Morality in Anthropology*, edited by Monica Heintz, 118–36. Oxford and New York: Berghan Publishers and the Max Planck Institute for Social Anthropology, Halle.

Rydstrom, Helle. 2012. "Gendered Corporeality and Bare Lives: Local Sacrifices and Sufferings During the Vietnam War." *Signs*, 37, no. 2: 275–301.

Rydstrom, Helle. 2016. "Vietnam Women's Union and the Politics of Representation." In *Gendered Citizenship and the Politics of Representation,* edited by Hilde Danielsen, Kari Jegerstedt, Ragnhild.L. Muriaas, Brita Ytre-Arne, 209–34. Basingstoke and New York: Palgrave Macmillan.

Rydstrom, Helle. 2017. "A Zone of Exception: Gendered Violences of Family 'Happiness' in Vietnam." *Gender, Place and Culture* 24, no. 7: 1051–70.

Rydstrom, Helle. 2019. "Disasters, Ruins, and Crises: Masculinity and Ramifications of Storms in Vietnam." *Ethnos, Journal of Anthropology*, published online before print. https://doi.org/10.1080/00141844.2018.1561490

Sandgren, Steve. 2009. "'Masculine Domination': Desire and Chinese Patriliny." *Critique of Anthropology* 29, no. 3: 255–78.

Schrock, Douglas, and Michael Schwalbe. 2009. "Men, Masculinity, and Manhood Acts." *Annual Review of Sociology* 35: 277–95.

Steffen, Will, Paul J. Crutzen, and John R. McNeill. 2007. "The Anthropocene: Are Humans Now Overwhelming the great Forces of Nature?" *Royal Swedish Academy of Sciences, Ambio* 36, no. 8: 614–21.

Sternberg, Troy, ed. 2019. *Climate Hazards Crises in Asian Societies and Environments.* London, New York: Routledge.

Stoler, Ann. 2013. *Imperial Debris: Reflections on Ruins*. Durham: Duke University Press.

The Sydney Morning Herald. 2017. "Vietnam Evacuates Thousands as Typhoon Tembin Nears." *The Sydney Morning Herald*, December 26. www.smh.com.au/world/vietnam-evacuates-thousands-as-typhoon-tembin-nears-20171226-h0a2wu.html.

Trinh, Tai Quang. 2008. "Marital Conflicts and Violence against Women." In *Rural Families in Transitional Vietnam,* edited by Trinh D. Luan, Helle Rydstrom, and Wil Burghoorn, 371–99. Hanoi: Social Sciences Publishing House.

True, Jacqui. 2013. "Gendered Violence In Natural Disasters: Learning from New Orleans, Haiti and Christchurch." *Aotearoa New Zealand Social Work* 25, no. 2: 78–89.

United Nations Asian Disaster Preparedness Center (UNADPC). 2010. "Disaster Proofing the Millennium Development Goals (MDGs)." UN Millennium Campaign and the Asian Disaster Preparedness Centre. www.preventionweb.net/publications/view/16098.

United Nations Development Programme (UNDP). 2016. "Overview of Linkages between Gender and Climate Change." Report Authored by Habtezion, Senay et al. www.undp.org/content/dam/undp/library/gender/Gender%20and%20Environment/PB1-AP-Overview-Gender-and-climate-change.pdf.

United Nations Economic and Social Commission for Asia and the Pacific (UNESCAP), Asian Development Bank (ADB), and United Nations Development Programme (UNDP). 2010. "Paths to 2015: MDG Priorities in Asia and the Pacific MDG Report 2010/11." www.unescap.org/resources/asia-pacific-mdg-report-201011-paths-2015-mdg-priorities-asia-and-pacific.

United Nations International Strategy for Disaster Reduction (UNISDR). 2009. *UNISDR Terminology on Disaster Risk Reduction*. Geneva: Switzerland. www.unisdr.org/we/inform/publications/7817.

Vigh, Henrik. 2008. "Crisis and Chronicity: Anthropological Perspectives on Continuous Conflict and Decline." *Ethnos* 73, no. 1: 5–24.

Walby, Sylvia. 2015. *Crisis*. Cambridge: Polity Press.

Weather Underground, 2013, November 10. "Category 1 Typhoon Haiyan Hitting Vietnam; Extreme Damage in the Philippines". https://www.wunderground.com/blog/Jeff Masters/category-1-typhoon-haiyan-hitting-vietnam-extreme-damage-in-the-phili.html

Wisner, Ben. 2016. "Vulnerability as Concept, Model, Metric, and Tool." *Natural Hazard Science*. DOI: 10.1093/acrefore/9780199389407.013.25

Wisner, Ben, Piers Blaikie, Terry Cannon, and Ian Davis. 2003. *At Risk: Natural Hazards, People's Vulnerability and Disasters*. Routledge and UNDP follow up to the Hyogo Framework for Action 2005.

Wisner, Ben, J. C. Gaillard, and Ilan Kelman. 2012. "Framing Disaster: Theories and Stories Seeking to Understand Hazards, Vulnerability and Risk." In *The Routledge Handbook of Hazards and Disaster Risk Reduction*, edited by Ben Wisner, J. C. Gaillard, and Ilan Kelman, 18–33. London: Routledge.

Young, Iris Marion. 1997. "Asymmetrical Reciprocity: On Moral Respect, Wonder and Enlarged Thought." In *Intersecting Voices Dilemmas of Gender, Political Philosophy and Policy*, edited by Iris Marion Young. Princeton: Princeton University Press.

Yount, Kathryn M., Eilidh M. Higgins, Kristin E. Vander Ende, Kathleen H. Krause, Tran Hung Minh, Sidney Ruth Schuler, and Hoang Tu Anh. 2015. "Men's Perpetration of Intimate Partner Violence in Vietnam: Gendered Social Learning and the Challenges of Masculinity." *Men and Masculinities* 19, no. 1: 64–84.

12 In the wake of Haiyan

An ethnographic study on gendered vulnerability and resilience as a result of climatic catastrophes in the Philippines

Huong Thu Nguyen

Introduction

Thursday late in the afternoon, on 7 November 2013, Maza,[1] a housewife and native of Tacloban, was on her way home after visiting a family relative in Samar. She heard about the weather warning while she was still in Samar but did not take it seriously. She said to her cousin: "Well, just another typhoon. It will come and go". On the road, Maza saw people trying to harvest their crops. Spatters of rain alternated with patches of sunshine. Some harvested paddy was left lying on the fields, soaked wet. People did not want to bring the crops home just yet. Life seemed absolutely normal before disaster struck Tacloban, the largest city and regional centre of the Eastern Visayas of the Philippines. Three days later, as Maza recalled, "I looked at the sky and saw dozens of military planes taking off or landing. The clouds were low and the air was dense with helicopter smoke. On the streets there were lots of foreign troops. At night we heard lots of shooting".

On 8 November 2013, super Typhoon Haiyan made its first landfall on Guian, a municipality of 47,000 located at the South Eastern tip of the province of Eastern Samar. It was one of the strongest storms ever recorded and climate change was suspected to have facilitated the extreme weather conditions. As of 17 April 2014, the National Disaster Risk Reduction and Management Council (NDRRMC) officially placed the death toll of Haiyan at 6,300 with 1,061 missing. Almost two years after the deadly typhoon ravaged Tacloban and nearby islands, many survivors continued to live in uncertainty. With almost 100,000 families still living in temporary houses, accelerating the construction of the government-funded 56,000 permanent houses and accompanying facilities appeared to be the most challenging task (Pham 2015). Displaced families were living in *bunkhouses* (transitional shelters), where they had little access to livelihood and basic needs such as water and proper sanitation, and relocation sites where some women feared for their safety (Catada 2015).

Information on the ramifications of Haiyan has been abundant in both domestic and international media, while academic work on the typhoon has been limited. The topics which have been covered include the vagaries of meteorological warning (Ahmed et al. 2015; Esteban et al. 2015; Lagmay et al. 2015; Leelawat et al.

2014; Toda et al. 2015), preliminary evaluations of relief and rehabilitation efforts from paediatric field hospital perspectives (Albukrek, Mendlovic, and Marom 2014; Yamada and Galat 2014), and public health needs of pregnant women (Sato et al. 2016). Yet, only little is known about the gendered dimensions of climate disasters and the specific hazards and vulnerabilities with which particular groups struggle in catastrophic times (McSherry et al. 2015; Nguyen 2018). Even though it is beyond the scope of this chapter to undertake a detailed inquiry into the Anthropocene and its conceptualisation, used as an expression for the geological Age of Man, and a 'driver' of global environmental change (Crutzen and Stoermer 2000; Palsson et al. 2013), it encourages recognition of the quotidian and normatively accepted practices and policies that "contribute to conditions of compounded risk and precarity" (Douglass and Miller 2018, 271). This perspective appears to be particularly relevant in anthropological definitions of disaster, which, for instance, are seen as "the end result of historical processes by which human practices enhance the materially destructive and socially disruptive capacities of geophysical phenomena, technological malfunctions, and communicable diseases and inequitably distribute disaster risk according to lines of gender, race, class, and ethnicity" (Douglass and Miller 2018, 272). Considered from a gender perspective, women are particularly vulnerable to destructive ramifications of climate disasters, one of these being male abuse (Nguyen and Rydstrom 2018).

Based on in-depth interviews with some 42 typhoon survivors and representatives of both civil society organisations and governmental agencies in my fieldwork in the Philippines during the summer of 2015 and previous trips in 2014, this chapter brings to the fore ethnographic narratives on the lived experiences of women survivors, who were directly affected by Haiyan when it hit the provinces of Leyte and Eastern Samar (Eastern Visayas region). Specifically, I focus on these women's so far untold experiences of suffering, their great losses, and their frustration regarding what they saw as a failure of the authorities to provide adequate and timely relief aid and rehabilitation support. I also examine how relief and rehabilitation efforts were beset by conflicting interests among international and national stakeholders and by the dynamics of metropolitan and provincial politics set against the background of pre-existing armed conflict. All of these factors combined, I suggest, render these people, figuratively and literally, into a state of *ruination* (Stoler 2008).

Ruination is "an act perpetrated, a condition to which one is subjected, and a cause of loss. These three senses may overlap in effect but they are not similar as each has its own temporality. Each identifies different durations and moments of exposure to a range of violences and degradations that may be immediate or delayed, subcutaneous or visible, prolonged or instant, diffuse or direct" (Fanon 1967, 195–6; see also Rydstrom this volume). In this regard, the metaphorical meaning of 'ruination' aptly reflects the precariousness of those simultaneously subjected to gendered violence and climate disaster consequences. It is the social of ruination that brings about severe impairment to one's health and honour. The notion of *ruination* appears to be helpful in understanding "a complex web of societal hierarchies, powers, and privileges which conditions female insecurity and precariousness" (Nguyen and Rydstrom 2018, 57). Ruination is an active process of violence that infuses both the psycho-social and material dynamics of everyday life. As Ann Laura Stoler (2008) argues, imperial formations and the processes of ruination they embrace are also defined by and might energise racialised hierarchies. In discussing violence, Stoler (2008, 194) refers to 'ruination' as "a corrosive process that weighs on the future and shapes the present".

As processes of ongoing ruination, imperial projects materially and socially shape the present. Therefore "asking how people live in and with ruins, both as a material and social force and a sustained political project, is critical to understand a history of the present" (Shalhoub-Kevorkian 2015, 19). As Stoler (2013, 11) reminds us, "ruin is both the claim about the state of a thing and a process affecting it. It serves as both noun and verb". Such an allegory is useful when attempting to create a connection between an associated 'culture of precarity' (Ortner 2016, 57) with the legacies of western colonialism and neo-colonial practice, as it helps us to link power in the past and present (Paprocki 2018). Such experiences and memories, Stoler (2008) refers to as 'ruins' of an imperial past or, in the words of Achille Mbembe (2001), 'traces and fragments' of colonial violence and excessive abuses (see Rydstrom 2012).

Relating an exploration of ruins to violence against women, on which I focus in this chapter, such violence can be understood as ruining 'unsayability' (Mertens 2016). Metaphorically, as Esther Terry (2016, 166) argues, rape and other forms of sexual violence perpetrated against the female body, such as in the Democratic Republic of Congo, becomes a raping of the land itself; "an act of defiance waged not just over women's bodies, but over the ruined body of the Congo herself" (Gener 2010, 122). Terry affirms that the female body bears the brunt of 'double' colonisation, and the body is seen as a site for gendered readings of postcolonial subjectivity (2016, 13). This is all the more poignant, since Eastern Visayas is the very location where, in the words of Bronwyn Winter (2012, 82), "militarization, globalization, post colonialism and monotheistic religion intersect at precisely the point where women experience the most profound and enduring violence". In such a setting, a climate disaster may fuel gender-based imbalances in differing ways, related to other factors such as age, ethnicity, sexuality and financial status (see Nguyen and Rydstrom 2018).

My discussion of the intersections between climate disasters and violence is epitomised by the experiences of Maza (38 years) from Tacloban, introduced above. She has been married for 12 years and used to have one daughter and three sons but her daughter and one of her sons were tragically swept away during Haiyan. At that time, Maza was five months pregnant with her now 13-month-old son. "The tidal waves came in so fast", she remembered. Maza thus witnessed how two of her children were struggling in the water without being able to do anything to save them, as her two other children were hanging on to her to avoid being swept away. In the aftermath, her son's body was recovered and buried in a mass grave but her daughter was never found. Maza used to do laundry for a living (150–200 pesos per day, some days only 80 pesos). Her husband is a *pedicab* driver and a heavy drinker. He often beat Maza as well as the couple's children;

even the youngest son was slapped on his legs. Whenever he starts hitting, Maza wants to escape but there is no shelter in her area for battered women, so abuse has become part of her daily life.

In the aftermath of a climate catastrophe, abuse against women and girls tends to increase due to factors such as lack of safety in resettlement areas and shelters, the collapse of a society's socio-cultural infrastructure and safety systems, male abuse inflicted by a partner or a relative, and the commodification of women's bodies (see Introduction; Nguyen and Rydstrom 2018; UN and Oxfam 2012). Due to general inequalities between female and male populations, women and girls are susceptible to the consequences of climate changes (Bradshaw and Fordham 2015; Warner et al. 2010). These inequalities include scarce resources and dependency upon livelihoods which might be sensitive to climate catastrophes, limited access to media information, and little influence on family decisions and finances at home and in society more generally, as well as being mainly responsible for the household and raising children. In the Philippines, as in other places, cultural biases tend to deprive the female population of the opportunity to learn survival skills such as swimming (see Introduction; Nguyen and Rydstrom 2018).

As ruination might facilitate resistance it could, at another level, create space for a more just future under climate change (Paprocki 2018), particularly in terms of collective efforts to overcome the condition of *being ruined*. Since disasters tend to loosen existing socio-economic and political hierarchies, they offer opportunities for social mobility (Oliver-Smith 1979, 100). This is largely in line with the current discourse on expressions of collective agency for effecting transformative change, such as the just distribution of resources to support recovery and social resilience (Douglass and Miller 2018). Therefore, "the exceptional character of a post-disaster scenario can inspire the public to take stock and critically negotiate the acceptable ethics of deliberation given particular contexts" (Curato 2015, 10). One possibility is that the affected populations are "able to speak for themselves as a collective [. . .] instead of being spoken for" (2015, 11).

Drawing on research in post-tsunami Southern Thailand, Jonathan Rigg and associates (2008) argue that local residents often successfully negotiate their shattered lives with the resources available and are optimistic about their future, despite current hardships. These studies indicate that scholars can think and write in terms of disaster survivors without "collapsing them into a wider sense of victimhood" (Rigg et al. 2008, 151). It is worth noting that sometimes major disasters can bring about dramatic political changes in different ways to affected countries that have endured pre-existing armed conflicts (Nel and Righarts 2008). For example, fighting escalated in Sri Lanka within a year of the tsunami (Billon and Waizenegger 2007), whereas the increased international presence in Aceh, Indonesia, was seen as a contributing factor to the ending of the conflict eight months after the disaster struck the area (Smirl 2008). Given the possibilities of resistance and expectations of social mobility following a disaster, it is necessary to look at the larger context of the disaster because, as David Alexander argues (1997, 299–300), this "offers a basis for analysing disaster in relation to current developments and preoccupations, and with respect to their variation from place to place".

In noting the importance of cross-societal and cross-cultural differences, disaster researchers will be able to see things that "we do not see at present, being somewhat a prisoner of our own culture" (Quarantelli, Lagadec, and Boin 2007, 39). Along this line Doug Henry contends that an 'expanded horizon' is necessary for examining the complex relationships between humans, culture and their environment; and issues such as vulnerability and perceived risk, individual and social responses and coping strategies (2007, 111). Such approaches go beyond "a habitual way of looking at disaster phenomena" (Quarantelli 1982, 453), and inspires researchers to adopt a more "nuanced, international, and comparative perspective" (Enarson, Fothergill, and Peek 2007). Research has been conducted on relief efforts and recovery processes in the Global North (Peacock, Morrow, and Gladwin 1997), as well as in developing countries such as Bangladesh (Haque and Zaman 1993) and Peru (Oliver-Smith 1994). Following fieldwork in Honduras and the United States, Anthony Oliver-Smith (2009), for example, presents similarities in the ways political economic behaviours and policies of governing elites are linked with common people's experiences of disasters following the 1992 Hurricane Andrew and the 1998 Hurricane Mitch.

Haiyan survivors *en masse* have demanded immediate relief and rehabilitation and have criticised the central government for alleged 'criminal negligence' months after the calamity (People Surge 2014). Oliver-Smith points out that while disasters diminish the agency of people, turning them into passive victims, people can reclaim agency and restore a sense of meaning and direction through the reconstruction process (2009, 26). As Sherry Ortner (2016, 60) reminds us, "it is important to look at the caring and ethical dimensions of human life, for what is the point of opposing neoliberalism if we cannot imagine better ways of living and better futures? How can we be both realistic about the ugly realities of the world today and hopeful about the possibilities of changing them?".

The empirical context

The Philippines, a densely populated country of 97 million people, is one of the most climate disaster-prone countries in the world. It experiences an average of 20 typhoons per year (Oxfam 2013), due to its location along the typhoon belt in the Western Pacific. In addition, the country sits astride the earth's 'Ring of Fire', being exposed to periodic volcanic eruptions and earthquakes.

From geographical and meteorological perspectives, Tacloban is a danger spot in itself because it is below sea level. In the north and the west, the city is bound by mountains, which often means landslides, especially when wind and rain are ferocious as during Haiyan (BPI Foundation and World Wide Fund for Nature 2014). Moreover, the V-shaped geography of the area serves to funnel the mass of water, with the shallow depths of San Pedro Bay amplifying the surge heights (Esteban et al. 2015). Tacloban is a migratory city of Eastern Visayas, with a population of about 200,000. Despite the fact that the city's coastal districts are below sea level, many major buildings were built right by the sea including the city hall, schools, colleges and hospitals. Also, there is a big gap between the resident- and the commuter-adjusted population and commuters can nearly quadruple the local population during the daytime. This makes Tacloban less homogenous than its neighbour municipality of Palo. There are, furthermore, a rather large number of migrants working in informal economic activities in Tacloban.

Weather warnings: a loss in translation

The storm surges of Haiyan were predicted two days in advance with a complete list of the highest predicted storm surges (Lagmay et al. 2014). Unfortunately, these advanced warnings were not translated into appropriate action in every coastal village in the region of the Central Philippines. The general public was not familiar with certain terminologies, such as 'storm surge' and its consequences. And there was a problem of interpretation: while Tagalog is the official language of the Philippines, Waray is mostly spoken on the islands of Leyte and Samar. Some people said that if it had been referred to as a tsunami, people may have been more alert because they knew what it was, being familiar with what happened in the Indian Ocean in 2004. However, government officials argued that if it had been called a tsunami it would have triggered undue fear and panic among local residents.[2] Some blame officials and the public alike for their complacency in underestimating the grave threat of Haiyan, noting that in a country like the Philippines where typhoons are so commonplace, disasters are perceived as a way of life (Bankoff 2003). Ryan, a non-governmental organisation practitioner, noted how there was a massive movement of residents a week before the typhoon because when the storm reached signal #3 it would be serious enough to start evacuating people.[3] But no one expected such deadly storm surges and a number of local residents decided to stay behind because they either wanted to protect their property or thought their concrete houses were secure enough to withstand the oncoming typhoon. This is illuminated in the story of Carla, a social worker whose parents-in-law died in Haiyan:

> We warned them [i.e., Carla's parents-in-law] about the typhoon, but they said: "well, we've been living here for years, typhoons come and go. This is just another of the thousands that hit the area before." And you know, they are senior to us, they are more experienced. When they spoke like that, what could you do? So, they stayed in the house, and as a result they both died.

Obviously, the risk perception of Carla's parents-in-law and other local residents was based on historical experience that did not match the forecast as most people did not expect the wind generated by the typhoon would lead to a surge in the sea level (Ahmed et al. 2015). This lack of understanding of the phenomenon led to inappropriate responses to the warnings, a topic to be discussed further in the following section.

Post Haiyan relief efforts: a 'second-order typhoon'

After Haiyan's onslaught in November 2013, the local administration was completely paralysed; even the Tacloban mayor himself was a victim of the typhoon. A Manila-based senior international non-governmental organisation officer called Ning told me that there were complaints about the local authorities' poor response, but added: "How could they respond when they were themselves affected by the typhoon? They were also desperate to get food and other basic necessities". As Tacloban native, high school female teacher cum activist Jo recalls:

> I think the government's response was just too slow. We got the first relief aid from the government only in March 2014 [almost five months after Haiyan]. We got aid from my friends in both Kabalogan and Kabayo cities. My friends in Manila managed to send us a bus full of emergency goods on November 19. We received twenty kilos of rice and shared them with everybody. Evidently the ravaging storm spared no one, ordinary citizens and political elites, rich or poor. However, the rescue efforts from the central government were thought to be too slow for our fellow villagers.

The delay of the central government in coming to the rescue of Tacloban residents resulted in a chaotic situation in administering post-disaster relief. Marie, a *sari-sari* (convenience store) female shop owner whose seven-year old boy was drowned in the floodwaters, narrates:

> There was *waray namon* (no light), *waray pagkaon* (no food), *waray fado* (no clothes). Local residents grabbed whatever at hand like *bukojuice* (coconut water), *ubod* (bamboo shoot) as lots of them were scattered on the ground during the storm. No drinking water. No *bunkhouse* (transitional shelter). After days we looked at each other and asked ourselves "Oh my God what was happening here?" We women could barely cover our bodies. Our clothes were tattered. We had no underwear. We picked pieces of clothing from muddy waters, let them dry and used them as tampons for our periods. Medical care was non-existent. We used leaves and herbs to treat our wounds. We took *malunggay* leaves and bark, mixed and crushed them to use as medicines to treat the wounds. They have antiseptic qualities. At least fifty people died from injuries or starvation in our *barangay* (village). We had no water to wash ourselves. We stank terribly, flies and insects swarmed around us. Our skin was full of rashes and sores. We looted stores like Seven Eleven for food and medicines. Men, women, children, old people all did it. My fourteen-year old daughter joined in the *loloting* (looting).

Marie indicates complex emergencies typified by direct physical injury, inadequate access to medical intervention, food and water. These pre-existing factors, in combination with situations of humanitarian emergencies, underlie the structural vulnerabilities to which women were exposed and which form the normal backdrop to women's lives. In sum, as a result of the complexity of pre-existing social inequities, local women and children faced singular challenges to their health and well-being in situations of complex humanitarian emergency. The teacher activist Jo shares her observations about the situation in the wake of Haiyan's destruction:

> I think natural calamities are not disasters per se. It would only become a disaster when the affected people cannot go beyond the catastrophe situations. Most of them are poor, only a small proportion of populations can cope with the aftermath of natural calamities because they are rich. In the bunk houses the limited space can be seen as a factor contributing to the risk of women being sexually harassed, and abused. We heard stories of women trying to limit their visits to the common room during their monthly periods, or complaining about their bladder pains from holding urine too long because they were afraid of running to the common room which is indecently small and located separately from the bunkhouse in the dark evening. In that situation people did not even care about hygiene issues. They just put on clothes which they grabbed from dead bodies on the street because their own clothes were torn and almost gone while struggling to escape from the typhoon.

Talking to some residents of Guian, I learnt that three days after the typhoon had hit, they received canned fish and rice rations from the local Department of Social Welfare and Development (DSWD). The death toll recorded at 60 *barangays* of Guian municipality was 101 persons. Evalyn, a female university lecturer, explained to me that Guian had a lower death toll, mainly because there are less residential housing buildings along the shore in Guian in comparison with Tacloban. According to press reports (O'Keeffe and Baylis 2013), when President Aquino arrived in Guian on 17 November 2013, Mr Gonzales, the mayor, with his hand bandaged due to storm-related injuries, welcomed the President with an elaborate report about the typhoon's impact on the town and its people. Government aid efforts in Guian were facilitated to some extent by access to the local airport which was still operational. This was not the case with Tacloban, whose airport was severely damaged. Relief aids were airlifted to Cebu international airport before being transported by road to Tacloban. My local interlocutors indicated that transportation played a vital role in relief operations. During my fieldwork in Leyte, I also heard complaints about malpractice in the distribution of DSWD allowances and foreign aid reliefs.

One thing stood out quite clearly; the early arrival of international non-governmental organisation workers on the scene to join forces with the government and local authorities in carrying out relief work. For example, the International Red Cross, Catholic Relief Services and Caritas arrived in Guian and Palo three days after Haiyan struck. International aid was coordinated by the Philippine government through the central DSWD. As the international non-governmental organisation representative Ning told me, "international non-governmental organizations did not want to interfere with foreign aid distribution and let the government deal with this matter". But DSWD, according to Carla, "is loyal to the yellow (Aquino) government so that they are not really helpful. Given the patronage system in the government bureaucracy, they are generally very slow in responding to the aftermaths of natural disasters". The question is, however, whether the ways relief operations were carried out might be politically motivated since the head of the central government (i.e., the former President) was the leading member of a political clan diagonally opposed to another powerful political clan, the Romualdez, whose leading member happened to be the current mayor of Tacloban, Alfred Romualdez. Obviously political patronage does play a role in government responses to the aftermath of natural disasters, having an impact on relief activities at the local level. A native of Leyte government employee, Joseph, shared with me his observation:

> All aids landed in Cebu (airport) because Tacloban airport was out of service at that time. My sister lived in Tacloban. Two weeks after I went there to look for my two nieces, I had to fly to Cebu first. Near the airport, I saw three warehouses full of goods for relief but there was nobody there to pack them. You know, when DSWD received the goods from international organisations they will pack for example foods into small portions to be distributed to the household level. But the aid goods were just there for display. They (DSWD) might say that there was not enough manpower to do that. But I think the bottom line is that the government did not mobilise people to do so. That's why there were a lot of anecdotes about rotten aid foods.

Stories about aid foods being left to rot or making their way into private grocery stores were plenty, as Mark, a local male activist, told me:

> I heard from someone inside (meaning DSWD) that they had to bury two truckloads of aid foods. But all the foods were rotten. Plus the fact that part of the relief goods turned up at the grocery stores. People bought cans of food with labels saying these are donations from the People of the America.

While distribution tasks were handled by the central government via DSWD, relief work on the scene and rehabilitation efforts were undertaken by international organisations relying on the support of the local administration, namely the *Barangay Kagawat* (Village Councillor) and the *Barangay Punong* (Village Captain). There were clear cases of blatant misuse of aid as told by Melissa, a non-governmental organisation practitioner in Eastern Samar:

> Speaking of emergencies, if you had been there you would have seen that most of the goods were delivered to the *Barangay Punong* and his relatives. That's how the *Barangay Punong* got a new house. After Yolanda there were lots of construction materials given to the village such as wood, cement, etc.

Where did all these aid goods go? You should come to see the house of the *Punong*, the houses of his children, the house of his siblings.

In addition, the ways relief organisations operated did not always meet the real needs of the victims as Melissa remarked:

There is something that international organisation should be aware of. The way they often do when they come to a community and try to determine who got help from International Organisation for Migration, who got aid from the United States Agency for International Development, then to check out those who did not get anything. But the fact is, that those who [. . .] received aid [were not necessarily offered] sufficient aid and [thus] would no longer need support. As one organisation may give them just one bag of rice, another organisation may hand out some canned foods.

Amid the vagaries of bureaucratic wrangling and political mistrust, some voluntary organisations had gained the admiration of the local population for their good deeds. The work of the Taiwan-based Buddhist charity Tzu Chi was a case in point. My local interlocutors said that this Buddhist Foundation was the first organisation to come to Tacloban and its surrounding areas immediately after the typhoon. They provided local residents with cash without any conditions. Each household received about 15,000 Pesos, and 8,000 Pesos for single-headed households. Tzu Chi also gave local residents rice, chocolate (for kids), canned foods, and kitchen ware/utensils. They coordinated with *Barangay Punong* but did not rely on the *Barangay Punong* for aid distribution. They really made sure that all needy people would receive the money in a timely manner. All of my interviewed residents in Leyte expressed their appreciation for Tzu Chi, whereas other organisations appeared to them as having lots of policies and protocols regarding aid distribution procedures.

At the regional level, it should be mentioned that the relief and rehabilitation work in the wake of Haiyan was carried out against the backdrop of a long drawn-out armed conflict between government forces (i.e., the Armed Forces of the Philippines) and the Communist-led New People Army (NPA) in Eastern Visayas. In the immediate aftermath of the disaster, the Communist Party announced a 10-day ceasefire, later extended to two months in order to facilitate humanitarian work. There were subsequent charges and counter charges from both sides, accusing one another of taking advantage of the situation to make political and military gains. During the fieldwork I collected the following story from Jennifer, a Manila based sympathiser of the NPA:

In some areas where there are organised farmer organisations, local residents pool resources together to overcome Yolanda's aftermath, but they are looked upon with suspicion by the central government. The suspicion was that the local residents could not get organised without support from NPA. That's why the Government sent military troops to the area. Schools were disrupted, women were harassed by militants, many guerrilla fighters were arrested. One local civilian was killed from military operations.

Nevertheless, some government employees blamed the NPA for the lack of security after Haiyan, as in the words of Camille – a Manila based police officer:

> The locals are also affected by armed communists in some parts of the island (Leyte). Just imagine, they are displaced by the disaster but some of them could not go home because their villages are under control of armed communists.

Cataclysmic events often create an environment leading to a *brevity of peace* (Nel and Righarts 2008). As Jennifer Hyndman (2009) argues, each conflict has distinct historical and geopolitical antecedents, resulting in different post-disaster political scenarios and challenges. Damien Kingsbury (2014) points out that the 2004 tsunami created 'organisational, economic and political pressure' in both Aceh and Sri Lanka. In Aceh, facing the immense scale of destruction, both the Indonesian government and the separatist Free Aceh Movement (*Gerakan Aceh Merdeka*) saw reconstruction as the priority task and, under international pressure, made mutual concessions for a negotiated settlement: limited local autonomy for the Free Aceh Movement within the Indonesian state which had just begun its own democratising process, thus bringing an end to the 30-year armed conflict. In Sri Lanka, momentum for a compromise between the government and the Tamil Tigers was lacking. The tsunami weakened Tamil infrastructures while the government emerged rather unscathed in the aftermath. Squabbles about aid distribution in Tamil-controlled areas and mutual distrust led to renewed hostilities. The 40-year conflict ended in 2009 with the government troops defeating the Tamil Tigers, causing many deaths and leaving the bitter ethnic animosities largely unresolved.

In the case of the Philippines, Haiyan did bring about some sort of a brief truce to the pre-existing armed conflict in Eastern Visayas between the Philippine army and the NPA. However, it did not have the catalysing effect of ending the war as in Aceh. The conflict remained a stalemate militarily as both sides were weakened after the disaster and had no incentives to disrupt the relief efforts. Besides there was no momentum for negotiation prior to Haiyan to build on, not to mention the fact that the government in Manila had to deal with internal conflicts elsewhere, notably the one with Muslim separatists in Mindanao. The following section deals with how variations in figures of the reported death toll affected people whose family members were among those considered as missing or unaccounted for, as seen from spiritual, emotional and institutional perspectives.

The number games: 'the loss of all losses'

As of 17 April 2014, the National Disaster Risk Reduction and Management Council (NDRRMC) officially placed the death toll of Haiyan at 6,300 with over a thousand people missing (1,061). The figure differs markedly from other estimates which were much higher. A Roman Catholic priest from Tacloban told churchgoers that more than 15,000 people had died during the onslaught of the Super Typhoon (Sison and Felipe 2015). All my local interlocutors agreed that the official death toll was far too low compared to the actual deaths during the super typhoon. Jo tells me:

> Personally I think the death toll must be as high as 20,000. Take a district like San Jose, the worst hit during Yolanda due to its geographical position. There are ten to fifteen *barangay* in the district [so] the number of victims must have been considerably high. A friend of mine saw with his own eyes heavy equipment were used to remove many bodies in Palo City two days after the typhoon had blown over.

Melissa describes the burial scene:

> In Negros, bodies of victims from the same family were put in the same coffin which was so small that the corpses were pressed against one another.

From qualitative interviews with local residents, there were suggestions that the government's low figures were based on a desire to control the relief operations in the sense that it wanted total control of the flow of international aid. Ning opines:

> The rule of the thumb is that if the death toll reaches 10,000 relief efforts would be taken over by the United Nations. The Government did not want to hand over the entire relief program to the United Nations. Local residents were aware of that. They were not happy about that.

One of the most tragic aspects of the typhoon was the number of children swept away by the tidal surge, especially in the *barangays* along the sea shore. Carla recalls:

> Parents brought their kids to the school as a site of evacuation because they thought it was a safe place. Then they went home to have a look at their belongings, small things like televisions, radios. Just to make sure that their property was secure. They wouldn't have known the waves could reach the school. It was really a tragedy.

During my fieldwork in the *barangays* along the coasts of Tacloban and Palo, almost every woman I interviewed had lost at least one child under different circumstances. Take the case of Celia, for instance, aged 45 and separated from her husband for 8 years, who was mourning the loss of her two children: a daughter who drowned after being evacuated to a high school thought to be safe from the incoming storm, and her youngest daughter who died from untreated wounds three days after the typhoon. Angelina, one of my local guides once pointed to the mass grave at the head of the road leading to her *barangay,* telling me: "My five-year-old son lies in there". What makes such personal losses all the more unbearable for these surviving mothers, and many others, is that the bodies of their loved ones, or what remained of these bodies, were buried in mass graves scattered around Tacloban and Palo. The biggest is the mass grave of 3,000 typhoon victims at the Holy Cross Cemetery in Barangay Diit, Tacloban City. For those who have submitted DNA to the National Bureau of Investigation in the hope of recovering the bodies of their loved ones among those buried in the mass grave, the answers are yet to come. For the time being, they have to console themselves by visiting the mass grave covered with small white crosses on which they have labelled the names of their loved ones. These white crosses dotting the mass grave of the Holy Cross Cemetery symbolise the final resting place for the deceased; they mark the departure of the loved ones in the minds of surviving family members. In the Catholic tradition, as my Filipino friends and informants have often told me, prayers are very important to family members when a loved one dies. More importantly, families need to be in control of the body and to participate in preparation for viewing because this practice may help with 'closure' (Lobar, Youngblut, and Brooten 2006, 48).

This explains why Carla and her family were very relieved when the bodies of her parents-in-law were recovered, knowing that they would get a proper burial under the circumstances. But Carla is one of the fortunate few. The pain of loss is far greater if it is coupled with feelings of guilt when the loved one does not get a proper burial, particularly for parents of children swept away by the storm and whose bodies have not been found, as in the cases of Angelica and Celia. The death of a child is tremendously painful for families and the mourning and sorrow make healing from the loss difficult (Lobar, Youngblut, and Brooten 2006). My local interlocutors told me that it is important to remember the dead by gathering together or having a picnic at their grave site on All Saints Day and All Souls Day; the two dates of the Catholic calendar when family members and friends remember and pray for the dead. As Marie explains:

> For us Catholics, burying the dead is a very important matter. If we could not give them a proper burial, their spirits will haunt us – the living. This is the problem for those whose family members are missing or unaccounted for. They are resentful of the fact that these are not acknowledged as dead in the list of actual death toll. It is an affront to their souls as they are not accounted for. Especially for those who have lost their entire family and are now only on their own. Their losses were not named. Historically speaking their death would not be recorded, I found it a great offense.

For those like Sasa whose family members are listed as missing two years after Haiyan, the likelihood of their survival is practically nil. In this connection, the death toll is itself a subject of controversy, since there seems to be a lack of interest in acknowledging the real extent of the death figures.

Conclusion: 'people surge' and the possibility of resistance

From the depths of despair and facing what seems to be inept government post-disaster response, on 20 January 2014, 12,000 Haiyan survivors gathered at the Eastern Visayas State University in Tacloban City demanding immediate relief and rehabilitation for typhoon survivors and criticising the central government for alleged 'criminal negligence' months after the calamity (People Surge 2014). Thousands of participants coming from various towns devastated by Haiyan were farmers and ordinary people including people from the academe and religious organisations. Out of this protest, starting out as a group of disaster survivors in Eastern Visayas demanding government response and accountability, People Surge came into being. The name 'People Surge', or *Duluk han Katawhan,* is an allusion to the two historical People Power Revolutions in the Philippines that toppled two regimes through collective action.[4]

It also stands as a metaphor of the tidal surge that devastated the central Philippines in early November 2013. Since the launch of the People Surge Alliance in January 2014, the group movement spread to other municipalities, villages and even provinces. About one month later, in February 2014, a group of this people's organisation went to Malacanang Palace, carrying a list of their demands to the government signed by 17,000 individuals. One of their immediate demands on the government was the financial assistance of 40,000 pesos for each family of around 1.4 million families whose houses were totally devastated by the typhoon. It was reported that the President would not allow financial assistance, and did not even read the petition (Sabillo 2014). Nevertheless, as strongly indicated in the People Surge statement on the second anniversary of the Haiyan tragedy (People Surge 2015), People Surge believes that "we must intensify our demands for justice holding the responsible ultimately accountable". On 6 November 2015 – just two days before the typhoon's second anniversary, thousands of people led by People Surge marched in the streets of Tacloban City while carrying placards to protest the pace of governmental reconstruction efforts (Reuters 2015). People Surge eventually made its voice heard beyond the shores of Eastern Visayas, linking up with global discourses of climate change governance by taking part in the People's Climate March held in New York in September 2014.

The case of People Surge provides indications that "if we make the world through social practice, we can unmake and remake the world through social practice" (Ortner 2016, 63). Put differently, understanding the efforts of Haiyan survivors and People Surge social movements through the lens of ruination helps to elaborate on both the webs of power in which we live as well as the possibilities of alternative futures, or in other words 'transformative possibilities' (Ortner 2016, 63). In this regard, gendered climate injustice calls for consideration of agency. As Helena Valdés points out, "[d]isaster reduction policies and measures need to be implemented with a twofold aim: to enable societies to be resilient to natural hazards, while ensuring that development efforts decrease the vulnerability to these hazards. Sustainable development is not possible without taking into account multi-hazard risk assessments in planning daily life, and as such is an issue that impacts on the lives of both women and men" (2009, 19). In this spirit, I find it productive to approach agency as a possibility for action rather than as a capacity for action; the capacity for action is inherently embodied in any human being whereas possibilities for action can be impeded due to climate changes, for instance, which could truncate the horizon of action (Brown and Westaway 2011).

The possibilities for agency might be circumvented due to a climate catastrophe but human resilience might at the same time lead to the reclaiming of action and restoration of a collective sense of direction (Oliver-Smith 1979; True 2016). In post-tsunami Southern Thailand, local residents were able to manage their shattered lives with the resources available to overcome immediate hardship, thus indicating that survivors of a disaster do not collapse into passivity and victimhood in a wider sense (Rigg et al. 2008). The possibilities for agency also carry a positive message to the effect that cultural, social and institutional factors that drive gendered injustice against women are not inevitable and can be transformed, making this a key moment to implement the intervention. Furthermore, this conceptualisation of agency as a possibility for action seems to be in line with the participatory community-based approach in many initiatives aimed at promoting actual gender equality in the contexts of the Philippines. It is thus fitting to end my chapter with a reference to what Arjun Appadurai calls an 'ethic of possibility', which should be grounded in "those ways of thinking, feeling, and acting that increase the horizons of hope" (2013, 295).

Acknowledgements

This chapter is the result of a current research project on "Climate Disasters and Gendered Violence in Asia: A Study on the Vulnerability and (In)Security of Women and Girls in the Aftermath of Recent Catastrophes in Pakistan, the Philippines, and Vietnam" funded by the Swedish Research Council (*Vetenskapsrådet*). Project participants include Helle Rydstrom (PI), Department of Gender Studies, Lund University, Sweden; Catarina Kinnvall Department of Political Science, Lund University; and Huong Nguyen, Department of Gender Studies, Lund University and Department of Anthropology, Hanoi University, Vietnam. I would like to thank the Commission on Filipinos Overseas, the Department of Anthropology and the Department of Political Science at the University of the Philippines (Diliman), the Department of Political Science at Ateneo de Manila University, in particular Imelda Nicolas, Janet Ramos, Nestor Castro, Jean Encinas-Franco and Lourdes Rallonza for their valuable support in facilitating data collection in the Philippines. Special thanks go to Aida Santos-Maranan, Tish Vito Cruz, Melinda Sagoullas and Carmela Bastes for their unfailing support in the field. I also appreciate the kind assistance of Flor desi Tumlos, Arvin Cordeta and Rowen Petilla. I am grateful to colleagues at the Department of Gender Studies, Lund University, Sweden who provided useful comments on an earlier version of the paper, as did Jonathan Rigg at the Asia Research Institute, National University of Singapore. Last but not least thanks are also due to Helle Rydstrom and Catarina Kinnvall, the two editors of this volume who carefully read my manuscript and provided me with insightful suggestions and comments. Parts of the chapter resemble a text I published in *Violence Against Women* (2018) under the title "Gendered Vulnerabilities in Time of Natural Disasters: Male-to-Female Violence in the Philippines in the Aftermath of Super Typhoon Haiyan."

Notes

1 All personal names in this paper are pseudonymous for reasons of protection.
2 There is an ongoing debate on Tagalog terms for storm surges. Experts and historians suggest terms like *daluyong* or *humbak*. While the first has long been used to describe big waves, and can also refer to tsunami, the latter is used in several provinces today to indicate the swell at sea (Pedrasa 2013).
3 In the Philippines, public storm warnings are categorised into four levels. The maximum is level 4 which indicates a situation in which all people should have completely evacuated (Leelawat et al. 2014).
4 The first People Power is also known as the EDSA Revolution in 1986 – which was named after the main road Epifanio Delos Santos Avenue in Manila, whereat nearly 2 million people were protesting at one point. The second People Power marked the forced resignation of President Joseph Estrada in 2001.

References

Ahmed, Atiq, Ardito Kodijat, Mayfourth Luneta, and Krishna Krishnamurthy. 2015. "Typhoon Haiyan, an Extraordinary Event: A Commentary on the Complexities of Early Warning, Disaster Risk Management Societal Responses to the Typhoon." *Asian Disaster Management News* 21: 20–25.

Albukrek, Dov, Joseph Mendlovic, and Tal Marom. 2014. "Typhoon Haiyan Disaster in the Philippines: Paediatric Field Hospital Perspectives." *Journal of Emergency Medicine* 31, no. 12: 951–53.

Alexander, David. 1997. "The Study of Natural Disasters, 1977–97: Some Reflections on a Changing Field of Knowledge." *Disasters* 21, no. 4: 284–304.

Appadurai, Arjun. 2013. *The Future as Cultural Fact: Essays on the Global Condition.* London and New York: Verso.

Bank of the Philippine Islands Foundation and World Wide Fund For Nature. 2014. "Business Risk Assessment and the Management of Climate Change Impacts: Sixteen Philippines Cities." www.bpifoundation.org/environment/bpi-teams-up-with-wwf-on-16-city-climate-change-study.

Bankoff, Greg. 2003. *Cultures of Disaster: Society and Natural Hazards in the Philippines.* London, New York: RoutledgeCurzon.

Billon, Philippe, and Arno Waizenegger. 2007. "Peace in the Wake of Disaster? Secessionist Conflicts and the 2004 Indian Ocean Tsunami." *Transactions of the Institute of British Geographers* 32, no. 3: 411–27.

Bradshaw, Sarah, and Maureen Fordham. 2015. "Double Disaster: Disaster through a Gender Lens." In *Hazards, Risks and Disasters in Society,* edited by Andrew E. Collins, Samantha Jones, Bernard Manyena, and Janaka Jayawickrama, 233–51. Amsterdam: Elsevier.

Brown, Katrina and Elizabeth Westaway. 2011. "Agency, Capacity, and Resilience to Environmental Change: Lessons from Human Development, Well-being, and Disasters." *Annual Review of Environment and Resources* 36: 321–42.

Catada, Rhea. 2015. *Three Faces of Women in the Resettlement*. Oxfam in the Philippines.

Crutzen, Paul J., and Eugene F. Stoermer. 2000. "Global Change Newsletter." *The Anthropocene* 41: 17–18.

Curato, Nicole. 2015. *Voice of Care and the Voice of Justice: Public Deliberation in the State of Exception*. University of Canberra: Working Paper.

Douglass, Mike, and Michelle Ann Miller. 2018. "Disaster Justice in Asia's Urbanising Anthropocene." *Environment and Planning E: Nature and Space* 1, no. 3: 271–87.

Enarson, Elaine, Alice Fothergill, and Lori Peek. 2007. "Gender and Disaster: Foundations and Directions." In *Handbook of Disaster Research*, edited by Rodriguez, Havidian, Enrico Louis Quarantelli, and Russell R. Dynes, 130–46. New York: Springer.

Esteban, Miguel, Ven Valenzuela, Nam Yun, Takahito Mikami, Tomoya Shibayama, Ryo Matsumaru, Hiroshi Takagi, Nguyen Danh Thao, Mario De Leon, Takahiro Oyama, and Ryota Nakamura. 2015. "Typhoon Haiyan 2013 Evacuation Preparations and Awareness." *International Journal of Sustainable Future for Human Security* 3, no. 1: 1–9.

Fanon, Frantz. 1967. *Black Skin, White Masks: The Experiences of a Black Man in a White World*. New York: Grove Press.

Gener, Randy. 2010. "In Defense of Ruined." *American Theatre* 27, no. 8: 118–22.

Haque, Emdad, and Muhammad Zaman. 1993. "Human Responses to Riverine Hazards in Bangladesh: A Proposal for Sustainable Floodplain Development." *World Development* 21, no. 1: 93–107.

Henry, Dough. 2007. "Anthropological Contributions to the Study of Disasters." In *Disciplines, Disasters and Emergency Management: The Convergence and Divergence of Concepts, Issues and Trends from the Research Literature*, edited by David McEntire, 111–23. Springfield, IL: Charles C. Thomas Publisher.

Hyndman, Jennifer. 2009. "Siting Conflict and Peace in Post-Tsunami Sri Lanka and Aceh, Indonesia." *Norsk Geografisk Tidsskrift-Norwegian Journal of Geography* 63, no. 1: 89–96.

Kingsbury, Damien. 2014. "2004 Indian Ocean Tsunami: How Aceh Recovered, and Sri Lanka Declined." *The Guardian*, December 29. www.theguardian.com/commentisfree/2014/dec/29/2004-indian-ocean-tsunami-how-aceh-recovered-and-sri-lanka-declined.

Lagmay, Alfredo, Rojelee Agaton, Mark Allen C. Bahala, Jo Brianne, and Louise T. Briones. 2015. "Devastating Storm Surges of Typhoon Haiyan." *International Journal of Disaster Risk Reduction* 11: 1–12.

Leelawat, Natt, Cheery Mateo, Sandy Gaspay, Anawat Suppasri, and Fumihilo Imamura. 2014. "'Filipinos' Views on the Disaster Information for the 2013 Super Typhoon Haiyan in the Philippines." *International Journal of Sustainable Future for Human Security* 2, no. 2: 16–28.

Lobar, Sandra, Joanne Youngblut, and Dorothy Brooten. 2006. "Cross-Cultural Beliefs, Ceremonies, and Rituals Surrounding Death of a Loved One." *Pediatric Nursing* 32, no. 1: 44–50.

Mbembe, Achille. 2001. *On the Postcolony*. Berkeley: University of California Press.

McSherry, Alice, Eric Julian Manalastas, JC Gaillard, and Soledad Natalia M. Dalisay. 2015. "From Deviant to Bakla, Strong to Stronger: Mainstreaming Sexual and Gender Minorities into Disaster Risk Reduction in the Philippines". *Forum for Development Studies* 42, no. 1: 27–40.

Mertens, Charlotte. 2016. "Sexual violence in the Congo free state: Archival Traces and Present Reconfigurations." *The Australasian Review of African Studies* 37, no. 1: 6–20.

National Disaster Risk Reduction and Management Council. 2014. *Updates Re Effects of Typhoon Yolanda*. Republic of the Philippines. https://web.archive.org/

web/20141006091212/www.ndrrmc.gov.ph/attachments/article/1177/Update%20 Effects%20TY%20YOLANDA%202017%20April%202014.pdf.

Nel, Philip, and Marjolein Righarts. 2008. "Natural Disasters and the Risk of Violent Civil Conflict." *International Studies Quarterly* 52, no. 1: 159–85.

Nguyen, Huong Thu. 2018. "Gendered Vulnerabilities in Times of Natural Disasters: Male-to-Female Violence in the Philippines in the Aftermath of Super Typhoon Haiyan." *Violence Against Women* 25, no. 4: 421–40. https://doi.org/10.1177/1077801218790701.

Nguyen, Huong Thu, and Helle Rydstrom. 2018. "Climate Disaster, Gender, and Violence: Men's Infliction of Harm Upon Women in the Philippines and Vietnam." *Women's Studies International Forum* 71: 56–62.

O'Keeffe, Kate, and Paul Baylis. 2013. "Small-town Mayor's Star Rises in Typhoon Haiyan's Aftermath." [Blog post] Indonesia Real Time. *The Wall Street Journal.* December2,http://blogs.wsj.com/indonesiarealtime/2013/12/02/small-town-mayors-star-rises-in-typhoon-hayians-aftermath.

Oliver-Smith, Anthony. 1979. "The Yungay Avalanche of 1970: Anthropological Perspectives on Disaster and Social Change." *Disasters* 3, no. 1: 95–101.

Oliver-Smith, Anthony. 1994. "Peru's Five Hundred-Year Earthquake: Vulnerability in Historical Context." In *Disasters, Development, and Environment*, edited by Ann Varley, 3–48. New York: John Wiley & Sons.

Oliver-Smith, Anthony. 2009. "Anthropology and the Political Economy of Disasters." In *The Political Economy of Hazards and Disasters*, edited by Eric Jones and Arthur Murphy, 11–28. Rowman Altamira.

Oxfam International. 2013. *Typhoon Haiyan. The Responses So Far and Vital Lessons for the Philippines*. Oxford: Oxfam. www.oxfam.org/en/research/typhoon-haiyan-response-so-far-and-vital-lessons-philippines-recovery.Palsson, Gisli, Bronislaw Szerszynski, Sverker Sörlin, John Marks, Bernard Avril, Carole Crumley, Heide Hackmann, Poul Holm, John Ingram, Alan Kirman, Mercedes Pardo Buendía, and Rifka Weehuizen. 2013. "Reconceptualizing the 'Anthropos' in the Anthropocene: Integrating the Social Sciences and Humanities in Global Environmental Change Research." *Environmental Science & Policy* 28: 3–13.Paprocki, Kasia. 2018. "All That Is Solid Melts into the Bay: Anticipatory Ruination and Climate Change Adaptation." *Antipode* 51, no.1: 295–315. https://doi.org/10.1111/anti.12421

Peacock, Walter Gillis, Betty Hearn Morrow, and Hugh Gladwin, eds. 1997. *Hurricane Andrew: Ethnicity, Gender, and the Sociology of Disasters*. New York and London: Routledge.

Pedrasa, Ira. 2013. "Finding a Filipino Word for Storm Surge: 'Daluyong' or 'Humbak'." *ABS-CBN News*. November 18. www.abs-cbnnews.com/focus/11/18/13/finding-filipino-word-storm-surge-daluyong-or-humbak.

People Surge. 2014. "12,000-Strong Storm Survivors March for Justice, Rights versus Aquino's Criminal Neglect." *Samar News*. www.samarnews.com/news_clips24/news510.htm.

People Surge. 2015. "PH Disaster Survivors to the World: Intensify our Demands for Justice, Our Survival Is Non-Negotiable." [Blog post]. https://peoplesurgephils.word-press.com/ph-disaster-survivors-to-the-world-intensify-our-demands-for-justice-our-survival-is-non-negotiable/.

Pham, Hang. 2015. Sendai Framework Applied Post-Haiyan. Report from the United Nations Office for Disaster Risk Reduction – Regional Office for Asia and Pacific.

Quarantelli, Enrico L. 1982. "General and Particular Observations on Sheltering and Housing in American Disasters." *Disasters* 6, no. 4: 277–81.

Quarantelli, Enrico L., Patrick Lagadec, and Arjen Boin. 2007. "A Heuristic Approach to Future Disasters and Crises: New, Old, and In-Between Types." In *Handbook of Disaster Research*, edited by Havidian Rodriguez, Enrico Louis Quarantelli, and Russell R. Dynes, 16–41. New York: Springer.

Reuters. 2015. "Typhoon Yolanda Survivors Protest Slow Pace of Reconstruction." *CNN Philippines*, November 7. http://cnnphilippines.com/regional/2015/11/07/Typhoon-Yolanda-survivors-protest-slow-pace-of-reconstruction.html.

Rigg, Jonathan, Carl Grundy-Warr, Lisa Law, and May Tan-Mullins. 2008. "Grounding a Natural Disaster: Thailand and the 2004 Tsunami." *Asia Pacific Viewpoint* 49, no. 2: 137–54.

Rydstrom, Helle. 2012. "Gendered Corporeality and Bare Lives: Local Sacrifices and Sufferings during the Vietnam War." *Signs* 37, no. 2: 275–99.

Sabillo, Kristine. 2014. "In the Know: What Is People Surge?" *Inquirer*, February 25. http://newsinfo.inquirer.net/580604/in-the-know-what-is-people-surge#ixzz4574b3Y3f.

Sato, Mari, Yasuka Nakamura, Fumi Atogami, Ribeka Horiguchi, Raita Tamaki, Toyoko Yoshizawa, and Hitoshi Oshitani. 2016. "Immediate Needs and Concerns among Pregnant Women During and after Typhoon Haiyan (Yolanda)." *PLoS Currents* no. 8. Edition 1. DOI: 10.1371/currents.dis.29e4c0c810db47d7fd8d0d1fb782892c.

Shalhoub-Kevorkian, Nadera. 2015. "Criminalizing Pain and the Political Work of Suffering: The Case of Palestinian 'Infiltrators.'" *Borderlands e-journal* 14, no. 1: 1–28.

Sherry B. Ortner. 2016. "Dark Anthropology and its Others: Theory since the Eighties." *Hau: Journal of Ethnographic Theory* 6, no. 1: 47–73.

Sison Jr., Bebot, and Cecillie Suerte Felipe. 2015. "Yolanda Death Toll as High as 15,000, Priest Says." *Philstar*, January 5. www.philstar.com/headlines/2015/01/05/1409522/yolanda-death-toll-high-15000-priest-says.

Smirl, Lisa. 2008. "Building the Other, Constructing Ourselves: Spatial Dimensions of International Humanitarian Response." *International Political Sociology* 2, no. 3: 236–53.

Stoler, Ann L. 2013. *Imperial Debris: On Ruins and Ruination.* Durham, NC: Duke University Press.

Stoler, Ann Laura. 2008. "Imperial Debris: Reflections on Ruins and Ruination." *Cultural Anthropology* 23, no. 2: 191–219.

Terry, Esther J. 2016. "Land Rights and Womb Rights." In *A Critical Companion to Lynn Nottage*, edited by Jocelyn L. Buckner, 161–79. London and New York: Routledge.

Toda, Luigi, Justine Orduña, Rodel Lasco, and Carlos Santos. 2015. "Geography of Social Vulnerability of Haiyan-Affected Areas to Climate-Related Hazards: Case Study of Tacloban City and Ormoc City, Leyte." The Oscar Lopez Center for Climate Change Adaptation and Disaster Risk Management Foundation, Inc., The Philippines.

True, Jacqui. 2016. "Gendered Violence in Natural Disasters: Learning from New Orleans, Haiti and Christchurch." *Aotearoa New Zealand Social Work* 25, no. 2: 78–89.

United Nations Viet Nam and Oxfam. 2012. Gender Equality in Climate Change Adaptation and Disaster Risk Reduction in Viet Nam. http://www.un.org.vn/en/publications/

Valdés, Helena M. 2009. "A Gender Perspective on Disaster Risk Reduction." In *Women, Gender and Disaster: Global Issues and Initiatives*, edited by Elaine Enarson and P. G. Dhar Chakarabarti, 18–28. New Delhi: Sage Publications.

Warner, Koko, Mo Hamza, Anthony Oliver-Smith, Fabrice Renaud, and Alex Julca. 2010. "Climate Change, Environmental Degradation and Migration." *Natural Hazards* 55, no. 3: 689–715. DOI: https://doi.org/10.1007/s11069-009-9419-7

Winter, Bronwyn. 2012. "Lily Pads and Leisure Meccas: The Gendered Political Economy of Post-base and Post-9/11 Philippines." In *Gender, Power, and Military Occupations; Asia Pacific and the Middle East since 1945*, edited by Christine De Matos and Rowena Ward, 89–107. London and New York, NY: Routledge.

Yamada, Seiji, and Absalan Galat. 2014. "Typhoon Yolanda/Haiyan and Climate Justice." *Disaster Med Public Health Preparedness* 8, no. 5: 432–35.

13 Accountability for state failures to prevent sexual assault in evacuation centres and temporary shelters

A human rights-based approach

Matthew Scott

Introduction

The gendered experience of disasters has been the subject of study since at least the 1990s (Alway, Belgrave, and Smith 1998; Enarson 1998) and academic as well as practitioner focus is increasing. Insights from, for example, Bradshaw and Fordham (2015), Fisher (2010), Hines (2007), and Neumayer and Plümper (2007) clearly demonstrate how gender roles that play out in everyday life continue to operate during disasters, often with negative impacts for marginalised groups, including women and girls and LGBTI+ persons (Dominey-Howes, Gorman-Murray, and McKinnon 2014). The risk of sexual assault,[1] one form of gender-based violence that is a feature of everyday life (Ellsberg et al. 2015), has been shown to increase in some situations of disaster (Delaney and Shrader 2000; IASC 2015).

There is evidence that the experience of displacement presents a particularly heightened risk of sexual assault for women and girls (Anastario, Shehab, and Lawry 2009; Picardo, Burton, and Naponick 2010). The range of proposals to address the phenomenon focus primarily on planning and operational measures that states and humanitarian actors can play to reduce exposure (Care International 2016; IFRC 2015; Seelinger, Thuy, and Freccero 2013). Yet, despite a host of principles, guidelines, recommendations and so forth from international, regional and national-level authorities, sexual assault in situations of disaster displacement remains pervasive.

Adopting a human rights-based approach, this chapter examines the role that legal actors can play in promoting accountability for, and thereby potentially contributing to the reduction in the incidence of sexual assault in situations of disaster displacement, particularly in evacuation centres and temporary shelters. Whereas much has already been written about the importance of having an effective justice sector to prosecute offenders and respond to the needs of victims (ICJ 2016; UN Women and the Advocates for Human Rights 2011), the question of the role justice-sector actors can play in reducing the incidence of sexual assault in evacuation centres and temporary shelters remains under-examined.

Drawing on the evolving principle of due diligence obligations of states to prevent sexual assault generally, this chapter argues that justice-sector actors can contribute to accountability for and a reduction in sexual assault in evacuation centres and temporary shelters by advancing legal arguments through litigation in domestic and regional courts, as well as in wider advocacy initiatives.

The chapter focuses narrowly on the question of accountability for failures to prevent sexual assault by non-state actors in evacuation centres and temporary shelters. A wide range of other issues relating to gender-based violence in the context of a changing climate could equally have been subjected to a human rights-based analysis, but this particular issue is sufficiently narrow to allow treatment within the constraints of the chapter, whilst also permitting new arguments relevant to a pervasive phenomenon to be developed. Although the due diligence standard has been widely discussed in relation to gender-based violence more generally (Liebowitz and Goldscheid 2015), how the standard applies in the context of disaster displacement has not been examined previously.[2] The chapter thus makes a specific contribution to the literature on both gender-based violence in disasters and disaster displacement by focusing on the due diligence argument in the context of evacuation centres and temporary shelters.

As this chapter appears in a volume where authors have been invited to reflect critically on gender-based violence in the context of climate change, and with a particular awareness of the newly proposed geological epoch of the Anthropocene, the following observations situate this contribution in context.

Disaster is the consequence of a changing climate when more frequent and intense natural hazard events interact with increasingly exposed and vulnerable social settings (Wisner et al. 2004). Physical exposure and social vulnerability are the consequence of social factors, with gender consistently recognised as a significant risk amplifier (Sendai Framework for Disaster Risk Reduction 2015–2030, paragraph 4). The Anthropocene emerges as the epoch in which more people live in cities than in rural areas, which in themselves tend to entail heightened exposure to natural hazard events, particularly for more marginalised groups living in informal urban settlements. Humans in urban areas rely heavily on national and municipal infrastructure, including in situations of disasters. The state, whilst increasingly seen as but one (multifaceted) actor in any social context, will continue to play a central role in disaster risk management and climate change adaptation as we move further into the Anthropocene.

At the same time as the growth of the municipality as the dominant habitat on the planet heralds the onset of the Anthropocene, the stability of climatic systems that characterised the Holocene erodes, bringing rising sea levels, decreased food production, heightened exposure to vector-borne disease and widespread decline in living standards. Of concern from a human rights perspective are the implications for the robustness of what may already be weak state institutions, including the judiciary and other actors within the justice-sector as discussed below. In short, the Anthropocene entails more concentrated and exposed populations, heightened risks and hazards, and weaker institutions, with significant implications for the prevention of gender-based violence in situations of disaster displacement.

A human rights-based approach to gender-based violence in the Anthropocene, set out more narrowly in this contribution relating to accountability for failures to prevent sexual assault in evacuation centres and temporary shelters, presumes the existence of a state with institutions more or less willing and able to perform their functions as established under domestic legislation. But it does not presume strict adherence or uniform interpretation of obligations. The persistence of torture, enforced disappearances, extrajudicial killings and indeed the crimes of genocide, war crimes and crimes against humanity reflect continued failures by states and the international community as a whole to guarantee rights reflected in international treaties, at least one of which every state on the planet has ratified. The same can be said regarding the persistence of hunger, inadequate shelter and clothing, poor working conditions and widespread, systemic discrimination against women, minority ethnic and religious groups, persons with disabilities and so forth. These challenges can only increase in the context of climate change and other pressures associated with the Anthropocene.

However, despite having roots in the Holocene, human rights speak clearly to the experience of people in the Anthropocene. Indeed, the rise of the human rights cities movement[3] speaks volumes in this connection. Rights are not only invoked in torture chambers and in the midst of armed conflict, but rather find reflection in volumes of legislation, judicial decisions, administrative circulars, classroom discussions and civil society initiatives from village to city to national, regional and international levels. The international legal framework for the protection of human rights, reflected also within regional and domestic legal instruments, provides a language and a range of mechanisms for addressing a wide variety of harms that the majority of countries in the world have agreed amongst themselves to protect people from. Violations remain widespread within and between societies, but so too does resistance, as well as concerted efforts by state actors themselves to address myriad rights-related challenges facing people within their jurisdiction.

Further, states themselves are far from monolithic, and different state and non-state actors within a society adopt varied positions in relation to human rights. For example, ministries, departments and agencies concerned with disaster risk management and climate change adaptation may see human rights as an effective and relevant lens for shaping interventions at local and even national levels. At the same time, other actors within the same state may routinely engage in torture, enforced disappearances and extrajudicial killings. Similarly, it is not uncommon to find the judiciary engaged in protracted struggles with the government on rights-related issues. A human rights-based approach to gender-based violence in the context of disasters and climate change does not, therefore, presume that articulation of legal principles brings about concrete improvements in lived experience. Rather, it provides a lens for analysing a situation, a language for articulating grievances as well as expectations, standards for measuring progress and mechanisms for holding responsible actors accountable for their failures.

The chapter has the following structure: First, the evidence that the everyday risk of exposure to sexual assault is heightened in evacuation centres and temporary shelters is presented. Then, guidelines from the international humanitarian and human rights communities concerning measures that should be taken by responsible actors to address the pervasive risk of sexual assault in evacuation centres and temporary shelters are summarised. Next, the question whether states have a legal obligation to protect persons from sexual assault in general is addressed, before considering the specific situation of evacuation centres and temporary shelters. Finally, strategies that justice-sector actors can adopt in order to contribute towards accountability are proposed and limitations identified and discussed. The chapter concludes with the recognition that the articulation of relevant legal arguments concerning state responsibility for the prevention of sexual assault in evacuation centres and temporary shelters may have more or less leverage depending on the particular context in which it is invoked, and that a reduction in sexual assault in evacuation centres and temporary shelters requires local actors and vernacular strategies.

The risk of sexual assault, a feature of everyday life is heightened in evacuation centres and temporary shelters

Sexual assault is widespread and endemic in many countries (Ellsberg et al. 2015). Multiple studies demonstrate that this already high risk of exposure to sexual assault is exacerbated in situations of disaster, with particular risks associated with situations of disaster displacement (Enarson 1998; IFRC 2018; UN Women 2016). The literature provides a number of interconnected explanations for the heightened risk of exposure to sexual assault in situations of disaster displacement. Some factors relate to the personal circumstances of individuals, both perpetrators of sexual assault as well as (potential) victims. Other studies focus on the conditions in shelters for persons displaced in this context.

Individual factors that increase exposure to sexual assault include separation from family members and wider social networks. Scarcity of food, water and shelter can increase the risk of forced and/or coerced prostitution, child and/ or forced marriage, trafficking for sexual exploitation and/or forced/domestic labour, amongst other forms of violence (IASC 2015). Intimate partner violence increases, with stress and predominantly male feelings of inadequacy, frustration and boredom generated by sometimes reversed gender roles identified as contributing factors (Nguyen and Rydstrom 2018). These pressures operate irrespective of whether displaced persons find themselves in formal, state-operated evacuation centres and temporary shelters, or whether they find shelter elsewhere, for example with friends, relatives or other persons in the private sector.

Particular risks arise in evacuation centres and temporary shelters

Although there are no formal legal definitions, an evacuation centre is generally considered to be a very short-term shelter for people in the period immediately preceding the onset of a natural hazard event, and during the emergency phase (Camp Coordination and Camp Management (CCCM) Global Cluster 2014; Government of Vanuatu 2016). Schools, stadiums, airports, open spaces amongst others may serve as evacuation centres in the immediacy of the hazard event. A temporary shelter can be used for somewhat longer periods until people are able to return or relocate (Sphere Project 2011, 245).

Factors that have been identified as contributing to the heightened risk of sexual assault in these settings include: the lack of accessible safe spaces for women, children and adolescent girls; overcrowding; lack of doors and partitions for sleeping and changing clothes; unsafe distribution of shelter-related non-food items, such as fuel and water, forcing women and girls to leave the shelter; poor lighting in general, and in particular at water, sanitation and hygiene (WASH) facilities, and; lack of safety audits and monitoring, in particular around WASH facilities (IASC 2015, 51–72)

The result of an unsafe evacuation centre or temporary shelter is that women and girls in particular are exposed to a heightened risk of sexual assault than in everyday life. Examples abound. In the aftermath of the Indian Ocean tsunami, the Asia Pacific Forum on Women, Law and Development recorded incidents of sexual violence in Indonesia, India and Sri Lanka, including in situations of displacement in shelters and camps (APWLD 2005). The Women and Media Collective Group reported that in Sri Lanka they had "received reports of incidents of rape, gang rape, molestation and physical abuse of women and girls in the course of unsupervised rescue operations and while resident in temporary shelters" (Pikul 2005).

In research conducted with almost 2,000 disaster-affected individuals in Indonesia, Lao PDR and the Philippines, the International Federation of Red Cross and Red Crescent Societies (IFRC) found that the risks of SGBV had been exacerbated during specific disaster situations in each country (IFRC 2018). Key factors identified in this and other studies included lack of lighting, lack of separate toilets for men and women and few safe spaces for women and children in temporary housing.

Additionally, reporting on field research conducted in relation to the Haitian earthquake, the IFRC notes how the breakdown of existing social protections in situations of disaster displacement can be a significant factor contributing to increased exposure to sexual assault:

> [A]fter the mass displacement caused by the earthquake, many Haitians found themselves either forced to move into communities where they knew no one, or to see their own communities suddenly filled with desperate unknown strangers, or worse, to move into a camp where no one knew anyone. The breakdown in this collective security seemed to create conditions making GBV more prevalent.
>
> (IFRC 2015, 21)

Similarly, Beyani (2015) recounts the common concern expressed by women and girls living in temporary 'bunkhouses' regarding their exposure to violence. Cramped living conditions in close proximity to strangers made parents, and particularly women heads of households, concerned about leaving their children in order to seek livelihood opportunities, as they feared they would be sexually assaulted. Similar findings are reported by the IFRC (2015), based on a study of gender-based violence in situations of disaster displacement in Haiti, Namibia, Malawi, Western Samoa, El Salvador and Romania.

Considering the initial response by national and international agencies in the aftermath of the 2004 Indian Ocean tsunami in the Maldives, Fulu (2007) notes that, in general, responses failed to adequately address the different situation of women and men. She describes a "male-dominated, technocratic, top-down" approach led by the Ministry of Defence, where a "tyranny of the urgent" resulted in an inadequate consideration of gender dynamics, resulting in cramped temporary housing for displaced persons disrupting "the normal protective function of the family" and "increas[ing] risks of violence and sexual abuse against women and adolescent girls". She quotes an OCHA official who worked with displaced persons in the aftermath of the tsunami:

> Most of the issues that we looked at we didn't consider the differential impact it might have between women and men . . . we didn't speak to women about the way the temporary housing structures were set up and what that could mean for family relations for example and it clearly did have a detrimental impact . . . causing serious stress between families, children and all sorts of protection issues coming out of that [. . .] but again this was something that was never really looked into in a thorough way by people who knew what they were doing.
>
> (Fulu 2007, 850)

In light of the foregoing, it is clear that sexual assault in situations of disaster displacement is widespread. Pittaway, Bartolomei and Rees (2007) argue that "[d] espite guidelines on a gendered response in disaster situations and a plethora of recommendations from UN agencies and government organisations, this outcome is common to most disaster responses". When sexual assault is a foreseeable consequence of disaster displacement, and when guidelines point to steps that responsible actors can take to addressing that risk, questions about legal responsibility for failure to take appropriate measures arise. In the following sections, the kinds of steps responsible actors can take are presented and the question of legal responsibility for failure to take such steps is considered.

Measures responsible actors should take to protect persons from sexual assault in evacuation and temporary shelters

Guidelines have been produced to assist states in this connection. The MEND Guide (2014), developed by the Camp Coordination Cluster in response to requests from states, provides only limited insights:

> Safe and appropriate structures and mechanisms for reporting, responding and preventing GBV need to be instituted in each evacuation centre.

Assistance should be provided for survivors of GBV, prevention activities must be put in place, and effective action to prevent and respond to GBV must be incorporated into all stages of the identification and management of evacuation centres.

(CCCM Global Cluster 2014, 95)

More detailed guidance is provided by the Inter Agency Standing Committee's Gender-Based Violence Guidelines (GBV Guidelines). Basic steps that should be taken to reduce the risk of gender-based violence in evacuation centres and temporary shelters include:

- Individual registration of married women, single women, single men, and girls and boys without family members registered
- Provision for registration of individuals with different gender identities in a safe and non-stigmatising way
- Promotion of active participation of women in all aspects of site governance and CCCM programming
- Ensuring awareness amongst staff of international standards, including the IASC Guidelines, for mainstreaming GBV prevention and mitigation strategies into their activities
- Adherence to Sphere standards
- Adequate lighting
- Provision and maintenance of women-, adolescent- and child-friendly spaces
- Enactment of a security strategy detailing when, where, how and by whom security patrols are conducted
- Addressing safety of water and distribution sites and whether they accommodate the specific needs of women, girls and other at-risk groups
- Considering the availability of 'safe shelters' for people at particular risk of GBV (IASC 2015)

The UN Population Fund (UNFPA) Minimum Standards for Prevention and Response to Gender-Based Violence in Emergencies (2015) expressly build on the GBV Guidelines by articulating a set of 18 inter-related and inter-dependent standards detailing "what should be achieved to prevent GBV and deliver multi-sector services to survivors in humanitarian settings".

Guidelines can also be found at the national level. For example, Vanuatu has produced National Guidelines for the Selection and Assessment of Evacuation Centres which includes guidelines and checklists that address a range of gender-specific security concerns, including around lighting, separate spaces, security and so forth (Vanuatu Government 2016). Guidelines such as these, whilst not legally binding, provide a good practice standard that states should, in the absence of compelling reasons, strive to follow. Failure to follow such standards may amount to a breach of the state's legal obligation to protect persons from sexual assault.

Accountability for failures to protect persons from sexual assault in evacuation centres and temporary shelters

The specific steps outlined above that responsible actors can take to protect persons from sexual assault in evacuation centres and temporary shelters are by no means exhaustive. Rather, they provide examples drawn from internationally recognised good practice that responsible actors can be expected to adhere to in fulfilment of their general obligation as states to protect persons within their jurisdiction.

In what follows, the scope of a state's legal obligation to protect persons from sexual assault in evacuation centres and temporary shelters is considered. First, the contours of a general obligation to take steps to protect persons from gender-based violence are sketched. Then, a specific argument concerning how the obligation applies in relation to sexual assault in evacuation centres and temporary shelters is set out. Finally, arguments against such a general obligation applying in evacuation centres and temporary shelters are considered.

The general obligation to take steps to protect persons from gender-based violence

The Convention on the Elimination of All Forms of Discrimination against Women (CEDAW) is the cornerstone of a gender-equal approach to international human rights law. The starting point for a human rights-based approach to gender-based violence is the prohibition of discrimination at Article 1 of the Convention. The CEDAW Committee, responsible under Part V for monitoring progress under the Convention, confirmed as long ago as 1992 that the non-discrimination provision at Article 1 includes a prohibition on violence against women (CEDAW 1992). In accordance with general principles of international human rights law, it is well-established that obligations have both negative and positive aspects. As a matter of general principle, the Human Rights Committee that monitors the International Covenant on Civil and Political Rights (ICCPR) confirms:

> [T]he positive obligations on States Parties to ensure Covenant rights will only be fully discharged if individuals are protected by the State, not just against violations of Covenant rights by its agents, but also against acts committed by private persons or entities that would impair the enjoyment of Covenant rights in so far as they are amenable to application between private persons or entities.
>
> (HRC 2004)

The content of the obligation to address gender-based violence has been developed by the CEDAW Committee and the Special Rapporteur on Violence against Women, Its Causes and Consequences, since the Committee first began systematically addressing violence against women in 1989 (CEDAW 1989). General Recommendation No 19 of 1992 identifies specific steps that states are expected to take to address sexual violence. These steps include both preventative and reha-bilitative measures. General Recommendation No 35 of 2017 further develops thinking around gender-based violence against women. In particular, General Recommendation No 35 further develops the notion of state responsibility for the acts or omissions of non-state actors, with reference to a 'due diligence' standard:

> That obligation, frequently referred to as an obligation of due diligence, underpins the Convention as a whole and accordingly States parties will be held responsible should they fail to take *all appropriate measures* to pre-vent, as well as to investigate, prosecute, punish and provide reparations for, acts or omissions by non-State actors that result in gender-based violence against women.
>
> (CEDAW 2017, emphasis added)

This perspective is highly relevant to the question of legal protection for women and girls exposed to sexual assault in evacuation centres and temporary shelters. Of course, states are required to operate a legal system that provides for sanctions for people who commit acts of violence against women and girls. It also follows that there is a particular responsibility when state agents, for example police or relief workers, sexually assault women and girls. There is also a basis for asserting a responsibility on the part of the state for sexual assault committed by non-state actors. The question in any case the state is accused of failing to fulfil its obliga-tions to protect persons from gender-based violence is whether 'all appropriate measures' were taken, in light of the particular facts of the case.

The due diligence standard is not limited to expression by the CEDAW Com-mittee, but is also articulated expressly in treaties including the 1994 Inter-American Convention on the Prevention, Punishment and Eradication of Violence against Women "Convention of Belem do Para" and the 2011 Council of Europe Convention on Preventing and Combating Violence against Women and Domes-tic Violence. It is important to note that, whilst this chapter focuses on principles under international human rights law, relevant legal action may also be framed within the context of domestic legal principles, such as civil liability for negli-gence (McLaren et al. 2001).

Jurisprudence concerning the due diligence standard relates predominantly to intimate partner violence in situations where the authorities knew or ought to have known of the risk of serious harm and failed to use measures available to them to prevent the violence.[4] Other strands of jurisprudence concern failures by the state to prosecute perpetrators of sexual assault,[5] and failures to provide redress to victims.[6] Myriad examples of the principle being applied in domestic jurisdictions worldwide can be found in the Due Diligence Project (Abdul Aziz and Moussa 2016). Academic engagement with the concept is also developing, with focus, for example, on its potential application in the context of 'honour related' violence (Grans 2018).

Although the standard has been found to apply primarily in situations where the risk related to a specific, known individual, it may also be found to apply in situations of more generalised violence. In *Kayak v Turkey*, for example, the Turkish authorities were found to be in breach of their obligation to protect the life of a 15-year-old boy, who was killed by a youth in the vicinity of the school following an altercation on the premises. In contrast to the intimate partner violence line of cases identified above, in this case neither the victim nor the perpetrator could have been identified beforehand. The failure related instead to the inadequate security within the school, in breach of the school's primary duty to protect students against any form of violence.

Of particular relevance in the *Kayak* case was the fact that the violence was foreseeable even if the parties to the violence were not identifiable. Further, and crucially, the fact that the violence took place within and in the vicinity of an institution charged with a particular duty of care was central to the judgment. This duty to protect from violence in state institutions has been found to extend across a range of institutions operated by the state, including prisons[7] and mental health institutions.[8]

This due diligence obligation, it is contended here, has application in situations where the state fails to take all appropriate measures to prevent sexual assault in evacuation centres and temporary shelters. Sexual assault is a foreseeable consequence of overcrowding, poor lighting, lack of security and so forth that, as Pittaway noted above, all too often characterise evacuation centres and temporary shelters. The fact that the victim and perpetrator are not necessarily identifiable in advance does not rule out the obligation of states to take all appropriate measures to protect persons within their jurisdiction, and particularly in settings where responsibility for security in a particular locality rests firmly with the state. The contours of the argument are developed below.

The due diligence standard in evacuation centres and temporary shelters

As a starting point, it warrants recalling that evacuation centres and temporary shelters house displaced persons, and that all of the human rights individuals are entitled to enjoy when living in their place of habitual residence also apply, without discrimination, in situations of internal displacement. This point is established most clearly under the 1998 Guiding Principles on Internal Displacement, which define internally displaced persons (IDPs) as:

> [P]ersons or groups of persons who have been forced or obliged to flee or to leave their homes or places of habitual residence, in particular as a result of or in order to avoid the effects of armed conflict, situations of generalized violence, violations of human rights or natural or human-made disasters, and who have not crossed an internationally recognized State border.

The Guidelines, which have been repeatedly endorsed at international and regional levels as a useful tool for addressing the phenomenon,[9] reaffirm at Principle 1 that "[i]nternally displaced persons shall enjoy, in full equality, the same rights and freedoms under international and domestic law as do other persons in their country".

This chapter argues that the 'appropriate measures' standard that applies generally to situations of gender-based violence applies in the same way in situations of internal displacement, including in particular in state-managed evacuation centres and temporary shelters. The 'appropriate measures' that states can be expected to take in order to protect persons from sexual assault within evacuation centres and temporary shelters are outlined in good practice guidelines described above. Due diligence in this context thus includes planning measures in advance of the disaster to ensure that, when displacement takes place and people are accommodated in evacuation centres and temporary shelters, appropriate mechanisms, such as adequate lighting, safe spaces, physical security and others are in place. More or less detailed guidelines relating to gender-based violence in emergency situations may have been developed in national and more local level law and policy providing an even more compelling case where failures to follow established guidelines contribute to the risk.

If authorities do not have a plan and the resources to implement such a plan in the context of a disaster, there is arguably a prima facie breach of the due diligence obligation. In situations of disaster, existing plans should be implemented to the fullest possible extent in light of all of the circumstances. The argument is particularly salient in shelters operated by the state, in light of the authorities identified above. However, as noted below, authorities can also be held accountable for sexual assault taking place in spontaneous informal shelters.

This general argument might be developed in a range of circumstances to promote accountability and contribute toward the prevention of sexual assault in evacuation centres and temporary shelters. For example, lawyers may assist an individual who has already been subjected to sexual assault to take legal action against the state for its failure to fulfil its due diligence obligations. Equally, the argument may be leveraged in more advocacy-based approaches, with civil society actors highlighting how existing measures to address sexual assault in evacuation centres and shelters fall short of the state's due diligence obligations. Strategies for leveraging the due diligence argument relating to prevention of sexual assault in evacuation centres and temporary shelters are considered towards the end of this chapter.

Limits to accountability for sexual assault in evacuation centres and temporary shelters

A range of arguments can be leveraged to counter the claim that a general obligation to prevent sexual assault applies in evacuation centres and temporary shelters. Four arguments are formulated below to highlight some legal challenges. First, the argument has been advanced that states ought not be held accountable for consequences flowing from uncontrollable acts of nature. Second, even if states may have positive obligations of some sort to protect persons from the adverse impacts of disasters, the obligation is curtailed by inevitable resource limitations. Third, even where good practice has been followed in preparation for a potential disaster, the challenges that ensue in the aftermath of the disaster may preclude effective implementation of planned response measures. Finally, domestic judicial authorities may reject interpretations of the extent of the state's legal obligations to prevent sexual assault in evacuation centres and temporary shelters that are not firmly grounded in domestic legal provisions. These challenges are considered in turn.

Disasters as acts of nature

It is common to see disasters represented as states of exception, and states as humanitarian actors making best efforts in extremely challenging circumstances. There is a kind of inherent attraction to this framing, as centuries of thinking about disasters either as acts of God or uncontrollable forces of nature continue to dominate popular approaches to disasters (Cedervall Lauta 2015). This stark epistemological division between human agency and the forces of nature remains convenient for authorities concerned to avoid responsibility for harm and damage arising in disasters. Gaillard, Liamzon, and Villanueva (2007), for example, noted in the context of their study of the social causes of deaths in a series of typhoons in the Philippines in 2004, that:

> The dominant view of Nature as the major agent of disaster in the Philippines is more than the persistence or the resurgence of an ancient scientific paradigm. Emphasizing the 'natural' dimension of catastrophic events here serves as an alibi and allows many stakeholders to escape from their responsibility.
> (Gaillard, Liamzon, and Villanueva (2007, 267)

This argument may be leveraged by actors interested in diverting attention away from questions about what steps should have been taken before the onset of the natural hazard event to prepare for and prevent foreseeable forms of harm. Clearly, the entire field of disaster risk reduction presents a challenge to the epistemology of disasters as merely the uncontrollable forces of nature. The Sendai Framework on Disaster Risk Reduction reflects the commitment of 187 states attending the Third UN World Conference on Disaster Risk Reduction (WCDRR) to take steps to address disaster risk.

The argument that nation-states cannot be held responsible for harm arising in the context of 'natural' disasters has failed before regional and domestic courts.

In the case of *Budayeva v Russian Federation*, the Russian Federation sought to avoid responsibility for the deaths of a number of inhabitants of the town of Tyrnauz, who died in the context of a landslide. The authorities expressly argued that the landslide was "an act of God" which could "neither be foreseen nor influenced" (para 117). The Court rejected this argument. Accepting that "an impossible or disproportionate burden must not be imposed on the authorities without consideration being given, in particular, to the operational choices which they must make in terms of priorities and resources", the Court nevertheless found that the authorities had failed to fulfil their positive obligation to protect people from what was a foreseeable disaster, noting that the scope of such positive obligations depends "on the origin of the threat and the extent to which one or the other risk is susceptible to mitigation".

States thus have a duty to take steps before the onset of a natural hazard event in order to prepare for its impacts. Such a duty arguably extends to following good practice guidelines relating to the prevention of sexual assault in evacuation centres and temporary shelters.

The resource implications of prevention

States may contend that, even if there might be a positive obligation to protect persons from sexual assault in evacuation centres and temporary shelters, resource constraints preclude adherence to good practice guidelines nationwide. Obstacles to effective disaster risk reduction and management may include the remoteness of some communities, insufficient human and financial resources, and difficulties recovering from recurrent disasters, particularly as the impacts of climate change become increasingly acute. Thus, a general argument that failure to follow good practice around preventing sexual assault in evacuation and temporary shelters is unlikely to be made out. States cannot be held to impossibly high standards, particularly when faced with the exigencies of disasters.

However, Farrior (2004) argues that recourse to the 'limited resources' argument is insufficient for avoiding accountability for due diligence failures. States must demonstrate that they have taken appropriate measures in light of all of the circumstances, which entails also an assessment of how resources have been allocated within the state. States can be expected to take steps to the maximum of available resources to ensure that, even in situations of disaster, people within their jurisdiction are not exposed to serious harm (CEDAW 2010).

In terms of the 'appropriate measures' test, this position recognises that resource constraints can indeed make it more difficult for states to prevent sexual assaults in evacuation centres and temporary shelters, but the onus lies on the state to explain why certain measures that could have been taken (for example, having adequate lighting, training shelter staff in measures to prevent sexual assaults, and so forth) were not taken.

The disaster seriously disrupts state capacity to implement even planned prevention measures

The International Law Commission's Draft Articles on the Protection of Persons in Situations of Disasters adopts a definition of disasters as:

> [A] calamitous event or series of events resulting in widespread loss of life, great human suffering and distress, mass displacement, or large-scale material or environmental damage, thereby seriously disrupting the functioning of society.
>
> (ILC 2016)

If one of the functions of society is to protect persons from sexual assault, and the disaster seriously disrupts this function, then surely states must be held to a lower standard in situations of disaster than in everyday life. This argument has some weight. First, particularly in large-scale disasters, actors typically responsible for protecting the public may themselves be affected by the disaster. They may have difficulty reaching evacuation centres. Communications may be down. Electricity may be down. Priorities may be directed to other security issues outside of the evacuation centre. There may not be time or capacity for individual registration and needs assessments, at least not in the initial phase of the disaster. These contingencies may limit a full implementation of a disaster risk management plan, including measures to prevent sexual assault in evacuation centres.

This argument would carry weight in an assessment of whether the measures that were actually taken were, in light of all of the facts of the case, 'appropriate', but probably only in the initial emergency phase.

Rejection by domestic legal authorities of international interpretations of relevant human rights instruments

Finally, irrespective of the apparent persuasiveness of the due diligence argument set out above, the legal reasoning may not carry weight in domestic jurisdictions. Although the committees established to monitor state party compliance with international human rights obligations have all developed the practice of issuing General Comments or General Recommendations, which clarify how the Committee interprets provisions of the relevant treaty, these interpretations do not bind states parties, and domestic courts will adopt different approaches to the weight they choose to attach to such documents when applying domestic law to domestic facts.

Even in the absence of domestic implementing legislation, however, courts may find a due diligence obligation with reference to domestic constitutional provisions, which very often contain non-discrimination clauses similar to those found in CEDAW, upon which, as noted above, the due diligence obligation to take measures to prevent gender-based violence is founded. For example, the Supreme Court of Bangladesh reasoned as follows in the case of *Bangladesh National Women Lawyers Association (BNWLA) vs. Government of Bangladesh, et al.*:

> The Fundamental Rights guaranteed in chapter III of the Constitution of Bangladesh are sufficient to embrace all the elements of gender equality including prevention of sexual harassment or abuse. Independence of judiciary is an integral part of our constitutional scheme. The international conventions and norms are to be read into the fundamental rights in the absence of any domestic law occupying the field when there is no inconsistency between them. It is now an accepted rule of judicial construction to interpret municipal law in conformity with international law and conventions when there is no inconsistency between them or there is a void in the domestic law.

Much, ultimately, depends on the disposition of the court, which will have a broad interpretative mandate where domestic law is silent or ambiguous. Still further arguments would inevitably arise in any situation in which legal actors sought to establish the accountability of the state for sexual assault in evacuation centres or temporary shelters. What this section has shown is that there is a due diligence standard that reflects the positive obligation of states to take 'appropriate measures' to prevent sexual assaults in evacuation centres and temporary shelters, and regional as well as domestic anchors upon which to build arguments in a particular case.

The chapter now turns to consider how justice-sector actors may leverage the due diligence argument to pursue state accountability for sexual assault in evacuation centres and temporary shelters, both in individual cases as well as in wider campaigns.

Strategies justice-sector actors may adopt to promote accountability and contribute towards reducing sexual assault in evacuation centres and temporary shelters

Action to promote the development of domestic law and policy

In light of the final challenge to legal action seeking to contribute to a reduction in instances of sexual assault in evacuation centres and temporary shelters, the domestic legal landscape needs to be surveyed. The results of the survey are not promising. Only 19 national law documents within the IFRC disaster law database contain the word 'violence' and in none of these is the word used in relation to gender-based violence. Eleven national-level policy documents contain the word 'violence', and only five address violence specifically in the context of gender-based violence. Engagement with gender-based violence at more local levels of government is even less evident, with only three provincial-level documents mentioning violence at all, none of which relate to gender-based violence, and no local level documents relating to violence in the context of disaster whatsoever. This is perhaps unsurprising given that the Sendai Framework itself does not make a single reference to gender-based violence.

Clearly, legal action to promote accountability and contribute to a reduction in sexual assault in evacuation centres and temporary shelters will have to encompass advocacy initiatives calling for the development of relevant domestic law and policy. Indeed, the absence of domestic law and policy may itself underpin a litigation and wider advocacy initiative. For example, in *Bangladesh National Women Lawyers Association (BNWLA) v Government of Bangladesh, et al.* the Supreme Court of Bangladesh criticised the failure by the state "to adopt guidelines, or policy or enact proper legislations to address the issue of abuse of sexual harassment for protecting and safeguarding the rights of the women and girl children at workplace, educational institutions/universities and other places wherever necessary". Recognising the fundamental importance both of domestic constitutional provisions as well as the 'great significance' of international conventions and norms for the development of guidelines to achieve protection from sexual harassment, the right to education and work with dignity, the Court took it upon itself to develop guidelines. In so doing, the Court was inspired by jurisprudence from courts in Canada and India, and was particularly inspired by the approach adopted by the Supreme Court of India in *Vishaka and Others v State of Rajasthan*, where similar guidelines were developed.

An advocacy initiative similar to that which gave rise to the BNWLA litigation in Bangladesh would have to come from below, and be based on whether local justice-sector actors consider the issue of sexual assault in evacuation centres and temporary shelters to be a relevant cause amongst many others. However, other forms of advocacy, including from international organisations, actors engaged in development cooperation, regional actors and others could also promote the development of national or more local guidelines, reflecting international good practice. Unfortunately, even where law and policy exists, the question of implementation is perennial.

Legal action to promote effective implementation

A second and potentially related approach entails actions promoting the effective implementation of existing domestic, regional and international law, policy and guidelines at the local level. Action may include litigation before domestic and/ or supranational judicial or monitoring bodies, preferably as part of a wider civil society campaign to increase action to address gender-based violence.

When litigation secures judicial endorsement of an important protection principle, such as the application of the due diligence obligation in situations of disasters, it can be a powerful tool for further legal action, including advocacy. However, even 'successful' litigation may fail to have the intended result for litigants, as implementation of judgements remains uneven at best. This dynamic is illustrated clearly in the case Women and Girls Residing in 22 Camps for Internally Displaced Persons in Port-au-Prince.

In this case, representatives of women and girls residing in 22 IDP camps that arose in the aftermath of the 2010 earthquake asserted that the government of Haiti was failing in its positive obligation to prevent sexual assault. On the basis of the information provided, the Inter-American Court of Human Rights (IACtHR) issued precautionary measures, requiring the Haitian government to take five specific steps, including in relation to the provision of appropriate medical assistance, implementation of effective security measures in the 22 camps, empowerment of public authorities charged to respond to incidents of sexual violence in the camps, creation of specialised units within the police and other relevant authorities for the investigation of sexual violence, and to provide guarantees of the active participation of women in the design of measures to address sexual violence in the camps.

Clearly, the decision of the IACtHR concerning the duty of Haiti to address widespread sexual assault in IDP camps is significant for its articulation of a due diligence standard in the context of disaster displacement. The principle can be used in a range of other scenarios, considered in more detail later in this section.

However, the litigation appears to have had no impact whatsoever on the lives of women living in the 22 camps, as the matter was subsequently brought before the Human Rights Council on the occasion of Haiti's Universal Periodic Review procedure in 2011. Here, representatives involved in the IACtHR case submitted a brief in parallel to the report submitted by the government of Haiti, in which it again highlighted the situation facing women in camps for displaced persons and called for the precautionary measures ordered by the Court to be implemented.[10] Five years after the earthquake, reports of ongoing sexual violence in camps for displaced persons remained widespread.[11] Thus, even when the argument that the state has a due diligence obligation to prevent sexual assault in situations of disaster displacement is accepted on the facts of a specific case, the impact on the ground may be very limited indeed.

Recognising the limited impact of the IACtHR litigation, Bookey (2011) provides an account of domestic challenges in Haiti to the implementation of the IACtHR judgment, including politicisation of the judiciary, lack of access to the judicial system as a result of public knowledge, public confidence, cost, limited facilities and corruption and attitudes of the legal system towards sexual assault cases generally. Highlighting the work of the Bureau des Avocats Internationaux, a public interest law firm working in Port-au-Prince, Bookey argues that:

> Following individual cases through the Haitian legal system will reinforce larger structural reforms and development projects that have, to date, produced only marginal results. It will also increase trust in the system from the bottom up, a foundation necessary for any system based on the rule of law.

The approach includes 'victim-centered legal advocacy', grassroots collaboration, including in litigation and work addressing the root causes of gender-based violence. The facts of the Haiti case are extreme considering the exceptionally difficult socio-economic conditions in the country and the inability of the state to effectively recover in the aftermath of an extremely powerful earthquake that brought massive destruction, thousands of fatalities and over 1.5 million displaced persons out of a population of approximately 10 million in 2010. Nevertheless, the difficulties Bookey identifies are not unique to the Haitian context, and the approach she advocates certainly contributes to accountability.

Advancing accountability for failures to prevent gender-based violence

Although no litigation seeking compensation for failure to prevent gender-based violence in the context of disasters was identified for this chapter, an indication of how such a case may proceed is provided by litigation currently underway in the Constitutional and Human Rights Division of the High Court of Kenya, which asserts that the widespread violence against women that took place in the context of unrest following elections in the country in 2007 reflect failures by Kenyan authorities to fulfil obligations under the domestic, sub-regional and international law, including the obligation to protect persons from sexual assault by non-state actors.[12] The case is not concluded at the time of writing,[13] but warrants monitoring given its potential to clarify the scope of 'appropriate measures' in situations of generalised violence.

Accountability has a value for persons who have been victims of human rights violations, as formal findings of failure tend to carry an obligation to make reparations. Strengthening this mechanism is one way of addressing failures to prevent sexual assault in evacuation centres and temporary shelters. However, accountability aims for more than an individual remedy in a particular case, and justice-sector actors are well-placed to advance claims relating to failures of the state to take 'appropriate measures' to prevent sexual assault in evacuation centres and temporary shelters to advocate for more systemic changes in law, policy and practice relating to disaster risk reduction and management at national and more local levels.

Parallel reports to treaty monitoring bodies

Human rights monitoring bodies, in addition to articulating general interpretations of the relevant international legal instruments, are also demonstrating an increasing concern with how human rights law applies in situations of disaster. The CEDAW (2018) Committee takes a leading role in this respect, and has developed General Recommendation No. 37 on gender-related dimensions of disaster risk reduction in the context of climate change.

The Committee also systematically raises the issue of gender-based violence with states, and is increasingly expressing concern about the differential impact of disasters and climate change on women, and the lack of gender mainstreaming in domestic legal frameworks relating to disaster risk reduction and management and climate change adaptation.[14] Legal actors concerned with the issue of gender-based violence in general, and sexual assault in evacuation centres and temporary shelters in particular, may contribute to raising the domestic profile of the issue through submitting parallel reports, as the justice-sector actors involved in the Haitian camps litigation described above did. Such interventions may lead the Committee to seek further information from states during the periodic review, and can result in more targeted recommendations. Of course, implementation of recommendations remains a perennial issue.

Engaging national human rights institutions

Finally, a role that has received only limited attention[15] in relation to its potential, is that which national human rights institutions can play in addressing gender-based violence in situations of disasters. National human rights institutions (NHRIs) are justice-sector actors established under law with a mandate to issue opinions, recommendations, proposals and reports on any matters concerning the promotion and protection of human rights. Under the Paris Principles (UNGA 1994), NHRIs may *inter alia* examine legislation, investigate human rights violations, contribute to reports to human rights monitoring bodies and assist in the formulation of programmes for teaching and research.

In May 2015, just one month after the April 2015 earthquake in Nepal, the National Human Rights Commission submitted a report (2015) on the initial response by the government. This critical rights-based report paid particular attention to the situation of women and girls, identifying a series of failures by the state to address particular needs and to take steps to reduce the risk of sexual assault, having taken note of five specific incidences of rape in one district. These cases were raised with the Chief District Officer and the chief of the District Police Office. The report also highlights the risk of trafficking to which women and girls living in camps were exposed. The Commission also developed a human rights monitoring checklist for use in evaluating conditions in the aftermath of the earthquake, which includes reference to sexual assault, but which does not provide guidelines for assessing the adequacy of evacuation centres and temporary shelters against international good practice. Integration of the guidelines described earlier in this chapter could support the articulation of concrete recommendations grounded in internationally recognised good practice.

Significantly, the Asia Pacific Forum of National Human Rights Institutions (APF) produced a manual in 2015 setting out how NHRIs can address gender-based dimensions of their work. The manual devotes considerable attention to gender-based violence, and includes a section explaining the due diligence obligation described above. Notably, the manual does not make any reference to disasters, thus missing an opportunity to link the general positive obligation to prevent sexual assault and other forms of gender-based violence to the specific situation of disaster displacement.

The manual further makes reference to the Amman Programme of Action, developed at the 11th International Coordinating Committee of National Institutions for the Promotion and Protection of Human Rights in 2012. The Programme of Action reflects the decision by NHRIs present to prioritise work around a series of themes, including violence against women, over the decade 2012–2022. Within the violence against women thematic area, priorities include building an evidence base around the nature, extent, causes and effects of all forms of gender-based violence, and on the effectiveness of measures to prevent and address gender-based violence, to promote and support the adoption of laws against gender-based violence, to support training of justice-sector actors, amongst others.

Hence, National Human Rights Institutions and their regional and international membership bodies have a distinctive potential to promote awareness of and accountability for gender-based violence in situations of disaster, including in evacuation centres and temporary shelters.

Conclusions

The positive obligation of states to take appropriate measures to prevent gender-based violence does not cease to operate in situations of disaster displacement. Rather, this flexible yet principled standard adjusts to the circumstances of each case and recognises that states cannot be held to impossible standards. Disasters present particular challenges that can overwhelm even well-laid plains. However, disasters are not unforeseeable acts of nature, and human rights law rejects claims that different standards should apply in what some may seek to describe as 'states of exception'. International good practice, at times incorporated into domestic and more local level law and policy, provides guidance on what steps states should take to reduce exposure to sexual assault in evacuation centres and temporary shelters. Failure to follow this good practice puts the burden on the state to explain why, and may amount to a breach of international as well as domestic law.

This chapter opened with a recognition that the human rights project entailed encounters between a range of actors at different scales and around different issues, with countless examples of widespread violations. With a host of state leaders disavowing central tenets of international human rights law there is cause to doubt the saliency of a human rights-based approach in an era of uncertainty, instability and widespread declines in living standards, with darker clouds on the horizon. At the same time, the human rights-based approach provides a language and mechanisms for different actors to use when describing and seeking to change lived experience. A legal right to be protected from sexual violence provides a platform for a host of actors to demand that states take steps, to the maximum of available resources, to fulfil their obligations towards people within their jurisdiction. Times may get harder, but the validity of this legal claim endures. How the claims are articulated and ultimately realised in practice depends on local actors and vernacular strategies.

Notes

1 Sexual assault is defined by the WHO as "A subcategory of sexual violence, sexual assault usually includes the use of physical or other force to obtain or attempt sexual penetration. It includes rape, defined as the physically forced or otherwise coerced penetration of the vulva or anus with a penis other body part, or object, although the legal definition of rape may vary and, in some cases, may also include oral penetration". See (WHO 2013, viii, citing Jewkes, Sen, and Garcia-Moreno 2002).
2 The application of the due diligence standard is referred to in Mireille Le-Ngoc, "Normative frameworks' role in addressing gender-based violence in disaster settings" Disaster Law Working Paper Series Paper No. 3, 2015 https://www.ifrc.org/Page Files/189264/Mireille%20Le%20Ngoc%20Working%20Paper%20GBV%20(final). pdf, but it is not developed in the context of disaster displacement, or at all
3 On human rights cities, see, for example, (Davis, Gammeltoft-Hansen, and Hanna 2017)
4 See, for example, *Osman v United Kingdom* [1998], *Yildirim v Austria* [2007], *Opuz v Turkey* [2009], *Valiuliene v Lithuania* [2013]
5 *Maria da Penha Maia Fernandes v Brazil* [2008]
6 *Jessica Lenahan (Gonzales) et al. v United States* (2011)
7 *Pantea v Romania* [2003]
8 *X and Y v The Netherlands* [1985]
9 See discussion with references to endorsements by *inter alia* the UN General Assembly, the UN Human Rights Commission, the Organization of African Unity (now African Union), IGAD, ECOWAS and OSCE in Kälin (2005)
10 www.madre.org/sites/default/files/PDFs/1302209987_UPR%20Submission%20 on%20Review%20of%20Haiti%20Final_0.pdf

11 See collection of reports at www.refworld.org/docid/58d539d04.html
12 www.opensocietyfoundations.org/sites/default/files/sgbv-kenya-20130219.pdf
13 See timeline here www.opensocietyfoundations.org/litigation/coalition-violence-against-women-and-others-v-attorney-general-kenya-and-others
14 See, for example, Concluding Observations on the Solomon Islands CEDAW/C/SLB/CO/1–3 (14 November 2014), Cambodia CEDAW/C/KHM/CO/4–5 (29 October 2013) and Bangladesh CEDAW/C/BGD/CO/8 (25 November 2016), amongst many others. Yet, treatment of the issue is inconsistent, with the 2014 Concluding Observations on China (CEDAW/C/CHN/CO/7–8 (14 November 2014)) silent on the issue despite the country being heavily exposed to natural hazard events, and related high levels of internal displacement.
15 The Brookings-Bern Project on IDPs has devoted some attention to the role of NHRIs in promoting and protecting human rights in situations of disaster. See Ferris (2008).

References

Abdul Aziz, Zarizana, and Janine Moussa. 2016. "Due Diligence Framework: Due Diligence Project's State Accountability Framework for Eliminating Violence against Women." International Human Rights Initiative, Inc. www.duediligenceproject.org/ewExternal Files/Due%20Diligence%20Framework%20Report%20final.pdf.

Alway, Joan, Linda Liska Belgrave, and Kenneth J. Smith. 1998. "Back to Normal: Gender and Disaster." *Symbolic Interaction* 21, no. 2: 175–95.

Anastario, Michael, Nadine Shehab, and Lynn Lawry. 2009. "Increased Gender-Based Violence among Women Internally Displaced in Mississippi 2 Years Post – Hurricane Katrina." *Disaster Medicine and Public Health Preparedness* 3, no. 1: 18–26.

Asia Pacific Forum of National Human Rights Institutions. 2015. "Promoting and Protecting the Human Rights of Women and Girls: A Manual for NHRIs." www.asiapacificforum.net/resources/manual-on-women-and-girls/.

Asia Pacific Forum on Women, Law and Development (APWLD). 2005. "Why are Women More Vulnerable during Disasters? Women's Human Rights in the Tsunami Aftermath." Bangkok: APWLD. http://iknowpolitics.org/sites/default/files/tsunami_report_oct2005.pdf.

Beyani, Chaloka. 2015. "Report of the Special Rapporteur on the Human Rights of Internally Displaced Persons on His Mission to the Philippines." A/HRC/32/35/Add.3. Geneva: United Nations. https://digitallibrary.un.org/record/842523.

Bookey, Blaine. 2011. "Enforcing the Right to Be Free from Sexual Violence and the Role of Lawyers in Post-Earthquake Haiti." *City University of New York Law Review* 14, no. 2.

Bradshaw, Sarah, and Maureen Fordham. 2015. "Double Disaster: Disaster through a Gender Lens." In *Hazards, Risks, and Disasters in Society*, edited by John F. Shroder, Andrew E. Collins, Samantha Jones, Bernard Manyena, and Janaka Jayawickrama, 233–51. Academic Press.

Budayeva and Others v Russia – 15339/02 [2008] ECHR 216.

Camp Coordination and Camp Management (CCCM) Global Cluster. 2014. "The MEND Guide: Comprehensive Guide for Planning Mass Evacuations in Natural Disasters." http://www.globalcccmcluster.org/system/files/publications/MEND_download.pdf

Care International UK. 2016. "GBV Risk Reduction in Shelter Programmes: Three Case Studies." www.sheltercluster.org/sites/default/files/docs/shelter-gbv_-_three_case_studies_-_december_2016.pdf.

CEDAW, Communication No. 6/2005, *Yildirim v Austria* [2007].

Cedervall Lauta, Kristian. 2015. *Disaster Law*. Oxford: Routledge.

Committee on the Elimination of Discrimination Against Women (CEDAW). 1989. "General Recommendation No. 12: Violence against Women." CEDAW.

Committee on the Elimination of Discrimination Against Women (CEDAW). 1992. "General Recommendation No 19 on Violence against Women."

Committee on the Elimination of Discrimination Against Women (CEDAW). 2010. "General Recommendation No. 28 on the Core Obligations of States Parties under Article 2 of the Convention on the Elimination of All Forms of Discrimination against Women." CEDAW/C/GC/28.

Committee on the Elimination of Discrimination Against Women (CEDAW). 2017. "General Recommendation No. 35 on Gender-Based Violence against Women, Updating General Recommendation No. 19." CEDAW/C/GC/35.

Committee on the Elimination of Discrimination Against Women (CEDAW). 2018. "General Recommendation No. 37 on Gender-Related Dimensions of Disaster Risk Reduction in the Context of Climate Change." CEDAW/C/GC/37.

Council of Europe. 2011. "Council of Europe Convention on Preventing and Combating Violence against Women and Domestic Violence." CETS No.210, Istanbul, 11.V.2011. Entry into force 01/08/2014. Available at: https://rm.coe.int/168008482e.

Davis, Martha F., Thomas Gammeltoft-Hansen, and Emily Hanna, eds. 2017. *Human Rights Cities and Regions: Swedish and International Perspectives*. Lund: Raoul Wallenberg Institute of Human Rights and Humanitarian Law.

Delaney, Patricia, and Elisabeth Shrader. 2000. *Gender and Post-Disaster Reconstruction: The Case of Hurricane Mitch in Honduras and Nicaragua*. Decision Review Draft. Washington, DC: LCSPG/LAC Gender Team. The World Bank.

Dominey-Howes, Dale, Andrew Gorman-Murray, and Scott McKinnon. 2014. "Queering Disasters: On the Need to Account for LGBTI Experiences in Natural Disaster Contexts." *Gender Place and Culture: A Journal of Feminist Geography* 21, no. 7: 905–18.

Ellsberg, Mary, Diana J. Arango, Matthew Morton, Floriza Gennari, Sveinung Kiplesund, Manuel Contreras, and Charlotte Watts. 2015. "Prevention of Violence against Women and Girls: What Does the Evidence Say?" *Lancet* 385, no. 9977: 1555–66.

Enarson, Elaine. 1998. "Through Women's Eyes: A Gendered Research Agenda for Disaster Social Science." *Disasters* 22, no. 2: 157–73.

Farrior, Stephanie. 2004. "The Due Diligence Standard and Violence Against Women." *Interights Bulletin* 14: 150.

Ferris, Elizabeth. 2008. "Natural Disasters, Human Rights, and the Role of National Human Rights Institutions," October 25. www.brookings.edu/on-the-record/natural-disasters-human-rights-and-the-role-of-national-human-rights-institutions/.

Fisher, Sarah. 2010. "Violence Against Women and Natural Disasters: Findings from Post-Tsunami Sri Lanka." *Violence Against Women* 16, no. 8: 902–18.

Fulu, Emma. 2007. "Gender, Vulnerability, and the Experts: Responding to the Maldives Tsunami." *Development and Change* 38, no. 5: 843–64.

Gaillard, Jean-Christophe, Catherine C. Liamzon, and Jessica D. Villanueva. 2007. "'Natural' Disaster? A Retrospect into the Causes of the Late-2004 Typhoon Disaster in Eastern Luzon, Philippines." *Environmental Hazards* 7: 257–70.

Global Protection Cluster. 2010. "Handbook for Coordinating Gender-Based Violence Interventions in Humanitarian Settings." www.unicef.org/protection/files/GBV_Handbook_Long_Version.pdf.

Government of Vanuatu. 2016. "Vanuatu National Guidelines for the Selection and Assessment of Evacuation Centres." https://nab.vu/document/republic-vanuatu-national-guidelines-selection-and-assessment-evacuation-centres.

Grans, Lisa. 2018. "The Concept of Due Diligence and the Positive Obligation to Prevent Honour-Related Violence: Beyond Deterrence." *The International Journal of Human Rights* 22, no. 5: 733–55.

Hines, Revathi I. 2007. "Natural Disasters and Gender Inequalities: The 2004 Tsunami and the Case of India." *Race, Gender & Class* 14, no. 1/2: 60–68.

Human Rights Committee (HRC). 2004. "General Comment No. 31: The Nature of the General Legal Obligation Imposed on States Parties to the Covenant." CCPR/C/21/ Rev.1/Add. 1326.

IAHRC, PM 340/10 – Women and Girls Residing in 22 Camps for Internally Displaced Persons in Port-au-Prince, Haiti.

Inter-Agency Standing Committee (IASC). 2015. "Guidelines for Integrating Gender-Based Violence Interventions in Humanitarian Action: Reducing Risk, Promoting Resilience and Aiding Recovery." https://gbvguidelines.org/en/.

Inter-American Convention on the Prevention, Punishment and Eradication of Violence against Women. 1994. "Convention of Belem do Para." www.oas.org/en/CIM/docs/ Belem-do-Para%5BEN%5D.pdf.

International Commission of Jurists. 2016. "Women's Access to Justice for Gender-Based Violence: A Practitioners' Guide." www.icj.org/wp-content/uploads/2016/03/Universal-Womens-accesss-to-justice-Publications-Practitioners-Guide-Series-2016-ENG.pdf.

International Federation of Red Cross and Red Crescent Societies (IFRC). 2018. "The Responsibility to Prevent and Respond to Sexual and Gender-Based Violence in Disasters and Crises: Research Results of Sexual and Gender-Based Violence (SGBV) Prevention and Response Before, During and After Disasters in Indonesia, Lao PDR and the Philippines." https://media.ifrc.org/ifrc/wp-content/uploads/sites/5/2018/07/17072018-SGBV-Report_Final.pdf.pdf.

International Federation of Red Cross and Red Crescent Societies (IFRC). 2015. "Unseen, Unheard: Gender-Based Violence in Disasters: Global Study." https://www.ifrc.org/ Global/Documents/Secretariat/201511/1297700_GBV_in_Disasters_EN_LR2.pdf

International Law Commission (ILC). 2016. "Draft Articles on the Protection of Persons in the Event of Disaster." Report of the Work of the 68th Session. A/71/10. http://legal. un.org/ilc/texts/instruments/english/draft_articles/6_3_2016.pdf.

Jewkes, Rachel, Purna Sen, and Claudia Garcia-Moreno. 2002. "Sexual Violence." In World Report on Violence and Health. *WHO Report*, edited by Etienne C. Krug, Linda L. Dahlberg, James A. Mercy, Anthony B. Zwi, and Rafael Lozano, 147–83. Geneva: World Health Organization (WHO).

Jordan National Centre for Human Rights (JNCHR), Office of the High Commissioner for Human Rights (OHCHR), Internationl Coordinating Committee of National Institutions for the Promotion and Protection of Human Rights (ICC). 2012. "Amman Declaration and Programme of Action." November 7. https://nhri.ohchr.org/EN/ICC/ InternationalConference/11IC/Background%20Information/Amman%20PoA%20 FINAL%20-%20EN.pdf.

Kälin, Walter. 2005. "The Guiding Principles on Internal Displacement as International Minimum Standard and Protection Tool." *Refugee Survey Quarterly* 24, no. 3: 27–36.

Lenahan (Gonzales) v United States of America, Case 12.626, Inter-Am. Comm'n. H. R., Report No. 80/11 [2011].

Le-Ngoc, Mireille. 2015. "Normative Frameworks Role in Addressing Gender-Based Violence in Disaster Settings." Disaster Law Working Paper Series Paper No. 3. www.ifrc. org/PageFiles/189264/Mireille%20Le%20Ngoc%20Working%20Paper%20GBV%20 (final).pdf.

Liebowitz, Debra J., and Julie Goldscheid. 2015. "Due Diligence and Gender Violence: Parsing its Power and its Perils." *Cornell International Law Journal* 48, no. 2: 301–45.

McLaren, John et al. 2001. "Civil Remedies for Sexual Assault: A Report Prepared for the British Columbia Law Institute by its Project Committee on Civil Remedies for Sexual Assault." British Columbia Law Institute (BCLI) Report No. 14. www.bcli.org/sites/default/files/CivilRemRep.pdf.

Nepal National Human Rights Commission. 2015. "Preliminary Report on Monitoring on the Overall Human Rights Situation of Earthquake Survivors, Loss of Lives and Properties including the Humanitarian Support Such as Rescue and Relief Distribution following the Massive Earthquake that Hit the Nation on 25th April, 2015." www.nhrcnepal. org/nhrc_new/doc/newsletter/Earthquake-Monitoring%20Pre-%20English-Report%20 2072.pdf.

Neumayer, Eric, and Thomas Plümper. 2007. "The Gendered Nature of Natural Disasters: The Impact of Catastrophic Events on the Gender Gap in Life Expectancy, 1981–2002." *Annals of the Association of American Geographers* 97, no. 3: 551–66.

Nguyen, Huong, and Helle Rydstrom. 2018. "Climate Disaster, Gender, and Violence: Men's Infliction of Harm Upon Women in the Philippines and Vietnam." *Women's Studies International Forum* 71: 56–62.

Opuz v Turkey – 33401/02 [2009] ECHR 870.

Osman v United Kingdom – 23452/94 [1998] ECHR 101.

Pantea v Romania – 33343/96 [2003] ECHR 266.

Pikul, Corrie. 2005. "As Tsunami Recedes, Women's Risks Appear." *Women's E News*, January 7. https://womensenews.org/2005/01/tsunami-recedes-womens-risks-appear/.

Picardo, C. W., S. Burton, and Naponick, J. 2010. "Physically and Sexually Violent Experiences of Reproductive-aged Women Displaced by Hurricane Katrina." *Journal of the Louisiana State Medical Society* 162, no. 5: 284–8.

Pittaway, Eileen, Linda Bartolomei, and Susan Rees. 2007. "Gendered Dimensions of the 2004 Tsunami and a Potential Social Work Response in Post-Disaster Situations." *International Social Work* 50, no. 3: 307–19.

Seelinger, Kim Thuy, and Julie Freccero. 2013. *Safe Haven: Sheltering Displaced Persons from Sexual and Gender-Based Violence.* Comparative Report. Human Rights Center Sexual Violence Programme, University of California Berkeley, School of Law. www. law.berkeley.edu/files/HRC/SS_Comparative_web.pdf.

Sphere Project. 2011. *Sphere Handbook: Humanitarian Charter and Minimum Standards in Disaster Response.* Rugby: Practical Action Publishing. www.ifrc.org/Page Files/95530/The-Sphere-Project-Handbook-20111.pdf.

Thomas, Cheryl, Laura Young, and Mary Ellingen. 2011. "Working with the Justice Sector to End Violence against Women and Girls." UN Women, and Advocates for Human Rights. www.ohchr.org/Documents/Issues/Women/SR/Shelters/UN%20Women%20 by%20Cheryl%20_%20Team_working%20with%20justice%20sector.pdf.

United Nations. 2015. "Sendai Framework on Disaster Risk Reduction 2015–2030." www. preventionweb.net/files/43291_sendaiframeworkfordrren.pdf.

United Nations General Assembly. 1994. "National Institutions for the Promotion and Protection of Human Rights." A/RES/48/134. March 4. https://documents-dds-ny.un.org/ doc/UNDOC/GEN/N94/116/24/PDF/N9411624.pdf?OpenElement.

United Nations Population Fund (UNFPA). 2015. "Minimum Standards for Prevention and Response to Gender-Based Violence in Emergencies." www.preventionweb.net/files/ submissions/47102_gbvie.minimum.standards.publication.final.eng.pdf.

Valiuliene v Lithuania – 33234/07 [2013] ECHR 240.

van Baaren, Ellie. 2016. "After a Devastating Cyclone, Fiji's Women Struggle to Rebuild Livelihoods." *UN Women,* February 29. http://asiapacific.unwomen.org/en/news-and-events/stories/2016/02/women-of-fiji-look-for-support-to-rebuild-their-livelihoods.

Vishaka and Others v State of Rajasthan AIR 1997 SC 3011.

World Health Organization (WHO). 2013. "Responding to Intimate Partner Violence and Sexual Violence against Women." *WHO Clinical and Policy Guidelines*. Geneva: World Health Organization (WHO). https://apps.who.int/iris/bitstream/handle/10665/85240/?sequence=1.

Wisner, Ben, Piers Blaikie, Terry Cannon, and Ian Davis. 2004. *At Risk: Natural Hazards, People's Vulnerability and Disasters*. Routledge: Oxford and New York.

X and Y v The Netherlands – 8978/80 [1985] ECHR 4.

14 Conclusions

Catarina Kinnvall and Helle Rydstrom

All chapters in this volume consider the societal dynamics involved in climate disasters and their gendered ramifications. Taken together, they discuss the relationship between community interactions and their environments and the possibility to cope, adapt and transform gender relations in regards to local knowledge in order to reduce gendered vulnerability and harm. In so doing, the authors touch upon the realities and critical dimensions of living in the Anthropocene, and how human influences have become agents of change in an era in which the Holocene has been left behind. Not only does the era of the Anthropocene raise a number of questions related to the relationship between nature and culture and between materiality and humanity, but it also underlines how climate change conceals any internal and external boundaries between them. In addition, most of the chapters in this volume are concerned with how international sustainable development planning agreements have been largely reluctant to incorporate gender into environmental protective strategies for developing more inclusive and sustainable climate strategies.

A focus on gender puts emphasis on its intersectional nature, but also on the parameters that define what is anticipated, endorsed and cherished in terms of expectations and power relations connected to being a man or a woman in any predefined context. As discussed in the introduction, gender and gendered relations are socially constructed and learnt through socialisation processes that are contextual, time-specific and changeable. Climate disasters, similar to other disaster experiences, thus enforce already existing gender asymmetries that prevail and define social and political realities. The fact that women and girls are often stereotypically defined in relation to nature tends to disguise how pre-existing societal inequalities make them disproportionally receptive to the negative impacts of catastrophes, making female life precarious in relation to climate disasters. All chapters discuss how such social inequalities have consequences for the everyday lives of women and girls and how institutional and cultural practices make them disadvantaged in terms of disaster preparedness.

However, taking a gendered to gender perspective does not exclude the experiences of boys and men, but rather reinforces predefined gender boundaries at times of disasters in which boys and men tend to become constructed as saviours of both material and physical livelihoods. Various notions of masculinity, especially hegemonic masculinities that value strength, rationality and independence, can render also men's and boys' lives precarious in emergency situations in which they are supposed to take on protective roles. Moreover, as discussed in some of the chapters, these predefined roles are not only explicitly related to gendered violence, but also to colonial and Western hegemonic definitions and power demonstrations associated with 'white men', as detailed in the chapter by Pulé and Hultman. In relation to international aid projects, this can easily result in other (non-western) men being stereotypically portrayed as aggressive, irrational and uncaring wife beaters in which the heterogeneity of men, maleness and masculinity in the Global South fulfil certain preconceptions of maleness.

Recognising internal differences in 'malehood' in relation to age, sexuality, ethnicity, class and bodyableness does not mean to disregard androcentric privileges, however. Regardless of context, we find that men's violence against women is a social reality for many females across the world and often intensified at times of disasters. Both physical and structural violence are at the core of experiences related to the Anthropocene, but they take on specific social, cultural and psychological dimensions in relation to the disaster experience and capitalise on already existing power relations and power asymmetries. At heart is a concern with the enhanced insecurities, risks and traumas related to the disaster that render gendered vulnerabilities, coping mechanisms and empowerment the focus of many of the chapters in this volume. More specifically is an emphasis on how a critical feminist perspective can uncover the gendered politics of climate change, as detailed in George's chapter, and how different feminist approaches are tackling the development discourse and its tendency to take women, rather than gender as the object of analysis. Hence, a number of chapters deal with the inadequacy of much research and policies on climate change to critically address gender as a conceptual tool for advancing our knowledge of adaptation, coping and mitigation strategies as related to disaster experiences and climate change.

The included chapters propose at least three ways in which to understand, analyse and describe the impact of living in an era defined by the Anthropocene. First, as eloquently put by Tanyag and True, the Anthropocene represents the fullest demonstration of human accumulation and concentration of power in which those mostly affected by climate change are those who have contributed the least to its creation. In this sense, the era of the Anthropocene defies any traditional conceptions of the state-society divide, but instead resists the predominance given to national borders. As discussed by George, certain spaces such as the Pacific Islands are not only particularly affected by climate change, but are actually at the forefront of the intensification of environmental dangers. The consequences for some places, compared to others, have been devastating for the livelihoods of people unable to leave. In gendered terms, as Wester and Lama point out, a number of post-disaster situations highlight how gendered vulnerability is likely to increase when men migrate in search of alternative means of livelihood, leaving women to fend for their families without a male head of the household. Impoverishment of women due to loss of productive assets in post-disaster situations can also force women into low-wage labour, adding to this vulnerability discourse. Any gender-responsive framework, as specifically delineated by Tanyag and True, must thus provide a more holistic understanding of climate change as it intersects with and is compounded by food insecurity, resource-based conflicts and gender-based violence.

Second, and related, the era of the Anthropocene puts issues of power, structures and agency at the fore of any debate on the realities of differentiated vulnerability on all scales of human society. As Raju discusses in his chapter on South Asia, it is of utmost importance to relate any discussion of vulnerability to everyday politics, and more specifically to political decision-making in relation to disaster risk reduction and climate change adaptation. The tendency to think in terms of technical solutions, as Oven and her colleagues point out, gives precedence to science and engineering as quick fixes to the consequences of climate change and climate disasters rather than exploring how gendered relations and processes of marginalisation intersect with contemporary transformations in economy and society. Merli's study of the aftermath of the 2004 Indian Ocean tsunami in Southern Thailand shows, for instance, how local epistemologies are often obliterated in favour of a scientific and policy-oriented discourse or paradigm which privileges, even after death, citizens of the Global North over those from the Global South. The importance of recovering local epistemologies and understandings in order to make sense of catastrophic events and for overcoming trauma should instead be emphasised as means to decolonise Western understandings and recapture the local context. As demonstrated in Wester's and Lama's comparison of the Global North and the Global South, there is a tendency in a number of disaster projects to recycle a certain kind of participation that adds women to climate change adaptation projects in order to support them in their attempts to improve livelihoods. However, this is often done without providing these women with access to positions of power or changing the context that discourages their participation in the first place. Hence, it is of immense value, as pointed out by both Wester and Lama and Tanyag and True, that gender-responsive climate change solutions take a comprehensive approach to security and different types of leadership to encompass the needs of the most marginalised groups of women and girls.

Third, embracing the Anthropocene as an idea means the possibility of reversing the trend of climate change and climate disasters as both physical and human experiences and realities. As Hansson and Kinnvall contend, it means treating humans not as insignificant observers of an assumed natural world but as central to its workings, as fundamental to resistance and change. Nguyen employs the concept of ruination in her analysis, arguing that this concept epitomises the rupturing caused by the super typhoon Haiyan. In the wake of a catastrophe, this rupture works as a violation of both the society and the surviving female population who have been the victims of abuse. Yet, the experiences and the analytical content of ruination, Nguyen suggests, also seem to capture the particular forms of resistance as collective efforts to overcome the condition of being ruined under horrifying and devastating circumstances. Unravelling the contextual temporalities, modalities and intensities of a crisis helps us, Rydstrom shows, to identify the gender specific and differentiated ruination which a climate disaster inflicts upon particular groups in the era of the Anthropocene. A crisis of emergency caused by climate disasters, such as typhoons, landslides and earthquakes, interacts with crisis conditions that were already framing daily life prior to the disruptive event. Depending on socio-economic resources shaped in relation to gendered and masculinised asymmetries, a crisis of emergency might for some be dealt with as an abrupt event, a bracketing of daily life that is possible to cope with, or may even offer a path to renewal and long-term positive change (i.e., *catharsis*). For others, though, a crisis of emergency might fuel, or even exacerbate, antecedent crises conditions, such as men's violence against women, and thereby transform the aftermath of a climate disaster into a crisis of chronicity which perpetually inflicts ruination and slow harm upon lifeworlds, livelihoods and environments.

As Matthew Scott's chapter makes evident, the Anthropocene entails more concentrated and exposed populations, heightened risks and hazards and weaker institutions, with significant implications for the prevention of gender-based violence in situations of disaster displacement. However, the Anthropocene also puts emphasis on the political will to change and prevent gendered violence if institutions can be built and held accountable for violations in relation to climate disasters and their aftermaths. Here Scott suggests a human rights perspective that puts human security at the forefront in all its different shapes and guises, and proposes a right-centred perspective that finds reflections in legislation, judicial decisions, administrative circulars, classroom discussions and civil society initiatives from village to city to national, regional and international levels.

Using diverse ethnographic approaches for studying local effects of living in an era of the Anthropocene, all the chapters provide a multitude of rich descriptions and analyses of pre- and post-responses to climate disasters and their gendered ramifications. While some give detailed empirical descriptions of how disaster narratives and practices are intrinsically gendered by highlighting issues of gendered vulnerabilities and gendered violences, others are more concerned with possible solutions in terms of narrative, practical and institutional change. The chapters also vary in their detailed discussions of Disaster Risk Reduction (DRR) strategies within the global field of policy planning and practice, policies that aim to minimise disaster risks and mitigate the adverse impacts of hazards. Here the main emphasis is on how many of these frameworks fail to take gendered vulnerabilities to disasters into account and the implications of this for effective and equitable risk reduction. In addition, the chapters are also divided in terms of their main focus being on women and gendered ramifications in relation to climate disasters or on notions of masculinity and their devastating consequences for gendered behaviour.

Drawing on the example of the Pacific Islands, George explicates for instance how a particular 'gendered architecture of entitlement' influences where and how women may participate in response to the environmental insecurity that afflicts their communities. Her discussion shows how strongly idealised representations of contemporary Oceanic femininity are often expected, and indeed, central to these political projects, and how Pacific women may face censure and obstruction when they choose not to embrace this ideal. This is similar to Raju's account of South Asia in which he points to the cultural and moral ramifications of gendered options for women and girls affected by the 2004 Indian Ocean tsunami in which gendered cultural codes of dress inhibited women's mobility, resulting in higher disproportionate mortality. As Hansson and Kinnvall remind us in the case of Pakistan, patriarchal control over women are exercised through institutionalised codes of behaviour, gender segregation and an ideology which associates family honour with female virtue. This implies that customary practices which aim to preserve the subjugation of women are often defended and sanctified as cultural traditions and given religious associations.

Similarly, Wester and Lama discuss how the increase of guesthouse tourism in the Maldives discourages local women from walking along the beaches of Maafushi since visitors usually wear light and revealing clothing or swim wear, something that local women should not be exposed to. This, they argue, has limited local women's participation in the livelihood activities of the island as a direct response to climate adaptation measures focused on tourism and fishery. Such developments are also frequently described in cultural terms as punishments from God. As discussed in Hansson and Kinnvall chapter, discourses of God's, or Allah's punishment, have aided religious extremists in their sectarian appeal at the same time as the disaster has been intimately linked to religious narratives. Similarly, Merli's account of the gendered effects of the 2004 tsunami in Thailand confirms a general conviction that the islands struck by the tsunami were being punished for the sexual sins related to mass tourist resorts. This idea of the West as an (immoral) neo-colonial power is even further explicated in Nguyen's chapter, in which she links the legacies of Western colonialism and neo-colonial practices to a 'culture of precarity'. This precarity is also described in terms of memories of colonial violence and sexual abuses, so-called ruins, in which the female body bears the brunt of 'double' colonisation, and where the body is seen as a site for gendered readings of postcolonial subjectivity.

In terms of solutions, Tanyag and True's chapter together with Scott's final chapter are explicit in this regard, although a number of other chapters also discuss possible ways to move beyond gendered ramifications of climate disasters. The suggestions by Tanyag and True to focus on: (1) the deliberation and democratisation of climate science; (2) gender-inclusive representation in climate governance, and; (3) the rethinking of norms and practices in climate governance, are all concerned with strategies for creating a gender-responsive framework to climate disasters. The first of these are meant to ensure that all forms and sources of knowledge are recognised as contributing to climate responses, and that knowledge production is translated into information, assistance and services that benefit rather than exclude the communities most affected by climate change. The second strategy refers to women's direct representation in different levels of governance, from the community to national decision-making bodies to global climate negotiations, while the third strategy involves re-visioning institutional pathways for linking these localised knowledges and gender-inclusive political representations at all levels of governance. This focus on re-visioning norms and practices in climate governance can be related to George's interview with Ursula Rakova from the Carteret islands in which she discusses how Rakova's response to climate change in the region, subverts rather than accommodates the gendered architecture of entitlement and how Rakova's defiance of male traditions destabilises the narrative of gendered virtue.

As strategies they are not entirely divorced from Scott's suggestions on how to: (1) promote the development of domestic law and policy; (2) the use of legal action to endorse effective implementation; (3) advancing accountability for failures to prevent gender-based violence; (4) parallel reports to treat monitoring bodies, and finally; (5) engaging national human rights institutions. Although an explicit legal focus, and perhaps more directly related to the chapters by Bondesson, Raju, and Oven and colleague's discussion and critique of current DRR approaches, the human rights-based approach suggested by Scott does provide a language and instruments for different actors to use when describing and seeking to change lived experience. As Scott maintains, a legal right to be protected from sexual violence provides a platform for a host of actors to demand that states take steps to fulfil their obligations towards people within their jurisdiction.

This is similar to Bondesson's claim that DRR policymaking must consider gendered vulnerabilities to disasters, since the lack of gender perspectives yields ineffective and inequitable risk reduction. In this, gendered mainstreaming is not enough. As argued by Raju in his chapter, while synergies may be drawn between climate change adaptation and disaster risk reduction strategies, both these processes are challenged as a result of gender inequalities on the ground. Here policy change must be seen in the light of power shifts in which the institutionalisation of gender becomes a key issue in larger development processes feeding into DRR and CCA (Climate Change Adaptation). Here it is interesting to note how Community Based Approaches to Disaster Risk Reduction (CBDRR) have become a focus of significant development funding in countries in the global South, reflecting a shift towards more bottom-up planning and community empowerment. This is explicated in Oven and her co-authors' chapter on CBDRR in Nepal, in which they find that despite the explicit aim to involve communities in disaster risk reduction, not much attention was given to women's strategic gender needs that emerged from wider issues of subordination. Instead the technocratic nature of the approach largely resulted in Nepalese women finding themselves attending rather than contributing to CBDRR, thus reflecting historical and inherited gendered relations and processes of marginalisation.

This focus on underlying gender norms, especially norms related to dominant versions of masculinity, is at the heart of many of the chapters in the volume. As a theme, it is consistently dealt with in Pulé and Hultman's chapter in which they locate hyper-masculinities at the heart of an ever-increasing climate emergency. Here they are particularly concerned with climate change deniers in the Global North, i.e., individuals and constituencies who are entangled with what they refer to as "fossil-fuel addicted industrialisation and corporatisation", and whose allegiances are emboldened by traditional socialisations of masculine identities referred to as 'industrial/breadwinner masculinities'. In their critique of hyper-masculine hegemonies they highlight how such notions of masculinity rely on hyper-separation, transcendence and dominance that engender a denial of environmental, as well as other crises. They emphasise how the relationship between being Western, white and male brings with it a heavy reliance on anthropocentric notions of natural resources as being up for grabs and how such presumed orders of entitlement follow racial divisions of privilege.

Masculinity is furthermore discussed in Hansson's and Kinnvall's chapter, which shows how a patriarchal order has become manifest in the Pakistani society in relation to ideas of religion and nationhood. By describing how a patriarchal order takes particular forms in Pakistani society, where control over women are exercised through institutionalised codes of behaviour, through gender segregation and through an ideology which associates family honour with female virtue, they show how the local and the national are connected. In their study of the Bahawalpur region they specifically discuss how masculinity and the patriarchal order are entwined with issues of mental health, gender-based and sectarian violence and how risky situations connected to local disasters have resulted in women being violated in the name of honour.

Rydstrom's chapter offers an ethnographic perspective on masculinity. By doing so, it shows that in-depth knowledge from the ground is critical for translating disaster mitigation frameworks into more context sensitive strategies. Masculinity and a patriarchal organisation of social life in central coastal Vietnam, are enmeshed with the livelihood of fishing, male sociality and gendered responsibilities. A gender specific division of labour, Rydstrom highlights, results in a differentiation of disaster-related hazards with mainly men being in charge of saving the boats, and thus their livelihood, and women taking care of the household. While men risk their lives when attempting to save the boats, the precariousness endured by women when a storm hits, should be seen against the backdrop of patriarchally induced hierarchies, powers, privileges and violences that pervade everyday life to understand why women and girls are rendered particularly vulnerable in times of climate disasters. A persistent problem which has been identified and examined in the chapters of this volume.

In sum, the overall contribution of this book can be condensed into three main points: First, international and national aid organisations are often quite unsuccessful in their attempts to mitigate gendered violence in disaster risk reduction policies. This is mainly due to the fact that most of these policies, such as the Sendai framework for action for instance, proceed from a fairly limited and problematic gender perspective. Second, in pre- and post-disaster contexts ingrained and powerful narratives of gender roles and gender norms continue to restrict and limit access to the broader public sphere and to the political community proper for a number of marginalised communities and more specifically for women and girls in these communities. And third and finally, there is a need to more thoroughly envision ways in which we can move beyond gendered ramifications of climate disaster. There is, in other words, a need for visions, strategies, policies and narratives that are able to mitigate the negative effects of insecurity, precariousness and various types of crises related to climate disasters. Such visions, strategies, policies and narratives are necessary in order to conceptualise notions of gender and masculinity that are sensitive to the realities experienced on the ground in various local contexts.

Index